BERNARD D. WOOD
Syracuse University

APPLICATIONS
OF THERMODYNAMICS

ADDISON-WESLEY PUBLISHING COMPANY

Reading, Massachusetts
Menlo Park, California · London · Amsterdam · Don Mills, Ontario · Sydney

This book is in the
ADDISON-WESLEY SERIES IN MECHANICS AND THERMODYNAMICS

Consulting Editors
HOWARD W. EMMONS   BERNARD BUDIANSKY

ISBN 0-201-08740-5
GHIJKLMN-HA-7987

To the memory of Robert C. Wiren, teacher and friend

PREFACE

The subject matter of this book is designated as applications of thermodynamics because the intention is to build on the foundation of a prerequisite first course in the fundamentals of thermodynamics in order to explain certain engineering systems. The content of that first course might be entirely classical (macroscopic) in point of view, or it might include some microscopic (statistical mechanical) theory to help explain laws and relationships. Either type of course should provide the background required. A course in fluid mechanics (now sometimes combined with thermodynamics) is also a prerequisite. Heat transfer, where treated as a separate subject, could be taken previously or concurrently.

The broad coverage of the book is indicated by the table of contents. There are available today individual books on the subject matter of each of the separate chapters and indeed whole books on many of the subheadings. In general, however, these books do not serve the reader who wishes to extend his knowledge of thermodynamics but is not ready to specialize. The specialized books, by their nature, tend to emphasize the peculiarities of each branch of this broad field rather than to illustrate the similarities and possible generalizations. Curriculum development some years ago tended to de-emphasize such arbitrary divisions as were imposed by separate courses, whether elective or required, in Internal Combustion Engines, Gas Turbines, Refrigeration, Air Conditioning, and so on. We would be repeating the errors of the past to present the applications of thermodynamics to students only in such separate courses as Direct Energy Conversion, Cryogenics, and Rocket Propulsion, without attempting to correlate their essentially similar problems.

The selection of topics for the book had to be arbitrary and inevitably reflects, to some extent, the personal interests of the author. While an attempt has been made to achieve a balanced course presentation, thermodynamics is much too broad a discipline for one to hope for a full coverage of its applications in one

v

book. Readers will undoubtedly be more interested in one topic than another; the various chapters can be read separately and in any order when allowance is made for the tendency of an author to refer back to a point already made. In fact, reference is repeatedly made to similar developments or formulas used elsewhere in the book, both earlier and later; that is basic to the character of the book.

Readers will note the absence of a separate chapter on nuclear power. A detailed consideration of the nuclear energy source is based on a body of knowledge essentially different from the subjects dealt with here. However, the various vapor and gas power cycles and certain of the direct energy conversion devices that are discussed have been used to harness the energy from nuclear sources.

While the arrangement of material in this book lends itself to lecture course presentation, the information also provides useful background material and broadening insights for thermodynamic laboratory and design problems for both students and practicing engineers. It is clear that much of the content can be dealt with in undergraduate courses. At the same time, the whole book provides a foundation for various specialized courses at the graduate level. For the narrower, more advanced courses, the instructor may be able to select additional textbooks or prepare reading assignments in the technical literature, but the interrelationships among the numerous application topics will not be overlooked if a unifying treatment is available for continuous reference.

My debt to many authors of books on the various applications of thermodynamics will be evident to any teacher or practicing engineer in this field. Those books that have influenced this writing and those that will be useful to others are listed as references at the end of each chapter. Despite care to cite sources of particular information or ideas identifiable as the contribution of one author, it is possible that credit may not have been given in a few instances; for this I apologize. I am grateful to Professors M. E. Barzelay, D. S. Dosanjh, E. E. Drucker, W. E. Gifford, R. W. Heimburg, and V. G. Korolenko, all of Syracuse University, and Dr. R. Longsworth of Cryomech., Inc., who read various portions of the manuscript in their particular fields of competence. Dr. Harold Smulyan of the Department of Medicine, Upstate Medical Center of the State University of New York, was extremely generous in spending time to discuss various physiological factors relevant to Chapter 6. Also, I am indebted to my students who pointed out errors and ambiguities where those were present in the printed class notes used while this manuscript was in preparation. The cooperation of several of the secretaries of the Department of Mechanical and Aerospace Engineering of Syracuse University was most helpful in the preparation of the manuscript.

Syracuse, New York
June 1968

B. D. W.

CONTENTS

FOREWORD TO THE STUDENT

Since the assumption has been made that the reader of this book is conversant with the fundamentals of engineering thermodynamics, the sequence of presentation of topics is not critical here. Nothing dictates a logical starting point in a discussion of applications of thermodynamics except pedagogy. Power cycles will be dealt with first because students are able to visualize these more readily than refrigeration cycles; the latter apparently seem more plausible once power cycles are understood. Direct energy conversion devices for both power and refrigeration are in a relatively early stage of industrial development. Their operation is therefore less familiar and, with no moving parts apparent, less obvious than the operation of the more conventional power and refrigeration systems. Consequently, the explanation of these direct conversion devices is left until last.

There is no separate review chapter for fundamentals of thermodynamics, fluid mechanics, and electricity. In the explanation of various systems in this book, the presentation often goes right back to first principles where that seems useful. On the other hand, numerous terms and relationships are merely used when necessary without proof; where a reader feels in need of some review or would like a formal proof or extended discussion, he is best served by whatever textbook or course notes he has already used or is using in the various background courses. The books on the fundamentals of thermodynamics and fluid mechanics that the author has found most useful as general references are listed at the end of this Foreword. Certainly the list could be much longer as there are many good books on these subjects.

Rigid definitions of many of the generally accepted terms of thermodynamics are not presented in this book. As an aid to the reader, important terms are set in **boldface type** when they are used for the first time in this book and often again when met for the first time in another section. Also, figure numbers are set in boldface where reference is first made to each figure. Letters set in boldface italic in the text correspond to circled letters representing thermodynamic state points on diagrams.

The potential usefulness or practical significance of a device or system will often be mentioned, but there will be no attempt to describe current practice in any detail; that would be beyond the scope of a course in thermodynamics, even one designated as "applications." Where it is specifically stated or somehow implied in this book that one system is not economically justified for a particular application, the term "impractical" will still be avoided. Present and future requirements for space travel, remote communication stations, underdeveloped regions of the world, and so on, remove many of the conventional restraints in the use of the word "practical."

All the diagrams in this book are schematic. The use of photographs has been purposely avoided for two reasons. First, such photographs seldom help explain thermodynamic processes; a much more simplified view is usually required by the reader who is not familiar with the particular apparatus. Second, the physical appearance of some piece of apparatus may change drastically while its basic principle of operation remains the same. The best illustrations of the physical arrangement and detail of engineering equipment are to be found in the various engineering journals, not only in the technical articles, but also in the advertisements.

A book on the applications of thermodynamics cannot be self-contained. It must make reference to other sources. Typical sources, both articles and specialized books, are indicated at the end of each chapter. The lists of references given for each chapter are relatively brief. There has been no attempt to provide a complete survey of the literature for any of the topics presented. Of course, those articles and books to which actual reference is made in the text have been included. In the same lists, which are arranged alphabetically by author, there are several items included for each chapter which provide either a useful broad survey of a field or expanded discussion of particular points. It has not always been necessary to cite these as references in the text, and their coverage is usually indicated sufficiently by their titles. The selection of these additional references has been guided to some extent by the desire to present a diversification of sources, though certain journals will predominate in their own fields.

Any attempt to use only one **system of units** throughout a book which covers such a wide range of topics would lead to eventual confusion rather than simplicity. The units chosen have generally been those most often encountered in the literature on a particular section of the work. The **English engineering system** (pound force, pound mass, foot, second) has been used through most of the book. In places, the **English gravitational system** (pound force, slug, second) seems appropriate. In Chapter 7, Direct Energy Conversion, the **mks** (meter, kilogram, second) and **cgs** (centimeter, gram, second) systems are used extensively. When Newton's second law of motion,

$$F = \frac{ma}{g_c}$$

is written, g_c has the value 1.0 in all the above systems except the first; in the English engineering system, g_c is 32.174 (pound mass/pound force)(ft/sec²). In many engineering books, pound force ($1b_f$) is carefully distinguished from pound mass ($1b_m$) by the subscript used with the abbreviation. Such distinction will be made here only where confusion might otherwise occur. Whether force or mass is intended will usually be obvious from the context. Conversion factors for basic and derived units are given in Appendix I.

Both **work** and **heat** will be dealt with extensively throughout the book. Both are forms of energy identified only in transit. There is little difficulty with the concept of **work** except that work very often is not immediately recognized as such. The term will be used wherever a transfer of energy actually displaces a force through a distance, or can be imagined to be convertible to the effect of raising a weight against the force of gravity in the surroundings of the thermodynamic system.

On the other hand, some confusion arises in the concept of **heat** because, in the long-discarded caloric theory, it was thought of as a massless substance contained in a body. The term heat now is only applicable to the energy transferred from one body to another in a process in which its quantity is established by the mathematical statement of the **first law of thermodynamics,**

$$Q = (U_2 - U_1) + W,$$

in which Q is the heat transfer, $(U_2 - U_1)$ is the change of internal energy (a thermodynamic property), and W is the work done in the process. All three terms are energy terms and must have the same units.

Although a body or system cannot "contain" heat, it will be useful in discussing many processes to speak of **heat received** or **heat rejected** so that the direction of heat transfer relative to the system is immediately obvious. This should not be construed as the treatment of heat as though it were a substance.

The numerical value of any thermodynamic property such as pressure, volume, temperature, internal energy, enthalpy, and entropy depends on the state point only and not on the path taken to reach that point. Small changes in property values can therefore be dealt with as **exact differentials,** designated by the letter "d" as for instance dT. On the other hand, the work and the heat transfer accomplished between two state points do depend on the paths followed. Small amounts of work and heat transfer will therefore be **inexact differentials** designated in this book by $đW$ and $đQ$, respectively.

In addition to the conversion tables for units already referred to, the appendices contain many tables and charts of thermodynamic properties of various substances. These will be useful in explaining certain thermodynamic effects as well as in numerical problems.

REFERENCE TEXTBOOKS IN THE FUNDAMENTALS OF THERMODYNAMICS AND FLUID MECHANICS

Binder, Raymond C., *Fluid Mechanics*. Englewood Cliffs: Prentice-Hall, 1962, 4th Edition.

Eskinazi, Salamon, *Principles of Fluid Mechanics*. Boston: Allyn and Bacon, 1962.

Hall, Newman A., *Thermodynamics of Fluid Flow*. New York: Prentice-Hall, 1951.

Hall, Newman A. and Warren E. Ibele, *Engineering Thermodynamics*. Englewood Cliffs: Prentice-Hall, 1960.

Hansen, Arthur G., *Fluid Mechanics*. New York: Wiley, 1967.

Hatsopoulos, George N., and Joseph H. Keenan, *Principles of General Thermodynamics*. New York: Wiley, 1965.

Keenan, Joseph H., *Thermodynamics*. New York: Wiley, 1941.

Kestin, J., *A Course in Thermodynamics*. Waltham, Mass.: Blaisdell, 1966.

Lee, John F., and Francis Weston Sears, *Thermodynamics*. Reading, Mass.: Addison-Wesley, 1963, 2nd Edition.

Lee, John F., Francis W. Sears, and Donald L. Turcotte, *Statistical Thermodynamics*, Reading, Mass.: Addison-Wesley, 1963.

Li, Wen-Hsiung, and Sau-Hai Lam, *Principles of Fluid Mechanics*. Reading, Mass.: Addison-Wesley, 1964.

Mooney, David A., *Mechanical Engineering Thermodynamics*. Englewood Cliffs: Prentice-Hall, 1953.

Obert, Edward F., *Concepts of Thermodynamics*. New York: McGraw-Hill, 1960.

Reynolds, William C., *Thermodynamics*. New York: McGraw-Hill, 1965.

Roberts, J. K. *Heat and Thermodynamics*. London: Blackie and Son, 1944..

Sonntag, Richard E., and Gordon J. Van Wylen, *Fundamentals of Statistical Thermodynamics*. New York: Wiley, 1966.

Van Wylen, Gordon J., and Richard E. Sonntag, *Fundamentals of Classical Thermodynamics*. New York: Wiley, 1965.

Zemansky, M. W. and H. C. Van Ness, *Basic Engineering Thermodynamics*. New York: McGraw-Hill, 1966.

CHAPTER 1

POWER CYCLES

1.1 INTRODUCTION

The material to be presented in this chapter is very broad in scope, but unified by the involvement of a **working fluid** utilized **for the production of power**. The power production, the efficiency, the limitations, and the potential improvements of various systems will be analyzed through the consideration of **thermodynamic cycles.** Such cycles can be represented or approximated on coordinates of thermodynamic properties. Involved in these cycles will be **processes** through each of which thermodynamic properties of the working fluid are changed; some of these processes where work is done on the fluid or by the fluid will be examined in some detail apart from the power cycles in which they might be used.

It will be usual to treat the working fluid as the **thermodynamic system** under consideration and all else will be the **surroundings.** In a completed cycle, the thermodynamic properties of the working fluid must be returned to their original values so that the starting point is reached again. In a continuous cycle, each element of the working fluid will be assumed to return to that point repeatedly. That is not to say that no net change will have taken place in the system plus its surroundings. Energy may be received from one or more **sources** and delivered to one or more **sinks,** and work may be done on the fluid by the surroundings or vice versa. A **closed system** does not exchange matter with the surroundings. In an **open cycle,** new fluid is continuously brought in at one set of conditions and released at another. No real "cycle" exists, but for purposes of analysis, it is often possible to imagine one or more processes that would bring the working fluid at the release conditions back to the conditions of the starting point and so to analyze the open cycle as though it were closed.

Frequently in the work to follow, ideal cycles will be discussed in which it will be assumed that certain processes are **reversible.** Usually our concern will be with the working fluids of such cycles, and what will be of interest to us will be whether the process is **internally reversible.** For instance, the transfer of heat cannot be accomplished without a temperature difference between two bodies or fluids, and the operation is therefore not reversible when both bodies are considered. The working fluid of the power cycle, however, can receive energy by heat transfer and then regain its original state point by rejecting heat energy along an identical process line. The larger system consisting of a heat source, the fluid, and a heat sink would undergo irreversible operations, but the process of the working fluid itself ideally could be internally reversible. It will usually be clear from the context whether the "system" implies only the working fluid or includes also its surroundings.

In this chapter, the emphasis will be on the production of useful power, although certain processes, notably compression, that are part of a power cycle will quite obviously absorb energy. In Chapters 4 and 5, cycles and working fluids will again be considered, but the purposes of those cycles will not be the production of power, but rather the useful transmission of heat from a low temperature source to a high temperature sink. It will be seen that power will have to be provided to those refrigeration and heat-pump cycles. In Chapter 7, the conversion of one form of energy to another will be considered with the emphasis on the conversion process which may or may not be part of a power or heat-pump cycle.

Practical and theoretical cycles will be presented schematically in the figures of this chapter. Physical details of components will be mentioned only where such details aid in the understanding of the theory or to explain why one type of component is preferred over another for other than thermodynamic reasons. The various components of a complex power plant will serve particular purposes regardless of what working fluid is used. Knowledge of the reasons for their use rather than the details of construction will provide the best basis for future, unpredictable applications. The conventional names of cycles and of components are used to help place things in categories as a convenience. It is far easier to say "ideal Carnot cycle" than to say "frictionless, reversible cycle composed of isentropic compression, isothermal heat addition, isentropic expansion, and isothermal heat rejection." With use, the names become familiar.

1.1.1 Efficiency and the Carnot Cycle

In this chapter, many different thermodynamic power cycles will be discussed. Real cycles will be compared with idealized cycles and idealized cycles will be compared with the **optimum cycle;** the optimum cycle is that which would produce the maximum thermal efficiency for the imposed conditions.

Thermal efficiency (η_{th}) is the ratio of the useful work output of the cycle to the energy that must be supplied to the cycle for that output:

$$\eta_{th} = \frac{\text{Useful work output}}{\text{Energy supplied}}.$$ (1.1)

Or, to consider rate of energy,

$$\eta_{th} = \frac{\text{Power output}}{\text{Rate of energy input}}. \tag{1.2}$$

We will find that the limitations to a power cycle are primarily those set by the temperature levels at which heat energy can be received and rejected. It is a **corollary of the second law of thermodynamics** that no cycle can be more efficient than a reversible cycle operating between the same temperature limits; also, any two reversible cycles receiving heat at one particular temperature and rejecting it at another particular temperature must have identical values of thermal efficiency. Thus it is almost always useful to compare the efficiency of the cycle being considered with the efficiency of a reversible cycle operating between the same two temperatures.

Any reversible cycle would be satisfactory for comparison, since all must have the same value of thermal efficiency between the same source and sink temperatures; however, for reasons that will become apparent, the **Carnot cycle** is usually the most convenient and will be discussed here. It is interesting to note that Carnot, in 1824, was the first to use a cycle in thermodynamic reasoning, and that the cycle that he conceived and that bears his name remains such a useful gauge of excellence.

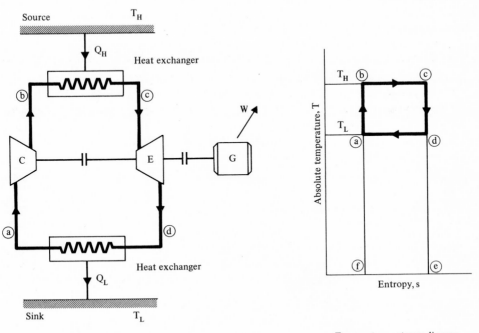

Schematic equipment   Temperature-entropy diagram

FIG. 1.1 The Carnot cycle.

The cycle is shown in **Fig. 1.1.** It is made up of two internally reversible **iso-thermal** processes and two internally reversible **adiabatic** (isentropic) processes. The whole heat exchange with surroundings must take place during the isothermal processes. The cycle is shown on temperature-entropy (*T-S*) coordinates so that the nature of the working fluid need not be specified; the shape of each process line is established by the nature of the process alone. A schematic arrangement of the equipment is shown also. Ideal, frictionless machines are assumed for the compression process from *a* to *b* and for the expansion process from *c* to *d*; neither process allows heat exchange with the surroundings. Ideal heat exchangers are assumed so that the working fluid can receive heat energy Q_H at the higher temperature T_H degrees absolute and reject heat energy Q_L at the lower temperature T_L with no temperature differences in either heat exchanger. The net work of the cycle *W* is absorbed by a device *G* that might be an electric generator. The units of energy for Q_H, Q_L, and *W* will be assumed the same, say, Btu per cycle.

The **first law of thermodynamics** stated mathematically is

$$đW = đQ + dU.$$

For a cycle, the working fluid returns to the same state point so that

$$\oint dU = 0.$$

Therefore the first law applied to a cycle gives us

$$\oint đW = \oint đQ,$$

$$W = Q_H - Q_L. \tag{1.3}$$

Thus the thermal efficiency is

$$\eta_{th} = \frac{W}{Q_H} \tag{1.4a}$$

$$= \frac{Q_H - Q_L}{Q_H}$$

$$= 1 - \frac{Q_L}{Q_H}. \tag{1.4b}$$

The **thermodynamic temperature scale** is defined in terms of the Carnot cycle by the relationship

$$\frac{T_H}{T_L} = \frac{Q_H}{Q_L},$$

where T_H and T_L are absolute values, i.e. degrees above absolute zero. Thus

$$\eta_{th} = 1 - \frac{T_L}{T_H} \tag{1.5a}$$

$$= \frac{T_H - T_L}{T_H} \tag{1.5b}$$

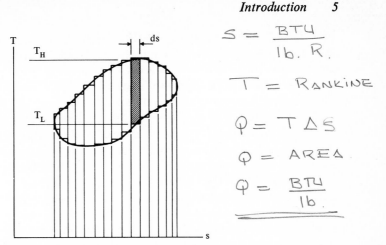

FIG. 1.2 General reversible cycle.

The same expressions can be derived immediately from the *T-s* diagram through the use of the definition of entropy:

$$dS = \left(\frac{dQ}{T}\right)_{\text{rev}}.$$

For the reversible isothermal processes at T_L and T_H,

$$Q = T\Delta S.$$

Thus the energy quantities Q_L and Q_H are represented by the areas under the lines *da* and *bc* respectively on the *T-s* diagram. Consequently,

$$\eta_{\text{th}} = \frac{\text{Area } \textbf{\textit{bcef}} - \text{Area } \textbf{\textit{adef}}}{\text{Area } \textbf{\textit{bcef}}} = \frac{T_H - T_L}{T_H}.$$

The net work of the cycle is represented by the cycle area *abcda*.

A generalization might be made on the basis of the dependence of Carnot cycle efficiency on temperature. Any reversible cycle can be thought of as a series of infinitesimally narrow Carnot cycles, as shown in **Fig. 1.2**, each cycle receiving heat and rejecting heat isothermally. Without the necessity of a rigid proof, it can be seen that the whole cycle thermal efficiency would be a composite of the individual cycle efficiencies. Thus the higher the average temperature at which heat is received and the lower the average at which heat is rejected, the greater the composite cycle efficiency.

If the abscissa of the temperature-entropy diagram is entropy per pound of working fluid (identified by the use of a lower case *s*), then areas delineated by reversible lines on the diagram represent energy per pound of fluid. In the usual engineering units, *T* would be degrees Rankine and *s* would be Btu per pound-degree *R*. Area would be Btu per pound. This type of diagram, familiar from the study of fundamentals of thermodynamics, will be employed extensively in this chapter and elsewhere in the book.

1.1.2 Relative Efficiency and Cycle Efficiency

The prime purpose of the preceding discussion about the Carnot cycle, which was taken as an example of a reversible cycle operating between two particular temperatures, was to provide a basis for comparison in the analysis of other reversible or actual cycles. For such comparison, ratios of efficiency values are useful.

The **relative efficiency** (η_r) of an ideal cycle (sometimes called reduced efficiency) is the ratio of the theoretical value of efficiency for the particular ideal cycle to the efficiency of a reversible cycle operating between the same principal temperature limits and receiving and rejecting heat energy only at those temperatures. While all reversible cycles receiving and rejecting heat at the same temperature levels will have the same values of thermal efficiency, other reversible cycles which receive and reject heat at varying temperatures will not have the same efficiency value. Furthermore, some "ideal" cycles are not even intended to be completely reversible and they would then not have efficiencies equal to those of ideal reversible cycles. The relative efficiency then is

$$\eta_r = \eta_{\text{th(ideal)}}/\eta_{\text{th(Carnot)}} \tag{1.6a}$$

and

$$\eta_{\text{th(ideal)}} = \eta_{\text{th(Carnot)}} \times \eta_r. \tag{1.6b}$$

Actual cycles always deviate in some way from ideal cycles and usually in such a way as to have lower values of thermal efficiency. The **cycle efficiency** (η_{cy}) is the ratio of the efficiency value for an actual cycle to the value for a particular ideal cycle with which it is being compared:

$$\eta_{\text{cy}} = \eta_{\text{th(actual)}}/\eta_{\text{th(ideal)}} \tag{1.7a}$$

When these two efficiency ratios are combined,

$$\eta_{\text{th(actual)}} = \eta_{\text{th(ideal)}} \times \eta_{\text{cy}} \tag{1.7b}$$

$$= \eta_{\text{th(Carnot)}} \times \eta_r \times \eta_{\text{cy}} \tag{1.7c}$$

1.2 VAPOR POWER CYCLES

The term vapor cycle will be reserved for those thermodynamic power cycles in which some of or all the working fluid undergoes a **change of phase** from vapor to liquid and back again. In the gas cycles of Section 1.3, the working fluid will remain always in the one phase, regardless of whether or not it approaches the saturated vapor conditions where its equation of state would deviate greatly from the perfect gas laws.

Vapor power cycles of the basic types described here can be used whether the source of energy is combustion of fossil fuel (coal, gas, and petroleum products), some other chemical reaction, solar energy, or nuclear fusion or fission. Over three-quarters of the installed capacity of stationary power plants in the United States today is made up of systems using vapor power cycles.

Largely because of its availability and safety, water is the working fluid used in almost all these power plants. Consequently, steam will be used often as an example of a working fluid. Nevertheless, our studies of steam cycles will reveal some disadvantages resulting from the thermodynamic properties of that particular fluid. Alternative fluids will suggest themselves by their superior properties, particularly from the point of view of cycle efficiency. While some of these other fluids may or may not find application in the large capacity stationary plant, they will undoubtedly merit consideration for innumerable situations such as sealed systems for space vehicles and transportation, and any other place where high delivery per unit of weight or maximum efficiency coupled with simplicity are prime governing factors.

1.2.1 Vapor Tables

The thermodynamic properties of many substances have been published in the form of tables. Most follow the familiar format of the "Thermodynamic Properties of Steam" by Keenan and Keyes [B. 11] and of the "1967 ASME Steam Tables" [B. 22].* Tables and charts from the latter are given in abridged form in the appendix. Tables 1 and 2 of that volume give the properties of **saturated liquid** (liquid in equilibrium with vapor) and **saturated vapor** (vapor in equilibrium with liquid) from 32F, at which the enthalpy and entropy of liquid are arbitrarily taken as zero, up to the **critical point**, at which the enthalpy of evaporation is zero. The subscript designations of those tables will be used throughout this text, the subscript f designating saturated liquid, g, saturated vapor, and fg, the change from one to the other. The symbols v, h, and s stand for specific values of volume, enthalpy, and entropy respectively. Thus

$$v_f + v_{fg} = v_g \text{ cubic feet per pound,}$$

$$h_f + h_{fg} = h_g \text{ Btu per pound,}$$

$$s_f + s_{fg} = s_g \text{ Btu per pound--degree Rankine.}$$

For calculations of values in the **wet-vapor** region where the fluid considered is a mixture of liquid phase and vapor phase fluid in equilibrium with each other, the symbol x will represent the **dryness fraction** (often called **quality**), which is the fraction of the total that is in the vapor phase. For equilibrium at a given pressure, the specific volume, enthalpy, and entropy change in proportion to the dryness fraction. Therefore in the liquid-vapor region,

$$v = v_f + xv_{fg}, \qquad h = h_f + xh_{fg}, \qquad s = s_f + xs_{fg}.$$

A particular use for these relationships will be found in the determination of endpoint values after isentropic expansion from a known state point into the wet-vapor region. Through the known specific entropy, the dryness fraction can be determined at the final pressure.

Table 3 of the ASME Steam Tables gives properties of **superheated vapor,** which is vapor at a temperature above the saturation temperature for its pressure, and of **compressed liquid,** which is liquid at a pressure higher than the saturation

---

*These references can be found at the end of each chapter.

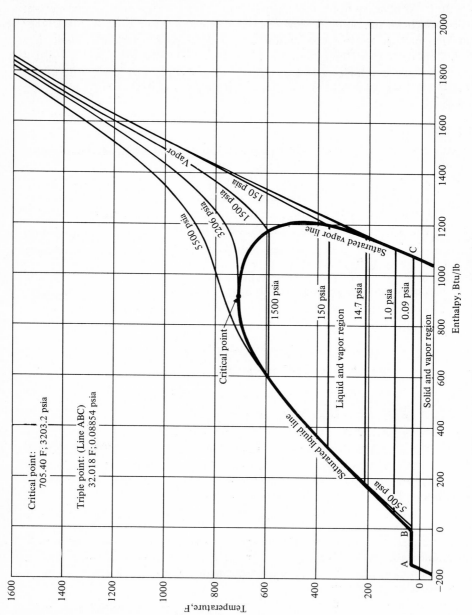

FIG. 1.3 Temperature-enthalpy diagram for water.

pressure for its temperature. Double interpolation between bracketing temperatures and between bracketing pressures may be necessary in this table. Other tables
of liquid and vapor properties will be referred to where required in this text.

It should be emphasized that dealing with the tabulated properties of one
particular substance, in this instance, water, provides an insight into the properties
of many other substances. Later in this section, for instance, the properties of mercury vapor will be mentioned, and it will be found that they are presented in the
same way. In Chapter 4, it will be found that the tables of thermodynamics properties of various refrigerants follow the same format as the steam tables except
that the arbitrary datum for enthalpy and entropy is usually at a lower temperature.

1.2.2 *Charts of Thermodynamic Properties*

The thermodynamic properties of substances can be presented on charts with
various pairs of coordinates. Perhaps the chart most easily visualized, because of
our familiarity with the boiling of water and other liquids at constant pressures,
is a chart on temperature-enthalpy coordinates, as shown for water in **Fig. 1.3**.
The saturated liquid and saturated vapor lines, coming together at the critical point,
enclose a region above 32 F in which liquid and vapor can exist in equilibrium.
This is sometimes referred to as the **wet-vapor dome** or saturated liquid-vapor dome.
Below 32 F, solid and vapor exist together in equilibrium, solid **sublimating** directly

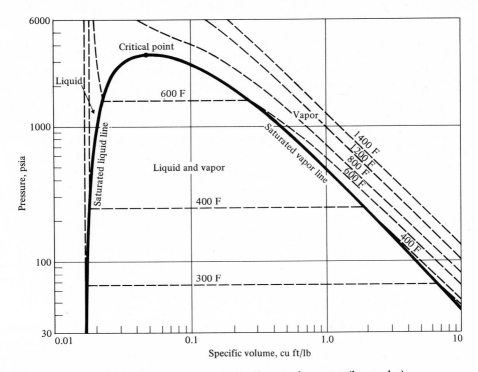

FIG. 1.4 Pressure-specific volume diagram for water (log scales).

to vapor in this region. The three phases can exist in equilibrium only at the **triple point temperature** which for water is 0.01 C (by definition) or 32.018 F at 0.1803 in. Hg vapor pressure.

The thermodynamic properties presented on pressure-volume coordinates are quite familiar. The presentation in **Fig. 1.4** using logarithmic scales indicates, in one way, the deviation of the vapor from the equation of state for an ideal gas, which is

$$pv = RT.$$

A constant temperature line following this relationship would be straight on logarithmic scales of p and v.

Thermodynamic Properties on T-s Coordinates   The most useful pair of coordinates for comparing power cycles on one plane is temperature and entropy, which were used already for the discussion of the Carnot cycle. **Figure 1.5** is an outline of a *T-s* chart for the thermodynamic properties of water. A more detailed chart above 32F is given by Keenan and Keyes [B.11], and is reproduced in the appendix. The shape of the liquid-vapor dome is not typical for all substances, except that the slope of the saturated liquid line is always positive. The slope must become zero at the critical point. For water, the slope of the saturated vapor line remains negative. Some other fluids exhibit a reversal of slope to positive over some part of the temperature range. The effect of the shape of the liquid-vapor dome on a power cycle will be discussed later.

On the *T-s* diagram, a different indication is apparent for the deviation of the vapor properties from those of an ideal gas. The enthalpy of an ideal gas would be a function of temperature only, not of pressure. It can be seen that this is approached at the lower pressures and higher temperatures.

Thermodynamic Properties on h-s Coordinates   While a temperature-entropy (*T-s*) diagram is extremely useful in the comparison of various thermodynamic cycles, an enthalpy-entropy (*h-s*) diagram is more frequently used for actual power calculations. Its main advantage is that energy transfers that result in changes of enthalpy in the working fluid are represented by vertical, linear distances rather than by areas.

The representation of thermodynamic properties on *h-s* coordinates is often referred to as a **Mollier diagram,** though that term is also applied by some to other coordinate systems such as enthalpy-concentration diagrams for solutions. The shape of the *h-s* diagram for water is outlined in **Fig. 1.6**. A chart with more detail of the upper part of this diagram, reproduced from Reference B. 23, is given in the appendix.

1.2.3 The Simple Rankine Cycle

The major thermodynamic advantage of a vapor cycle over a gas cycle is that isothermal heat transfer is a real possibility. Heat transfer processes similar to those of the Carnot cycle can be achieved within the wet-vapor region of thermodynamic properties where a change of enthalpy of the working fluid results in evaporation

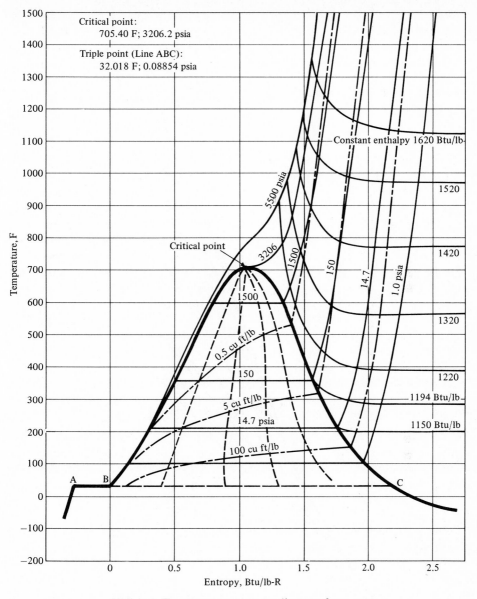

FIG. 1.5 Temperature-entropy diagram for water.

or condensation, but not in a temperature change. The temperature is governed only by the vapor pressure of the fluid. Isentropic expansion can be approached reasonably closely by the fluid in the vapor phase or when the dryness fraction is high. On the other hand, the approximation of isentropic compression in a very wet vapor is difficult if not completely impractical. Thus it is primarily in the deletion

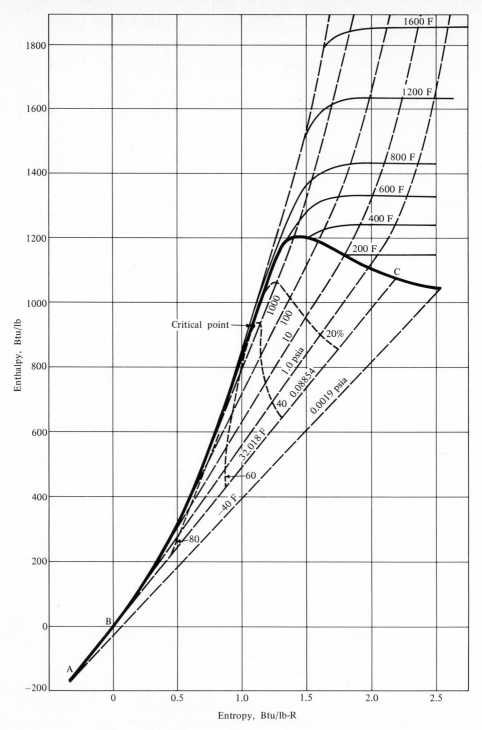

FIG. 1.6 Enthalpy-entropy (Mollier) diagram for water.

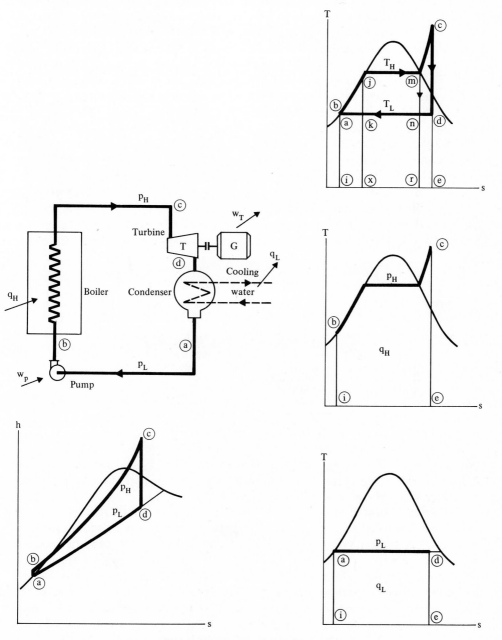

FIG. 1.7 The simple Rankine cycle.

of this compression process that the simple **Rankine cycle** is different from the Carnot cycle. **Figure 1.7** shows a schematic arrangement of equipment for the simple cycle and also the changes of state through the cycle on T-s and h-s coordinates.

The pump work to raise the pressure of the working fluid in the liquid phase from p_L to p_H will be much less than the work of compression for a vapor-liquid mixture over the same pressure range. Point *b* at the pump delivery conditions is only slightly above point *a* on the saturated liquid line. The addition of heat energy to this cycle is not entirely at constant temperature as it was for the Carnot cycle. The addition of heat energy is at constant pressure p_H. If it is stopped at point *m* on the saturated vapor line, then for water vapor, isentropic expansion from *m* at p_H to *n* at p_L puts the state point *n* in the wet-vapor region. This is not so important for a reciprocating expansion engine, although heat losses might be increased compared to losses from a dry vapor, but for a high-speed steam turbine, the inclusion of liquid-phase water in the expanding fluid introduces the mechanical problems of cavitation and possibly unbalanced forces. To avoid this, heat addition at p_H may be continued to some point *c* which is far enough into the superheated region to allow isentropic expansion to *d* at which point the vapor is still slightly superheated or only slightly wet. For any cycle, the thermodynamic efficiency is defined by Eqs. (1.1) and (1.2). When one pound of working fluid is considered,

$$\eta_{\text{th}} = \frac{\text{Useful work output per pound of fluid}}{\text{Energy supplied per pound of fluid}}.$$

From Eq. (1.3), this becomes

$$\eta_{\text{th}} = \frac{\text{Heat supplied} - \text{Heat rejected}}{\text{Heat supplied}}$$

$$= \frac{q_H - q_L}{q_H} \tag{1.8a}$$

$$= 1 - \frac{q_L}{q_H}. \tag{1.8b}$$

Both simple Rankine cycles *abmna* and *abcda* operate between heat absorption at p_H for which the temperature of vaporization is T_H and heat rejection at p_L for which the temperature of condensation is T_L. A large part of the heat transfer is isothermal simply because a change of phase is involved.

The **work output** from the cycle at the turbine can be calculated in terms of fluid properties by an application of the first law of thermodynamics to the turbine as an open stationary system. The energy of the fluid at *c* plus the work energy and heat energy added to the fluid between *c* and *d* must be equal to the energy of the fluid at *d*. It is recalled that for a flowing fluid, the product (pv) is an energy term sometimes called the flow energy. Let consistent units be used so that all energy terms are expressed as Btu per pound of flowing fluid, it being assumed that the equivalency factor, J, is applied where necessary to convert foot pounds to Btu. The term g_c is the dimensional constant for Newton's second law of motion as discussed in the foreword. Then

$$\left[u_c + p_c v_c + \frac{\bar{V}_c^2}{2g_c} + z_c \right] + q = \left[u_d + p_d v_d + \frac{\bar{V}_c^2}{2g_c} + z_d \right] + w_T \text{ Btu/lb.}$$

$$(1.9a)$$

By definition, $(u + pv)$ is the **specific enthalpy** (h). The **change of elevation** (z) through a machine such as the turbine is usually insignificant. No **heat energy** (q) is added or removed. Thus

$$h_c + \frac{\bar{V}_c^2}{2g_c} = h_d + \frac{\bar{V}_d^2}{2g_c} + w_T. \qquad (1.9b)$$

Under some circumstances, and usually for a stationary power plant, the change in kinetic energy from inlet to outlet can be ignored. Where that is not true, the reduction of Eq. (1.9b) to only the work term and a change in enthalpy can be accomplished by the use of the term **stagnation** enthalpy (h_0). It is defined as

$$h_0 = h + \frac{\bar{V}^2}{2g_c}.$$

This term will be found particularly useful in the consideration of certain gas cycles, and will be dealt with further in Section 1.3.9.

Where change in kinetic energy is negligible, or where stagnation enthalpy is used for enthalpy, the turbine work is

$$w_T = +(h_c - h_d) \text{ Btu/lb.} \qquad (1.10)$$

The **positive** sign, by our arbitrary convention, designates work **out** of the system.

The **gross work** of the turbine then is not affected by the left-hand side of the diagram, but only by the state points c and d. The **net work** of the cycle is less than the gross work by the amount of energy required by the pump. Equation (1.10) applied between points a and b with the same qualifications noted as for the case of the turbine shows that pump work is

$$w_P = -(h_b - h_a).$$

The **negative** sign designates work done **on** the working fluid. The **net work output** of the cycle is the difference between the work out and the work in. Thus

$$w_N = (w_T + w_P)$$
$$= (h_c - h_d) - (h_b - h_a). \qquad (1.11)$$

It is interesting to note that this is the same as

$$w_N = (h_c - h_b) - (h_d - h_a). \qquad (1.12)$$

Since heat addition to the cycle and heat rejection from the cycle are assumed to take place in constant pressure, reversible processes, the heat transferred in each case is equivalent to the difference in enthalpies of the endpoints of the process. In other words, Eq. (1.12) is the same as Eq. (1.3):

$$w_N = (\text{Heat supplied}) - (\text{Heat rejected}).$$

This was derived from the first law of thermodynamics applied to the system of the working fluid completing a whole cycle.

The heat added, q_H, and the heat rejected, q_L, being equivalent to differences in enthalpy of the working fluid, are represented by areas under the two constant pressure lines on *T-s* coordinates, as shown on the separate diagrams of Fig. 1.7. Thus the net work of the cycle is represented by the difference between the area *ibce* and the area *iade*. The net work of the cycle is therefore represented by the enclosed area **abcda**.

The purpose of most steam power plants is the production of electric power through a generator. Therefore a practical inverse indication of the cycle efficiency is the **heat rate,** usually expressed in Btu per kilowatt-hour (kwhr). Where the output of the plant is shaft work, the heat rate might be expressed in Btu per horsepower-hour (hphr). Care must be taken to specify whether or not the engine (or turbine) and the generator (or machine) efficiencies have been taken into account in the calculation of the power-plant heat rate. Where they have not, the value is sometimes called the **indicated heat rate**. The heat rate is easily calculated from the energy equivalent of a kilowatt-hour or a horsepower-hour:

$$\text{Heat rate} = 3413/\eta_{\text{th}} \text{ Btu/kwhr} = 2544/\eta_{\text{th}} \text{ Btu/hphr.}$$

Plant performance expressed in these terms has more significance to plant operators than cycle efficiency expressed as a percent of heat supplied. Where the **heating value** of a fuel is known and a **boiler efficiency** is known or assumed, fuel rates are easily approximated from heat rates.

The **liquid pump** work for any fluid is very small compared to the other energy quantities of the cycle. For water, it can be determined from Table 3 of the ASME steam tables. A value close enough to indicate the order of magnitude is obtained by assuming a constant specific volume (incompressible fluid) and finding the change in the flow energy term (pv product) over the pressure range desired.

The ideal cycle assumed, for which each process is quasistatic and reversible, is a reversible cycle. But it is immediately obvious from the *T-s* diagram that the Rankine cycle **abjmna** is less efficient than the Carnot cycle **kjmnk**. The heat rejected, which must be considered a loss for the simple power cycle, is a larger fraction of the heat supplied in the Rankine cycle than in the Carnot cycle, because from **b** to **j** , the heat is supplied at an average temperature below T_H; the ratio of the resulting trapezoidal work area to the heat rejected area **ianr** is less than it would be for a rectangular work area of the same height. Also from the diagram, it is apparent that if **c** were high enough, cycle **abcda** could be more efficient than the Carnot cycle **kjnmk**, but not more efficient than a Carnot cycle between the temperatures T_L and T_c.

Clearly, the efficiency of the simple Rankine cycle for a particular working fluid is affected by the pressures of vaporization and condensation and the number of degrees of superheat at point **c** . Pressure and temperature at **c** are referred to as **throttle conditions,** that is, the conditions at which the vapor is supplied to the turbine or engine through a throttling control valve. (In the ideal cycle, no losses

are assumed in the supply lines and valves). It would seem that the throttle temperature and pressure should both be as high as possible to increase the cycle efficiency.
Figure 1.8, with a limited throttle temperature, shows the effect of pressure alone
on two simple Rankine cycles. The condensation pressure and temperature are the
same for both. For water vapor, the diagram shows that the work areas remain
practically the same for an upper pressure of p_H or p_{HH}. However, the heat rejected
is obviously less for the higher pressure cycle, and therefore its thermal efficiency
will be higher. One would expect the thermal efficiency to be better for the higher
pressure cycle, since the average temperature of heat supply has been raised.

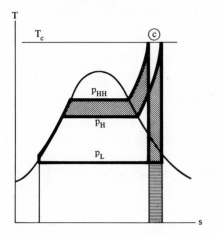

FIG. 1.8 The effect of pressure on a
simple Rankine cycle.

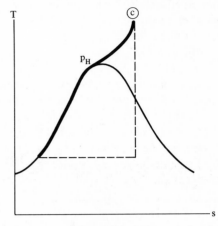

FIG. 1.9 Supercritical addition of heat.

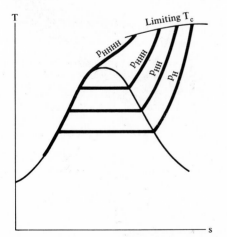

FIG. 1.10 The effect of throttle pressure on temperature limit.

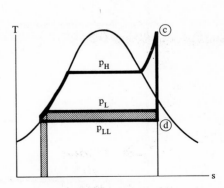

FIG. 1.11 The effect of condenser pressure on a simple Rankine cycle.

Purely from the point of view of cycle efficiency, the throttle pressure should be as high as possible. In fact, today, some steam plants do operate on **supercritical** cycles, which means that the pressure line **bc** is above the critical pressure for steam (3206 psia). Such a cycle, with expansion directly to exhaust pressure, is shown in **Fig. 1.9.** The disadvantages of a high-pressure cycle are (a) that expansion directly to exhaust pressure produces very low quality steam in the last stages of expansion, and (b) that the strength of the high-pressure elements in the system must be greater, which will increase first cost.

The limitations of throttle pressure and throttle temperature for steam plants cannot be separated. Metal creep becomes significant at high temperatures, thus limiting stresses for given materials. The interdependence of the two is shown in **Fig. 1.10**, the range of permissible temperatures being about 1000 to 1100 F (1460 to 1560 R) at the present time.

The effect of a reduction in the condensation pressure from p_L to p_{LL} is shown in **Fig. 1.11**. The increase in the work area is very nearly the same as the increase in the heat-supply area, so that the ratio w_N/q_H (which is the thermal efficiency) is increased. The potential output of the working element of the cycle, the turbine or engine, is clearly increased by a reduction in p_L. At the same time, an increase in the specific volume of the fluid at **d** must be recognized, and the size or speed of the machine would have to be greater for the greater potential to be realized. The pressure at **d** may be atmospheric or above if the exhaust steam is to be used

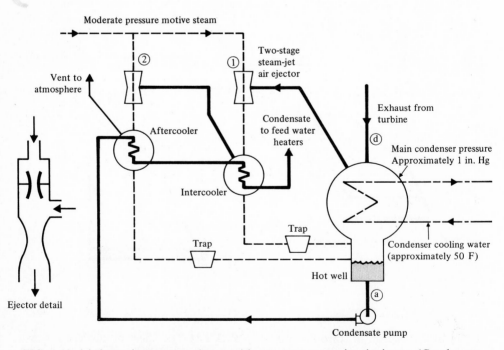

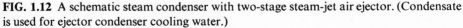

FIG. 1.12 A schematic steam condenser with two-stage steam-jet air ejector. (Condensate is used for ejector condenser cooling water.)

in some heating process. When that is not the case, the **condenser** pressure is kept as low as possible. Cooling water is passed through tubes within the condenser shell and the rapid condensation of the steam can produce a very high vacuum, the absolute pressure being the saturation pressure at the temperature of the leaving condensate. However, that vacuum can be maintained continuously only if non-condensable gases are removed. This is usually accomplished by the use of a **steam-jet air ejector** as shown in **Fig. 1.12**. For high vacuums, this must have at least two stages. The medium pressure **motive steam** or **extractive steam** for the ejector nozzles may be throttled from the high pressure supply or bled from the turbine. Rather than use additional cooling water, condensate from the main condensers can be utilized as shown for cooling water in closed heat exchangers. This condenses both the motive steam and the entrained vapor inevitably removed from the main condenser shell. The condensate from both the intercondenser and the aftercondenser are returned to the hotwell of the main condenser through traps. In place of steam-jet ejectors, rotary-type air pumps or else air-jet ejectors are sometimes used.

1.2.4 Turbine Efficiency

It should be recalled that, by the definition of entropy,

$$ds = \left(\frac{dq}{T}\right)_{rev}.$$

In the ideal Rankine cycle, the expansion process through the turbine or engine is reversible and adiabatic. It is therefore also **isentropic.** The Clausius inequality corollary of the second law of thermodynamics states that, in a cycle,

$$\oint\left(\frac{dq}{T}\right) \leq 0,$$

and from this it can be shown that, in a process from point ① to point ②,

$$\int_1^2 \frac{dq}{T} \leq s_2 - s_1.$$

Thus, for an adiabatic process,

$$s_2 \geq s_1.$$

This is true whether the work of the process is done on the fluid (work in) or by the fluid (work out).

Any irreversibility in the turbine expansion process therefore moves the state point to the right of the isentropic line on a *T-s* or an *h-s* diagram. The *T-s* diagram of **Fig. 1.13** shows that the heat rejection from a Rankine cycle is thereby increased, but it does not show directly that the work done is decreased by the same amount. This follows from Eq. (1.3):

Work done = Heat supplied — Heat rejected.

On the *h-s* diagram of Fig. 1.13, it is immediately clear that an increase in entropy during expansion must decrease the work output, since the change of enthalpy of the fluid is reduced. In this text, a symbol with a "prime" such as d' will indicate

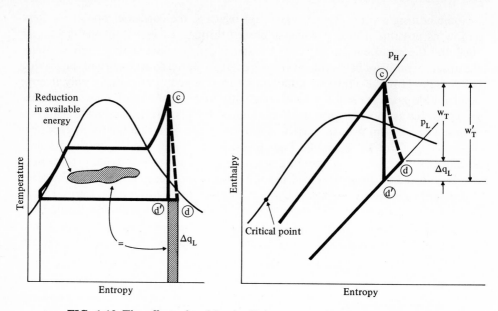

FIG. 1.13 The effect of turbine inefficiency on a simple Rankine cycle.

the completion of an ideal reversible process. Thus the expansion from **c** to **d'** is isentropic and $(h_c - h_{d'})$ represents the maximum possible work output for adiabatic expansion from point **c** to p_L. Point **d** is a possible state point after irreversible expansion from **c** to p_L. The change of enthalpy $(h_c - h_d)$ is less than the ideal enthalpy change $(h_c - h_{d'})$.

The **turbine efficiency** (without consideration of mechanical friction losses) is defined in terms of these enthalpy changes:

$$\eta_T = \frac{\text{Work from actual expansion process}}{\text{Work from isentropic expansion}}$$

$$= \frac{w_T}{w_{T'}} = \frac{h_c - h_d}{h_c - h_{d'}}. \tag{1.13}$$

On a Mollier diagram, the state point at the end of isentropic expansion to a given back pressure is easily located if the initial conditions are known. For a known turbine efficiency, the actual end conditions are readily found. If tables are to be used for greater accuracy or because an *h-s* diagram is not available for the fluid used, then s_d must be equated to s_c to find the dryness fraction, or superheat, and the enthalpy at the end of isentropic expansion.

In Fig. 1.13, the line **cd** on the *T-s* coordinates is shown as a broken line for two reasons: (1) the area bounded by an irreversible line on these coordinates has no significance, and (2) only the endpoints are known usually and not the actual path. On the *h-s* coordinates as well, the broken line is used because the path is not known.

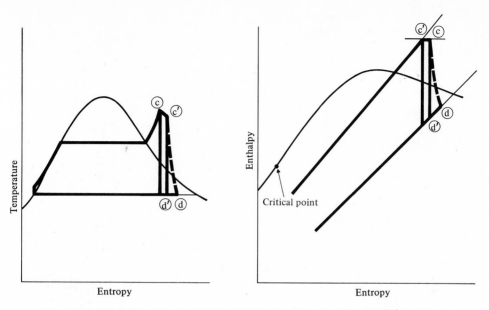

FIG. 1.14 The effect of piping friction loss on throttle conditions.

The difference between the actual turbine work (w_T) and the ideal turbine work ($w_{T'}$), as shown on the *h-s* diagram of Fig. 1.13, is sometimes referred to as a "loss." A particular definition of "loss" could justify this designation, but it seems better to avoid it. The process *cd* was designated as adiabatic, and even in a real turbine, the flow rate is usually so rapid that the heat lost per pound of fluid to the surroundings between *c* and *d* is not appreciable. The enthalpy difference between $h_{d'}$ and h_d was not converted to useful work, but still was not lost from the fluid. As far as the turbine is concerned, that amount of energy was not available. This is confirmed by the increase in area representing this amount of energy below the p_L line on the *T-s* diagram. The reason for the "loss" was principally fluid friction in which available energy was dissipated. In a more complex system, this energy or some of it may be reclaimed for purposes other than work at the shaft of the turbine shown. It is interesting to note that for steam, the inefficiency of the turbine has improved the vapor quality at *d* by raising the enthalpy at that point.

The **pressure drop** resulting from fluid friction **in the piping** from the boiler to the turbine reduces the available energy for expansion. As shown in **Fig. 1.14**, with perfect insulation and therefore constant enthalpy, the throttle condition is moved to the right and downward on the *T-s* diagram and to the right on the *h-s* diagram by the piping pressure drop.

1.2.5 Rankine Cycle with Reheat

The basic simple Rankine cycle can be greatly improved for a steam power plant. The improvements require additional equipment and piping and are generally used in relatively large installations. However, they may not be economically justifiable

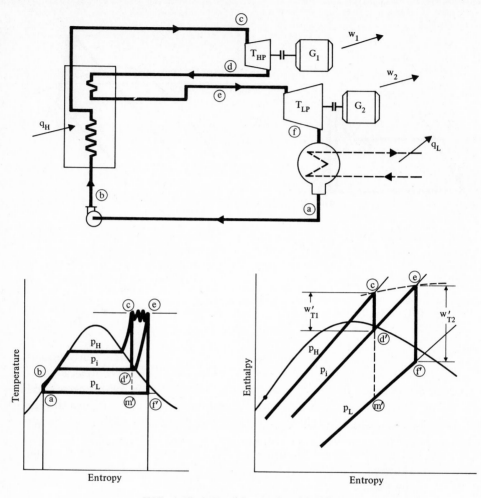

FIG. 1.15 A Rankine cycle with reheat.

in small plants, particularly where moderate top temperatures and pressures are used.

State points easily found on a Mollier (h-s) diagram for steam show that expansion, even with turbine inefficiency considered, from say 1800 psi or 2400 psi, and about 1000 F to an exhaust pressure less than 2 in. Hg abs. would produce wet vapor of very low quality. No more than 7 to 10 percent liquid can be tolerated in the last stages of a large steam turbine, because liquid droplets can cause rapid erosion of both the moving and the stationary blades, as well as reducing the internal efficiency of the turbine expansion process.

The low quality of wet vapor at the end of the expansion can be avoided by the **reheat cycle** shown in **Fig. 1.15**. In this cycle, the steam is expanded from *c* at p_H to **d** at an intermediate pressure p_i. At this pressure, the steam is taken from the turbine to the original boiler or a different heat source and reheated to a super-heat temperature T_e, which is approximately the same as T_c. An expansion to **f** at the final pressure P_L puts the final state point reasonably near a dry saturated vapor condition.

The effect of reheating on cycle efficiency is not immediately obvious from either the *T-s* or the *h-s* diagrams. Depending on conditions of operation, it could theoretically produce either a slight increase or a slight decrease in cycle efficiency. Nevertheless, in addition to the improved exhaust quality, another advantage is obvious: more work is delivered per pound of working fluid circulated. This is noted in the increased work area of the *T-s* diagram and in the greater enthalpy drop apparent on the *h-s* diagram because of the divergence of the constant pres-sure lines. For a given power rating of the plant, auxiliaries such as the pump and some pipe sizes can be smaller, although more complex piping is required.

A very large number of reheats would theoretically allow isothermal heat transfer to be approached between points **c** and **e** as shown by the saw-tooth line on the *T-s* diagram. Ideal cycle efficiency would obviously be improved, but losses would be increased. More than one reheat is seldom economical except in some nearly-critical or supercritical pressure systems.

Each major stage of expansion may actually require a multistage turbine. (See Section 3.4.3.) The two major stages of expansion can be accomplished in sepa-rate turbines, as shown, or with the two rotors on a common shaft. There are advantages to the separate shaft arrangement, because the two turbines can be designed for different operating speeds.

The reheat cycle is similar in some respects to two-stage compression discussed in Section 3.2.4, and to two-stage vapor refrigeration cycles discussed in Chapter 4, Section 2. Application of the gas laws to the problem of two-stage gas compression or expansion allows an optimum intermediate pressure value to be selected for two given pressures p_L and p_H; this will be shown in Section 3.2.4. The complexity of the property relationships for a vapor make such a generalization impossible for vapor power reheat cycles or two-stage vapor refrigeration cycles.

1.2.6 Rankine Cycle with Regeneration

In many thermodynamic cycles, heat can be transferred usefully from the working fluid at one part of the cycle to working fluid at a different part of the same cycle. This internal transfer of heat in a power cycle is called **regeneration,** and the heat transfer apparatus is a **regenerator.** The term **recuperator** is sometimes used, par-ticularly where it is desirable to distinguish between counter-flow and reversed-flow heat exchangers. (See Section 4.7.3.)

Regeneration can be added to a simple Rankine cycle to improve the cycle efficiency. In theory, the efficiency could be equal to that of the Carnot cycle. **Figure 1.16** shows a highly idealized system in which the feed water returning from the condenser is brought in contact with the turbine in such a way that there is enough heat transfer from the expanding vapor to preheat the feed water right up to T_H.

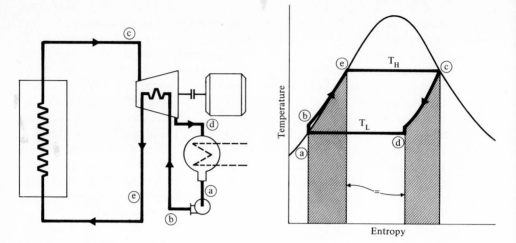

FIG. 1.16 Ideal regeneration in a Rankine cycle with saturated throttle conditions.

On the *T-s* diagram, the area under the line *cd* would be identical to the area under the line *ae*. The only external heat transfer would be along the line *ec* in the boiler and the line *da* in the condenser, and it would be isothermal in each. The work area *abecda* would be the same as the area of a rectangle of the same height on the base *ad*, and so the thermal efficiency would be that of the Carnot cycle between T_H and T_L.

Such perfect heat transfer with no terminal temperature difference would be impossible. Further, the extremely high moisture content of the vapor after full expansion would be undesirable for reasons already noted. However, the principle of feed water heating by regeneration can be applied through the process of **extraction.** In this procedure, a relatively small percentage of the vapor flowing through the turbine is bled off from the turbine at one or more intermediate pressures between the supply and exhaust pressures, and it is used to preheat feedwater in closed or open heat exchangers. The extraction rates are controlled by condensate levels. The enthalpy change in condensation of this steam raises the temperature of part of or all the feed water close to the saturation temperature of the extracted steam.

Figures 1.17 and **1.18** show two slightly different systems in which feed water heating is accomplished by extraction steam in two **closed heaters;** the condensing vapor is separated by tube walls from the pressurized feed water. In the **cascaded** system of Fig. 1.18, the condensate from the second heater at condition *m* is allowed to flow to the first heater through a **trap** which prevents vapor passage. Compared to the alternative system of Fig. 1.17, one pump is saved, but the *T-s* diagram is not improved quite so much.

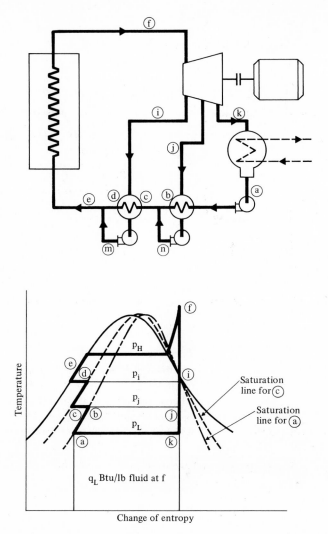

FIG. 1.17 A Rankine cycle with regeneration using two independent closed heaters.

Note that the amount of working fluid reaching *k* is less than at *j*, which in turn is less than the amount at *i* or *f*. An attempt is made to show this on the *T-s* diagrams through the superposition of saturation lines representing the appropriate quantities. The position of *k* relative to the saturation lines does show its quality correctly. However, the change of enthalpy in the condenser per pound of fluid originally supplied at *f* is not represented by the area under the line *ky*. Because there is less fluid at *k* than at *f*, the area under the line *ka* represents the condenser heat transfer in Btu per pound of working fluid at *f*. The symbol *ṁ*

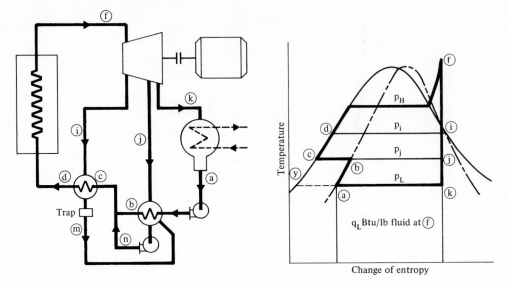

FIG. 1.18 A Rankine cycle with regeneration using two cascaded closed heaters.

will be used for mass rate of flow. The length of the line **ka** will be to that of **ky** as $\dot{m}_k$ is to $\dot{m}_f$. The composite *T-s* diagrams are drawn so that areas represent energy in Btu per pound of fluid at the throttle which is point **f**. The base is "Δs" rather than "s" above an arbitrary datum, because the saturation lines shown shift as the fluid quantities change. This is necessary so that the position of each point relative to the appropriate saturation lines will indicate the dryness fraction correctly.

The composite *T-s* diagrams show that much less heat is rejected from the cycle than from a comparable simple Rankine cycle. Although work per pound of steam circulated is reduced, efficiency is increased. In very large plants, as many as ten extraction points are used. If an infinite number could be used, the shape of the left side of the *T-s* diagram of Fig. 1.17 would approach that of the Carnot cycle.

The extraction quantities must be determined by mass and energy balances on the individual heat exchangers. For the system of **Fig. 1.18** only:

$$\dot{m}_d(h_d - h_c) = \dot{m}_i(h_i - h_m),$$
$$(\dot{m}_d - \dot{m}_i - \dot{m}_j)(h_b - h_a) = \dot{m}_j(h_j - h_n) + \dot{m}_i(h_m - h_n).$$

The fluid at points **a**, **b**, **c**, **d**, **m**, and **n** can be assumed to be saturated liquid at their respective pressures. The relatively small pump work has been ignored to simplify the diagrams and the calculations. Enthalpy values at **i** and **j** can be obtained from tables and calculations or from a Mollier diagram. With $\dot{m}_d$ (which is the same as $\dot{m}_f$) known in pounds per hour or assumed unity, only the two extraction rates $\dot{m}_i$ and $\dot{m}_j$ are unknown in the two equations.

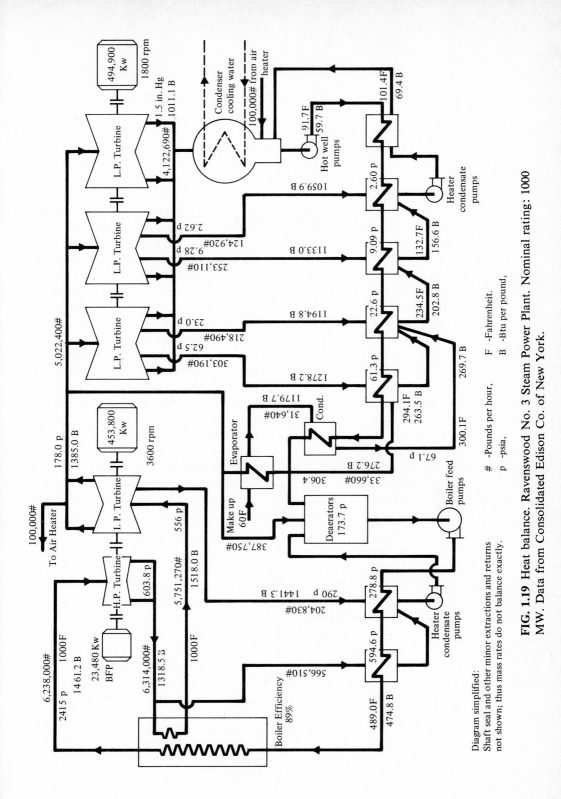

FIG. 1.19 Heat balance. Ravenswood No. 3 Steam Power Plant. Nominal rating: 1000 MW. Data from Consolidated Edison Co. of New York.

Diagram simplified:
Shaft seal and other minor extractions and returns not shown; thus mass rates do not balance exactly.

\# -Pounds per hour,   F -Fahrenheit.
p -psia,   B -Btu per pound.

The ideal thermal efficiency for the cycle of Fig. 1.18 would be

$$\eta_{\text{th}} = \frac{\dot{m}_f(h_f - h_i) + (\dot{m}_f - \dot{m}_i)(h_i - h_j) + (\dot{m}_f - \dot{m}_i - \dot{m}_j)(h_j - h_k)}{\dot{m}_f(h_f - h_d)}.$$

One or more **open heaters**, also called **direct contact heaters**, might be used for regeneration. In these, the condensing vapors and the feed water mix intimately at the same pressure so that heat transfer is much more effective. Also, an open heater can act as a **deaerator** when properly vented to allow the escape of previously dissolved and entrained gases. Oxygen and carbon dioxide in boiler water accelerate metal corrosion. The calculations for flow rates for an open heater are essentially the same as for a closed heater. All leaving fluids will be in equilibrium.

From the preceding discussions, it is clear that, for initial steam pressures over about 1200 psi, a reheat cycle is necessary to avoid unreasonably wet conditions during the expansion process, and for large power plants, regeneration is necessary for economy. **Figure 1.19** shows the vapor cycle for a very large power plant with a high-pressure turbine, reheat, an intermediate-pressure turbine, a low-pressure turbine with a separate shaft, and a total of seven points of steam extraction. All feed water heaters are of the closed type except for the deaerating heater. An **evaporator,** not previously mentioned, is used to vaporize **make-up** water that must replace leakage. The high-pressure and intermediate-pressure turbines are both balanced-flow types with the steam entering at the midpoint and expanding axially in both directions, and the low pressure unit is made up of three balanced-flow turbines on a common shaft, all discharging to the condenser. These split-flow turbines not only balance end thrust forces, but also reduce the extremely large radius necessary in later stages to accommodate the tremendous increase in specific volume as exhaust pressure is approached. Note that in this particular plant, the boiler feed pumps, which require a considerable amount of power, are driven by the output of a generator that is electrically independent of the main power generators. The whole diagram is highly schematic.

1.2.7 Alternative Working Fluids

The thermodynamic properties of water put limitations on the efficiency of a steam cycle even with the complexities of reheat and regeneration. For relatively small power plants, these complexities are not warranted, and at the same time, the cost of other fluids might not be prohibitive.

The greatest disadvantages of water as a working fluid are its high pressure and relatively low temperature at the critical point, the negative slope of the saturated vapor line on *T-s* coordinates, and its low molecular weight. Some other substances do have a vertical or positively sloped saturated vapor line. The slope is governed primarily by the number of atoms in the molecule [A.40], though that is not a completely reliable criterion.

The shapes of the saturation lines for several substances are shown on *T-s* coordinates in **Fig. 1.20**. Not all these are practical working fluids over wide ranges of temperature. For instance, the saturation pressure of mercury (Hg) becomes

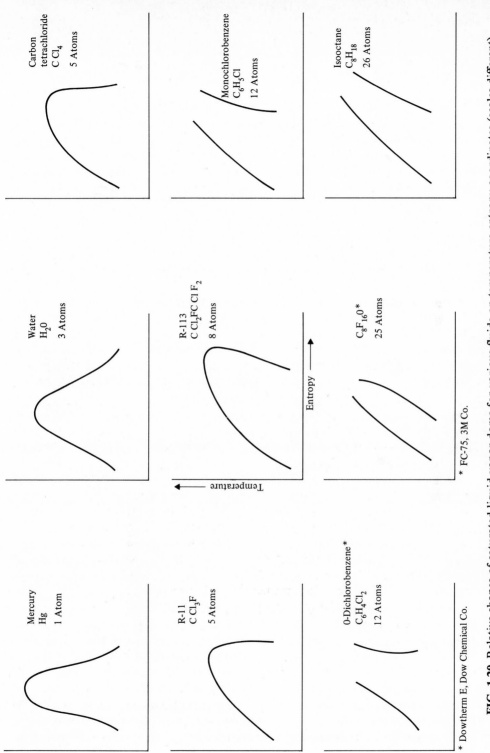

FIG. 1.20 Relative shapes of saturated liquid-vapor dome for various fluids on temperature-entropy coordinates (scales different).

Mercury
Hg
1 Atom

Water
H₂0
3 Atoms

Carbon
tetrachloride
C Cl₄
5 Atoms

R-11
C Cl₃F
5 Atoms

R-113
C Cl₂F C Cl F₂
8 Atoms

Monochlorobenzene
C₆H₅Cl
12 Atoms

0-Dichlorobenzene*
C₆H₄Cl₂
12 Atoms

C₈F₁₆0*
25 Atoms

Isooctane
C₈H₁₈
26 Atoms

Entropy ⟶

Temperature ⟵

* Dowtherm E, Dow Chemical Co.

* FC-75, 3M Co.

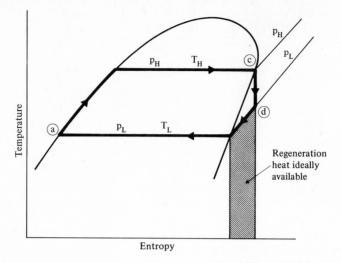

FIG. 1.21 The possibility of regeneration in a simple Rankine cycle if the saturated vapor line has a negative slope.

extremely small, and the specific volume becomes extremely large at normal ambient temperatures. Carbon tetrachloride (CCl_4) will decompose even below 500 F, a moderate temperature for reasonable cycle efficiency. Similarly, the halogenated hydrocarbons such as R-11 and R-113, developed particularly for refrigeration cycles (see Chapter 4), are unstable at the temperatures usually desired for power cycles, except that they may be quite satisfactory for relatively low temperature, waste heat recovery [A.22] and [A.28].

A nearly vertical saturated vapor line would mean that superheating above the saturation temperature at which the fluid is evaporated would not be necessary to avoid the wet vapor region at the end of isentropic expansion. While a positively sloped saturation line would guarantee dry vapor throughout the expansion process, **Fig. 1.21** shows that an increased heat rejection is unavoidable for the simple Rankine cycle. However, the temperature at the end of expansion (T_d) is higher than the temperature of condensation (T_L), so that some regeneration is possible without extraction if the condensate exchanges heat with the superheated exhaust vapor. Such regeneration is sometimes called "feed heating."

High **molecular weight** of the working fluid is desirable primarily to reduce the required blade speed of a turbine. It is possible to show that the nozzle velocity for a given temperature drop in one stage is approximately inversely proportional to the square root of the molecular weight of the fluid. The optimum blade speed is about one-half the nozzle velocity. Where a steam turbine usually requires several stages to keep the enthalpy drop per stage low, and thus keep the blade speed within the bounds of reasonable centrifugal force, a turbine using a high molecular-weight fluid might need only one stage. Although nozzle velocities are lower for high molecular-weight fluids, the pounds per minute circulated must be greater for a given output because enthalpy drop per pound is approximately inversely proportional to the molecular weight.

The search for alternative working fluids has been going on for a long time. In a book published in 1912 [B.5], W. D. Ennis discussed nine substances besides water that had been suggested and tried by that time. Mercury was not then considered. Only water had had wide commercial success. Since then, many factors have changed. First, much more is known of the thermodynamic properties of most natural substances, and many synthetic compounds have been extensively studied. Second, vastly improved technology has made sealed systems more reliable. Third, the thermodynamics and fluid mechanics of turbines are better understood so that they can be designed for particular fluids and conditions. Fourth, the conventional range of temperatures for a stationary steam power plant will not always apply to unusual situations such as utilization of solar energy, waste heat recovery, and space flight, and unforeseen requirements of the future. Finally, in many systems, particularly small ones, the first cost of working fluid and auxiliaries may be much less important than either cycle efficiency or specific output (kilowatts per pound or per cubic foot of the power plant).

1.2.8 Binary Vapor Cycles

The high critical pressure of water and the value of its critical temperature, which is below the permissible temperature limit for boilers, are undesirable characteristics on the high side of the cycle; its very low vapor pressure at usual condensation temperature is a disadvantage on the low side. No single fluid spans the maximum possible range without some undesirable features. For this reason, cycles have been proposed and some have been built that utilize more than one fluid, each over its most satisfactory range. A **binary** vapor power cycle is one that employs two different working fluids. This is a special case of the thermally interdependent systems described in Section 1.4.2. It is similar to the binary vapor refrigeration cycles described in Chapter 4.

The earliest proposals for binary vapor cycles, over 100 years ago [B.4], were for water on the high side and a more volatile fluid (lower atmospheric boiling point) such as ether on the low side. The maximum temperatures and pressures then possible were not high for the properties of water, but the maintenance of a high condenser vacuum remained a problem for some time. The water, condensing at a temperature over 200 F, vaporized the second fluid that could then be condensed at ambient temperature after expansion to a positive gauge pressure. The development in this century of highly effective steam jet air ejectors, described previously, has made this unnecessary.

The one working fluid that has been used in a binary system with water in a few large power plants in this country is mercury. Its critical temperature, about 1649 F at 2646 psia, is above permissible temperatures to date so that superheating is not required, and its very steep liquid saturation line (low specific heat of the liquid) makes regeneration unnecessary. **Figure 1.22** presents a schematic layout for

THROTTLING → AN ADIABATIC, IRREVERSIBLE PROCESS IN WHICH A GAS EXPANDS BY PASSING FROM ONE CHAMBER TO ANOTHER CHAMBER WHICH IS AT A LOWER PRESSURE THAN THE FIRST CHAMBER.

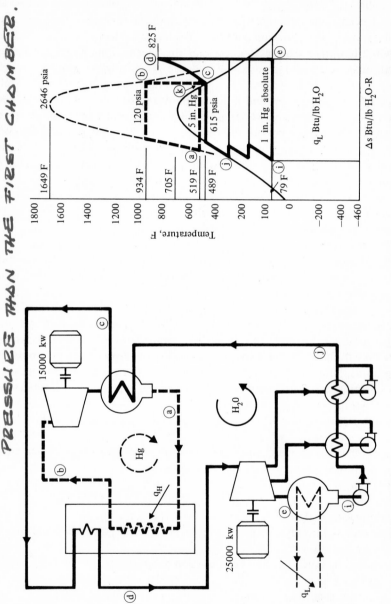

FIG. 1.22 Binary vapor cycle (mercury/water). Simplified schematic Schiller Station (1950).

a binary system of water and mercury. The condenser of the mercury cycle is the boiler of the steam cycle. The heat equivalent of the enthalpy change in condensation of the mercury from **k** to **a** is transferred to the water which is heated and evaporated from **j** to **c**. To obtain maximum efficiency in the lower cycle, the steam is superheated in the main boiler and extraction is used for regeneration. Temperature ranges can readily be shown on the accompanying *T-s* diagram with the mercury cycle superimposed on the steam cycle. However, the horizontal scale is different for the two substances, because the latent heat of vaporization for mercury is much less than that of water. About 13 lb of mercury must be circulated per pound of water to accomplish the heat transfer shown.

While the schematic diagram is a highly simplified presentation and the *T-s* diagram shows two ideal cycles, the temperatures and pressures noted are approximately those of an actual power plant, the Schiller Station of the Public Service Company, built in 1950 [A.13]. Two mercury turbines with separate electric generators are employed to give the gross ratings shown. A temperature difference of 20 to 30 F degrees is maintained in the mercury-condenser/steam-boiler so that high rates of heat transfer are possible. In some mercury cycles, the feed pump is done away with, as shown, since the static head of the heavy liquid is enough to maintain circulation if the condenser is placed above the boiler. Pressure loss between the boiler drum delivering saturated mercury and the turbine throttle produces several degrees of superheat.

The initial cost of mercury and the precautions necessary because of its extreme toxicity militate against its use unless decided economies are realized through the improved cycle efficiency. More recent improvements in steam-cycle efficiency through the use of high pressures and temperatures, and particularly through more effective regenerative heating have made the complexities of the mercury/steam binary cycle economically noncompetitive. Still, the thermodynamic advantages of a binary cycle under certain conditions cannot be denied. Heat is added only from **a** to **b** in the mercury cycle and from **c** to **d** in the steam cycle. Heat is rejected only in the steam cycle from **e** to **i**. With regeneration in the steam cycle, the shape of the diagram is reasonably close to that of the Carnot cycle. A great advantage of mercury is that, being an element, it is completely stable. Other substances have been suggested, notably diphenyloxide. Under various demands of the future, innumerable combinations may prove to be practical.

1.3 GAS CYCLES

In the introduction to this chapter, cycle efficiency and the Carnot cycle were discussed; the nature of the working fluid was purposely not specified. In Section 2 of the chapter, vapor cycles were examined in which a change of phase from liquid to vapor and back again took place. In this third Section of the chapter, gas cycles will be described. No change of phase will occur in the working fluid. The properties of real gases used for such cycles may vary only slightly from the ideal gas laws, or may vary greatly if the fluid is in a vapor state relatively close to the saturated vapor conditions. As a general rule, deviations will be greater at lower temperatures and higher pressures. (Ionization of gases at high temperatures will not be considered here.) The reader may wish to refer to a textbook on the fundamentals of thermodynamics, such as those listed at the end of the introduction to this book, for some discussion of the properties of a perfect gas as well as definitions of properties such as internal energy (u), enthalpy (h), and entropy (s). Certain relationships will be used here without a formal proof.

For the development of expressions for thermal efficiency of **ideal gas cycles**, an idealized working fluid will be assumed. This hypothetical gas will be **frictionless** (nonviscous) and will be assumed to follow the **perfect gas law**:

$$pv = RT. \qquad (1.14)$$

A definition of a perfect gas usually includes, as well, the stipulation that internal energy and enthalpy are functions of temperature only. The hypothetical gas that is assumed for the ideal gas cycles to be discussed is limited further by the assumption that the **specific heat at constant pressure** (c_p) does not vary with temperature. Now it can be shown that for a perfect gas,

$$c_p - c_v = R/J, \qquad (1.15)$$

where c_v is the **specific heat at constant volume** and R is the individual gas constant, ft-lb$_f$/lb$_m$-degree R. The **ratio** c_p/c_v will be designated by the symbol γ. With c_p constant, Eq. (1.15) shows that c_v is constant also, and therefore γ will be assumed constant for this hypothetical working fluid. Later, the effect of changes in specific heat and therefore in γ will be discussed.

From the perfect gas law, it can be seen that an **isothermal** process will follow the relationship

$$pv = \text{const.} \qquad (1.16)$$

It is readily shown that a reversible adiabatic or **isentropic** process for a perfect gas will follow:

$$pv^\gamma = \text{const.} \qquad (1.17)$$

Through these assumptions and relationships, expressions for heat gain, heat loss, work done, and efficiency are made quite simple for the various ideal gas cycles to be discussed.

Because the working fluid is so often air, a further assumption is sometimes made. The value of c_p can be taken as 0.24 Btu per pound per degree F which is the value for dry air at standard atmospheric pressure and temperature. The actual value of γ for standard air is 1.40 which is the ideal value for a diatomic gas. A cycle based on the hypothetical fluid which has properties similar to those of standard air and has constant c_p and γ values is called an "**air-standard cycle**."

1.3.1 *Constant Pressure and Constant Volume Lines on T-s or h-s Coordinates*

The definition of entropy is given by the expression

$$\Delta S = \int \left(\frac{dQ}{T}\right)_{rev}.$$

From the first law of thermodynamics, the heat added to the system, dQ, must be $dE + dW$. For a fluid for which the only work is the result of pressure forces

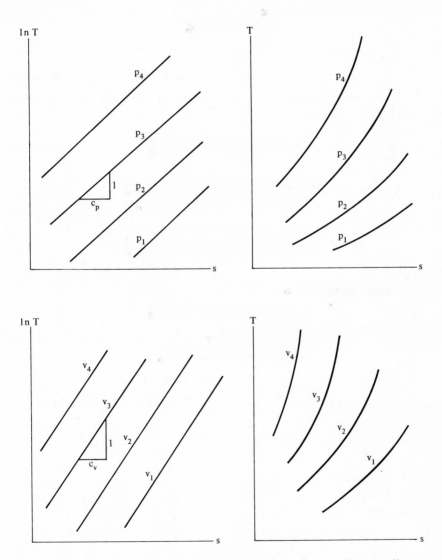

FIG. 1.23 Constant pressure and constant volume lines on *T-s* coordinates.

acting through a change of volume and the only energy change is internal energy change, the infinitesimal specific entropy change must be

$$ds = \frac{du + p\,dv}{T}.$$

The definition of enthalpy is given by the expression

$$h = u + pv$$

which, for a perfect gas, becomes

$$h = u + RT.$$

Internal energy and enthalpy are functions of temperature only for a perfect gas. Thus *T-s*, *h-s*, and *u-s* diagrams are essentially the same, and the choice of ordinate for an entropy abscissa is a matter of convenience. Diagrams on any of the three pairs of coordinates will be similar. This was not found to be true for a vapor, as shown by the striking difference in *T*-s and *h*-s diagrams of Figs. 1.5 and 1.6.

To continue with the implications of the perfect-gas relationships,

$$dh = du + p\,dv + v\,dp$$

so that

$$ds = \frac{dh - v\,dp}{T} = c_p \frac{dT}{T} - R\frac{dp}{p}.$$

And also,

$$ds = \frac{du + p\,dv}{T} = c_v \frac{dT}{T} + R\frac{dv}{v}.$$

Above a given datum at which entropy is zero, the value of specific entropy for a **constant pressure line** will be

$$(s)_p = c_p \ln T + \text{a constant.}$$

The value along a **constant volume line** will be

$$(s)_v = c_v \ln T + \text{a constant.}$$

Thus if log *T*, log *h*, or log *u* were the ordinate and *s* the abscissa, constant pressure or constant volume lines would be straight and parallel, the slope being given by the respective specific heats. This is shown in **Fig. 1.23** which also shows the divergence of these lines if linear scales are used. At any state point, the slope of the constant volume line will be steeper than the slope of the constant pressure line.

1.3.2 The Carnot Gas Cycle

If an ideal gas were the working fluid for the schematic Carnot cycle of Fig. 1.1, the expression for cycle efficiency could be put in another form by the use of Eqs. (1.14) and (1.17). These produce the following expression for isentropic compression or expansion:

$$\frac{T_2}{T_1} = \left(\frac{p_2}{p_1}\right)^{(\gamma-1)/\gamma} = \left(\frac{v_1}{v_2}\right)^{\gamma-1}. \tag{1.18}$$

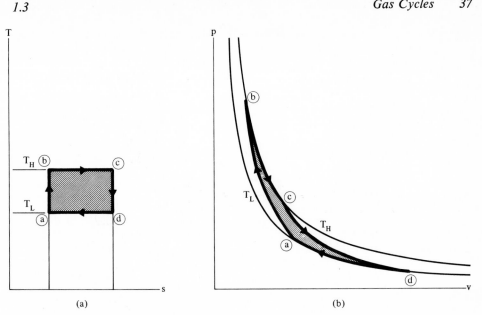

FIG. 1.24 The Carnot gas cycle.

The volume ratio for a given mass of working fluid is the same as the ratio of specific volumes. This volume ratio from the beginning to the end of an isentropic compression process will be called the **isentropic compression ratio** $(CR)_s$.

The efficiency of the Carnot cycle as given in Eq. (1.5a) can now be expressed in different ways:

$$\eta_{th} = 1 - \frac{T_L}{T_H} \tag{1.5a}$$

$$= 1 - \left(\frac{p_a}{p_b}\right)^{(\gamma-1)/\gamma} = 1 - \left(\frac{p_d}{p_c}\right)^{(\gamma-1)/\gamma} \tag{1.19a}$$

$$= 1 - \left(\frac{v_b}{v_a}\right)^{\gamma-1} = 1 - \left(\frac{v_c}{v_d}\right)^{\gamma-1} \tag{1.19b}$$

$$= 1 - \frac{1}{(CR)_s^{\gamma-1}}. \tag{1.19c}$$

Figure 1.24 repeats the T-s diagram of Fig. 1.1 for comparison with a p-v diagram of a Carnot gas cycle. The two isothermal and two isentropic lines form a very narrow p-v diagram for a gas cycle. This illustrates the very large volume changes necessary for even small work output per cycle. With a real gas having internal friction and with friction of moving parts of the machine taken into account, it is evident that the ideal Carnot cycle is not easily approached in practice. The various expressions for the thermal efficiency just presented, however, do provide a useful criterion for evaluation of other gas cycles as Eqs. (1.5a) and (1.5b) did for the vapor cycles.

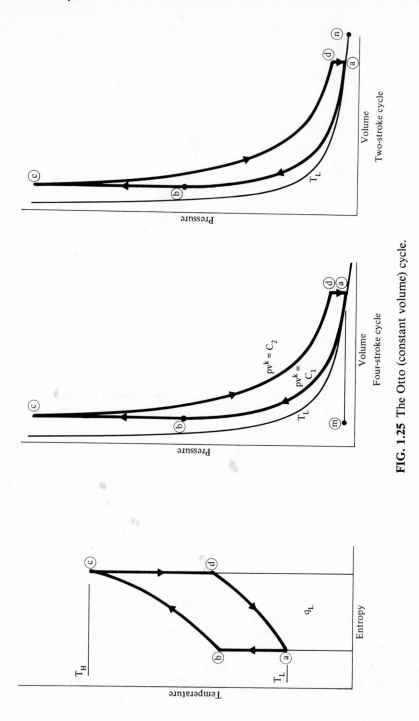

FIG. 1.25 The Otto (constant volume) cycle.

1.3.3 The Otto (Constant Volume) Cycle, the Diesel Cycle, and the Mixed Cycle

Perhaps the most familiar power cycle is the so-called **Otto** gas cycle, which is approached in most spark-ignition gasoline engines and some compression-ignition engines. It is known also as the **constant volume** cycle, because all heat addition and heat rejection take place at constant volumes, the other two processes being adiabatic compression and expansion. **Figure 1.25** shows the cycle on both *T-s* and *p-v* coordinates. Reciprocating engines in which this cycle is used will be described in Section 2 of Chapter 3, and it will be noted that either a **four-stroke** (two crankshaft revolutions) or a **two-stroke** (one crankshaft revolution) cycle can be employed. In the former, the exhaust and induction strokes are lines *am* and *ma* while in the latter these two functions are accomplished in an extension of the *p-v* diagram on the lines *an* and *na*. For a frictionless fluid, the induction and exhaust operations would require no pressure difference, and no energy expenditure would be involved. Because the cycle is not closed during these operations and the amount of working fluid is not constant, the lines *ama* and *ana* have no significance on the *T-s* diagram.

    With the assumptions stated for an ideal gas cycle, the efficiency of the Otto cycle is easily determined:

$$\eta_{\text{th}} = \frac{\text{Heat added} - \text{Heat rejected}}{\text{Heat added}}$$

$$= \frac{c_v(T_c - T_b) - c_v(T_d - T_a)}{c_v(T_c - T_b)}$$

$$= 1 - \frac{(T_d - T_a)}{(T_c - T_b)}$$

$$= 1 - \frac{T_a}{T_b}\frac{(T_d/T_a - 1)}{(T_c/T_b - 1)}. \tag{1.20a}$$

The isentropic compression from *a* to *b* is through the same volume ratio as the isentropic expansion from *c* to *d*. From the relationships of Eq. (1.18),

$$T_d/T_c = T_a/T_b,$$

and so

$$T_d/T_a = T_c/T_b.$$

The expression for η_{th} reduces to

$$\eta_{\text{th}} = 1 - (T_a/T_b) \tag{1.20b}$$

$$= 1 - (1/\text{CR})_s^{\gamma-1}. \tag{1.20c}$$

    Note that this is identical in form to the expression for the thermal efficiency of the Carnot gas cycle. While the forms are the same, two differences must be noted. The temperature at *b* in the Otto cycle is not the maximum temperature in the cycle; heat is added through the temperature range T_b to T_c, and so the efficiency of the Otto cycle will always be less than that of the Carnot cycle between

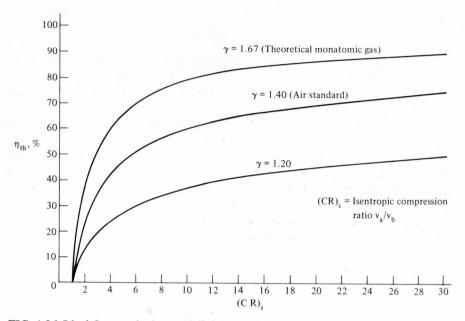

FIG. 1.26 Ideal Otto cycle thermal efficiency. Applicable also to the Carnot and Brayton cycles.

the same temperature limits from T_L to T_H. The compression ratio v_a/v_b for the Otto cycle is a ratio between the maximum and minimum specific volumes (or total volumes for a fixed mass of fluid) over the whole cycle; in the Carnot gas cycle, v_a/v_b is the isentropic compression ratio, but there is also isothermal compression from d to a. These two cycles will be compared with the Brayton cycle after that one has been analyzed.

If a reciprocating engine operates on an ideal Otto cycle, then its potential thermal efficiency is a function of only compression ratio $(CR)_s$ and γ. **Figure 1.26** shows the relationship between η_{th} and $(CR)_s$ for three different values of γ. The curve for the air-standard cycle, for which γ is 1.40, represents a theoretical limit for the conventional internal combustion engine with air as the major part of the working fluid. (Reasons for deviations from this will be discussed later). The air-standard cycle efficiency is by no means the maximum theoretically attainable. Monatomic gases, according to simplified kinetic theory, would have γ values equal to $1\frac{2}{3}$. Theoretically, at least, the Otto cycle and others could be operated with helium, H_e ($\gamma = 1.66$), argon, A ($\gamma = 1.67$), or neon, Ne ($\gamma = 1.64$). Figure 1.26 shows the great advantage of such monatomic gases. Those named have been proposed or actually used in some thermodynamic cycles. The value of γ for carbon dioxide, CO_2, is close to 1.3 and for ethane, C_2H_6, is close to 1.2; the same figure shows the effect of such low values on potential cycle efficiency.

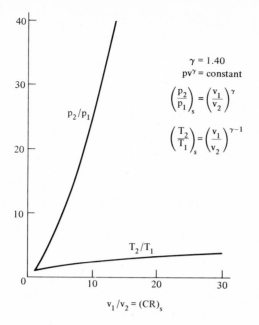

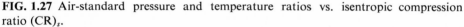

FIG. 1.27 Air-standard pressure and temperature ratios vs. isentropic compression ratio $(CR)_s$.

The three curves of Fig. 1.26 show that an increase in compression ratio from say 4:1 to 8:1 should yield substantial improvement in cycle efficiency, but that an increase from say 12:1 to 16:1 might not be worthwhile. Problems of structural strength, bearing reliability, and combustion must all be considered in the selection of compression ratio for a reciprocating Otto cycle engine.

Some appreciation of the temperature and pressure changes resulting from isentropic compression or expansion is obtained from **Fig. 1.27**. The pressure ratios and temperature ratios are plotted on a base of compression or expansion ratio from the relationships of Eqs. (1.17) and (1.18) for a gas with a constant value of γ equal to 1.40 (air standard). It must be kept in mind that absolute values of pressure and of temperature are used in the respective ratios. The perfect gas law, Eq. (1.14), dictates that, for the constant volume lines, pressure and temperature ratios are numerically equal; for isothermal lines (not used in the Otto cycle), pressure ratios are simply the inverse of volume ratios.

The Diesel Cycle The difference between the Otto cycle just discussed and the **Diesel** cycle is that heat addition is at constant pressure in the latter. For this reason, the Diesel cycle is sometimes referred to as the "constant pressure" cycle, but confusion with the Brayton cycle results. Pressure drop or blow-down at the end of expansion is still at constant specific volume if a fixed amount of working fluid

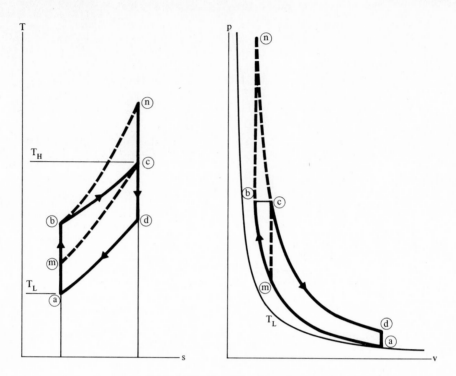

FIG. 1.28 The Diesel cycle.

is considered to be present. The Diesel cycle is shown in **Fig. 1.28**. For comparison, an Otto cycle with the same maximum pressure and one with the same compression ratio are both shown with broken lines. Note that the compression ratio and the expansion ratio are not equal for the Diesel cycle.

The thermal efficiency of the ideal Diesel cycle is

$$\eta_{th} = \frac{\text{Heat added} - \text{Heat rejected}}{\text{Heat added}}$$

$$= \frac{c_p(T_c - T_b) - c_v(T_d - T_a)}{c_p(T_c - T_b)}$$

$$= 1 - \frac{1}{\gamma}\left(\frac{T_d - T_a}{T_c - T_b}\right)$$

$$= 1 - \frac{T_a}{\gamma T_b}\left(\frac{T_d/T_a - 1}{T_c/T_b - 1}\right). \tag{1.21a}$$

Because the compression and expansion processes are both isentropic in the ideal cycle,

$$\frac{T_a}{T_b} = \left(\frac{v_b}{v_a}\right)^{\gamma-1} \quad \text{and} \quad \frac{T_d}{T_c} = \left(\frac{v_c}{v_d}\right)^{\gamma-1}.$$

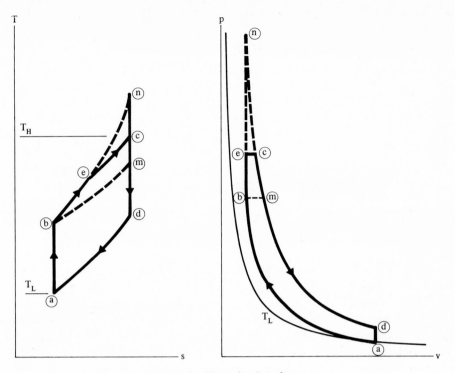

FIG. 1.29 The mixed cycle.

The heat addition from **b** to **c** is at constant pressure, so that

$$\frac{T_c}{T_b} = \frac{v_c}{v_b}.$$

Also,

$$v_d = v_a.$$

Thus

$$\frac{T_d}{T_a} = \frac{T_c}{T_b}\left(\frac{v_c/v_d}{v_b/v_a}\right)^{\gamma-1} = \frac{v_c}{v_b}\left(\frac{v_c}{v_b}\right)^{\gamma-1} = \left(\frac{v_c}{v_b}\right)^{\gamma}.$$

With these substitutions in Eq. (1.21a),

$$\eta_{\text{th}} = 1 - \frac{1}{\gamma(\text{CR})_s^{\gamma-1}}\left[\frac{(v_c/v_b)^{\gamma} - 1}{(v_c/v_b) - 1}\right]. \tag{1.21b}$$

A comparison of Eq. (1.20a) with Eq. (1.21a) and Eq. (1.20c) with Eq. (1.21b) may not immediately show which of the two cycles has the higher thermal efficiency. However, an inspection of the *T-s* diagram of Fig. 1.28 does allow a comparison. For the same isentropic compression ratio from **a** to **b**, more work is done in the Otto cycle than in the Diesel cycle while the same amount of heat energy is rejected; thus the efficiency of the Otto cycle must be greater, but it should be noted that the maximum cycle temperature is higher. For the same maximum temperature at point **c**, the Diesel cycle does more work, and the efficiency must be higher than for the Otto cycle, because the heat rejected is still the same.

The Mixed Cycle When a cycle with adiabatic compression and expansion has heat addition partly at constant volume (Otto) and partly at constant pressure (Diesel), it is called a **mixed** cycle. The terms **limited pressure** cycle and **dual** cycle are also used. The last can be confused with a dual-fuel cycle and should probably be avoided. The mixed cycle is shown in **Fig. 1.29**. The expression for thermal efficiency reduces to

$$\eta_{th} = 1 - \frac{1}{(CR)_s^{\gamma-1}}\left\{\frac{(p_e/p_b)(v_c/v_e)^\gamma - 1}{[(p_e/p_b) - 1] + \gamma(p_e/p_b)[(v_c/v_e) - 1]}\right\}. \tag{1.22}$$

This becomes the same as the expression for the Otto cycle as v_c/v_e approaches unity, and the same as for the Diesel cycle as p_e/p_b approaches unity.

1.3.4 The Stirling, the Ericsson, the Atkinson, and the Lenoir Cycles

After a considerable success in the nineteenth century, the **Stirling** cycle was later considered impractical or at least noncompetitive with the internal combustion engine. Recently, reciprocating engines have been developed that use it, and they will be described in some detail in Section 2 of Chapter 3. In the theoretical cycle, all heat transfer between the working fluid and the surroundings is achieved isothermally as in the Carnot cycle; pressure increase and decrease at constant volume are achieved by internal heat transfer between the two processes. The cycle is illustrated in **Fig. 1.30**. It is very much like the ideal regenerative Rankine vapor power cycle illustrated in Fig. 1.16. An ideal regenerator with 100 percent effectiveness would have to be employed to achieve the heat transfer from process **da** to process **bc** as shown. With the areas on the *T-s* diagram under the lines **da** and **bc** equal, the thermal efficiency is seen to be the same as that for a Carnot cycle between the same temperatures T_L and T_H, Eq. (1.5a). Indeed, this must be so since both are reversible cycles receiving heat and rejecting heat at two identical temperatures. An advantage of the Stirling cycle over the Carnot cycle is that the work area **abcda** is larger for the same change in specific volume. The Stirling cycle, like the Carnot cycle, can be used also for refrigeration when all processes are reversed. See Chapter 4, Section 4.4, and Chapter 5, Section 5.2.

Like the Stirling cycle, the **Ericsson** cycle was not important commercially for many years. However, a gas turbine power plant with intercooling, reheat, and regeneration (Fig. 1.39) is an approach to this cycle, and it is therefore of interest at the present time. Heat addition to and rejection from the working fluid are both isothermal as in the Stirling and Carnot cycles. In the Ericsson cycle shown in **Fig. 1.31**, the other two processes are constant pressure expansion and contraction in which the heat transferred out of the working fluid during contraction from **d** to **a** is equal to that transferred to the fluid during expansion from **b** to **c**. The **regenerative** operation is therefore similar to that of the Stirling cycle. The thermal efficiency of the Ericsson cycle for operation between T_L and T_H is

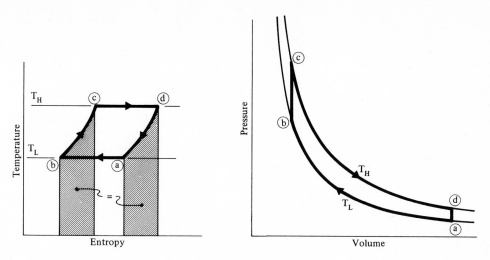

FIG. 1.30 The Stirling cycle.

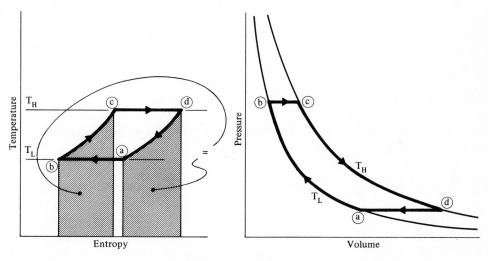

FIG. 1.31 The Ericsson cycle.

identical to that for the ideal Stirling and Carnot cycles between the same temperatures, Eq. (1.5a).

In **Fig. 1.32,** a comparison is made among the Carnot, the Stirling, and the Ericsson gas cycles all operating between the same source and sink temperatures, T_H and T_L, and with equal changes in specific volume. It can be seen that the work area is largest for the Stirling cycle and smallest for the Carnot cycle, and the work of the Ericsson cycle is between the two. Large work output per cycle is an important advantage in a practical engine. One advantage of the Ericsson cycle is its smaller pressure ratio for a given ratio of maximum to minimum specific volume.

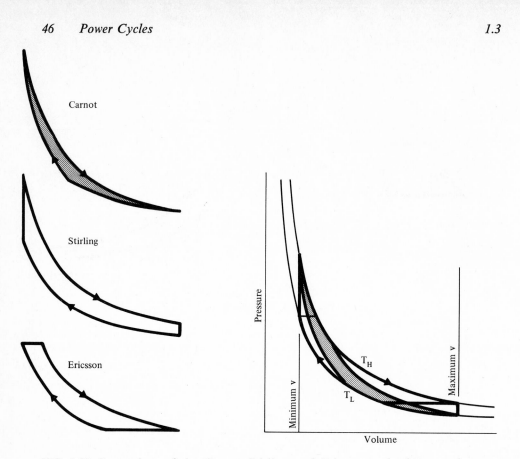

FIG. 1.32 Comparison of the Carnot, Stirling, and Ericsson cycles between the same temperatures and specific volume changes.

A limitation to the work output of the Otto cycle is that the expansion ratio can be no greater than the compression ratio. This is inherent in the operation of the simple reciprocating internal combustion engine. The gases at the end of the Otto cycle's isentropic expansion, however, could do more work if they were allowed to continue isentropic expansion to the lowest cycle pressure. The **Atkinson** cycle, which incorporates this additional expansion, is shown in **Fig. 1.33**. In comparison with the Otto cycle, operating between v_a and v_b and within the same temperature limits, the additional work of the Atkinson cycle is represented by the area **amda**. This cycle is approached in the commercially important combination of a free piston engine/compressor with a gas turbine. This will be described in Section 4.3 of this chapter.

The thermal efficiency of the ideal Atkinson cycle is

$$\eta_{\text{th}} = \frac{c_v(T_c - T_b) - c_p(T_d - T_a)}{c_v(T_c - T_b)}.$$

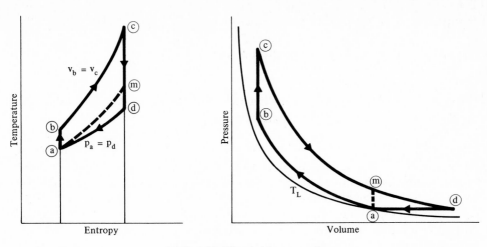

FIG. 1.33 The Atkinson cycle.

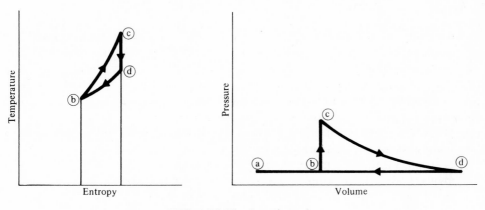

FIG. 1.34 The Lenoir cycle.

This reduces to

$$\eta_{\mathrm{th}} = 1 - \frac{\gamma}{(\mathrm{CR})_s^{\gamma-1}}\cdot\left[\frac{(p_c/p_b)^{1/\gamma} - 1}{(p_c/p_b) - 1}\right], \tag{1.23}$$

where p_c/p_b is the constant-volume pressure ratio and $(\mathrm{CR})_s$ is the isentropic compression ratio, v_a/v_b, as in previous expressions.

Before the significance of high-compression ratio was appreciated, an internal combustion engine was constructed to operate on the **Lenoir** cycle as shown (idealized) in **Fig. 1.34**. Induction of working fluid takes place from **a** to **b**. This is followed by heat addition (combustion) at constant volume from **b** to **c**. Work is accomplished in the isentropic expansion from **c** to **d**, and spent gases would be removed from **d** to **a**. This cycle is approximated in the intermittent firing duct or pulse-jet propulsion systems to be described in Section 7 of Chapter 2. For that open system or for a reciprocating engine where the quantity of fluid is varying from *a* to *b*, the line *ab* has no significance on the *T-s* diagram.

1.3.5 The Brayton Cycle

Comparable to the Otto cycle in importance is the **Brayton** cycle, also called the "**Joule**" or the "**constant-pressure**" cycle; the last term, however, might be confused with the Diesel cycle. It is most readily applied in a system that uses a separate compression machine and a separate expansion machine. Where the latter is a steady-flow gas turbine, the allowable maximum temperature is much less than the allowable maximum in a reciprocating type of engine, where the metal parts are exposed to that maximum for only a short period of time in each cycle. Steady-flow gas compressors have been relatively inefficient until quite recently. The practical application of the Brayton cycle, which was conceived many years ago, has awaited advancements in metallurgy and a better understanding of the fluid mechanics of both compressors and turbines. Now it is the basic cycle for most **gas turbine power plants** whether they are used in stationary installations, for automotive and locomotive propulsion, or for aircraft propulsion.

Variations on the cycle are illustrated in **Fig. 1.35**. The schematic diagrams show the compression and expansion processes taking place in separate components of the system. In the simple closed and open cycles shown, the **compressor,** the **turbine**, and the **load** all have a common shaft, which may be either directly coupled between components or geared to achieve optimum speed in each. The **net work** output of the cycle is the difference between the **gross work** of the turbine and the **back work** required by the compressor. The **free-power turbine** arrangement shown allows the net work and back work to be separated by the use of two turbines which are mechanically independent, but in series thermodynamically. As a result of this arrangement, the compressor speed can be matched to the mass rate of flow requirements while the free-power turbine is controlled to a speed dictated by the load requirements. Similarly, the jet propulsion cycle may have one or more compression and one or more expansion stages. Where two stages are advantageous, the low-pressure compressor stage can be driven by the low-pressure turbine, and the two high-pressure units can be coupled so that separate speed selection is possible, the smaller high-pressure units generally running faster. Constant-pressure heat addition to the working fluid may be accomplished by heat transfer in a heat exchanger, as shown for the **closed cycle**, or more directly from chemical energy in a combustion chamber between *b* and *c*, as shown for the **open cycles**. Heat rejection in a closed cycle must be accomplished by a heat exchanger. In an open cycle, the process *da* is imaginary since the same fluid is not recycled. For jet propulsion, only sufficient shaft work is taken from the turbine to drive the compressor and necessary auxiliaries; the net output of the cycle is derived from the utilization of the difference in energy of the working fluid between point *m* and point *d* in a propulsion nozzle. This will be discussed in greater depth in Chapter 2.

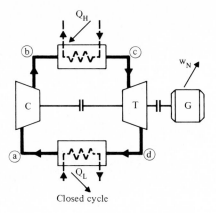

Closed cycle

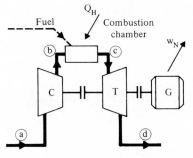

Simple open cycle

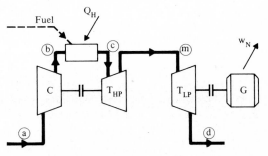

Open cycle with free-power turbine

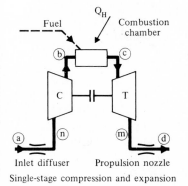

Inlet diffuser   Propulsion nozzle

Single-stage compression and expansion

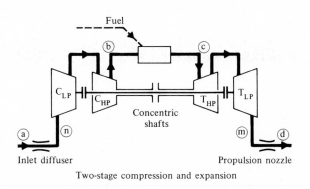

Inlet diffuser   Propulsion nozzle

Two-stage compression and expansion

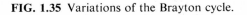

Turbo-jet propulsion cycles

FIG. 1.35 Variations of the Brayton cycle.

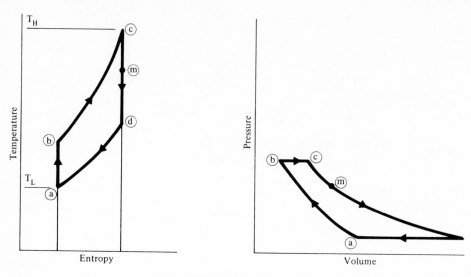

FIG. 1.36 The Brayton cycle.

Figure 1.36 shows the Brayton cycle on $T\text{-}s$ and on $p\text{-}v$ coordinates. It is composed of two isentropic processes, **ab** and **cd**, and two constant pressure processes **bc** and **da**. All heat is added between **b** and **c**, and all heat is rejected between **d** and **a** . Therefore the thermal efficiency of the ideal cycle is

$$\eta_{th} = \frac{c_p(T_c - T_b) - c_p(T_d - T_a)}{c_p(T_c - T_b)} = 1 - \frac{T_a}{T_b}\frac{(T_d/T_a - 1)}{(T_c/T_b - 1)}.$$

Because the expansion from **c** to **d** is between the same pressures as the compression from **a** to **b** ,

$$T_d/T_c = T_a/T_b.$$

So the expression for cycle efficiency reduces to

$$\eta_{th} = 1 - (T_a/T_b) \tag{1.24a}$$
$$= 1 - 1/(CR)_s^{\gamma-1} \tag{1.24b}$$
$$= 1 - 1/(p_b/p_a)^{(\gamma-1)/\gamma}. \tag{1.24c}$$

These expressions are essentially the same as Eqs. (1.5a), (1.19c), and (1.19a), which were developed for the thermal efficiency of the Carnot gas cycle. They are also identical in form to Eqs. (1.20b) and (1.20c) for the Otto cycle. The compression ratio, $(CR)_s$, in all three cycles is the isentropic compression ratio from **a** to **b** . However, this is the overall ratio of volumes for the Otto cycle only; the pressure is the overall range for the Brayton cycle only; and the temperature ratio is the minimum to the maximum for the Carnot cycle only.

The ideal efficiency curves of Fig. 1.26 apply as well to the Brayton and the Carnot cycles as to the Otto cycle. Figure 1.27 can be used to convert compression

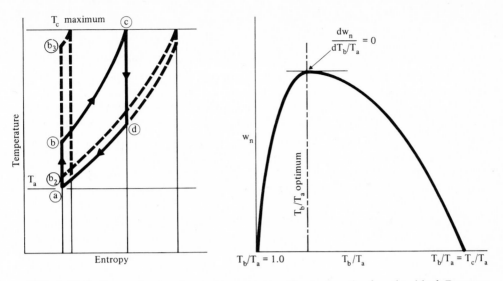

FIG. 1.37 Optimum isentropic - compression temperature ratio for the ideal Brayton cycle.

ratio to temperature ratio (perhaps more useful for consideration of the Carnot cycle) or to pressure ratio (in general more useful for consideration of the Brayton cycle).

Despite the fact that the expressions for cycle efficiency are the same for the Brayton and Otto cycles, actual engines attempting to approximate one or the other are limited by different factors. The compression ratio of a spark-ignition internal-combustion engine is limited by the phenomenon of end-gas detonation or "knock" (see Section 3.2.6). A compression-ignition engine usually must have a higher compression ratio than a spark-ignition engine to ensure proper combustion, but is limited eventually by the allowable maximum pressure for structural and vibration reasons; thus to achieve higher specific outputs in terms of horsepower per cubic inch of piston displacement, the *c-i* engine will operate on a mixed cycle; still, its isentropic compression ratio and therefore its thermal efficiency will be higher than those for a spark-ignition engine.   The Brayton cycle, as used in gas turbines, is limited primarily by the allowable temperature at the turbine inlet, which is governed by the materials available for nozzle and turbine blades; with T_c limited, the work per cycle would be low if the compression ratio were too high.

The last point concerning possible work from a Brayton cycle will be considered in more detail. The *T-s* diagram of **Fig. 1.37** shows three ideal Brayton cycles operating between a fixed minimum temperature, T_a, and a fixed maximum temperature, T_c. The net work of the ideal cycle is the difference between the heat added and the heat supplied:

$$w_N = c_p(T_c - T_b) - c_p(T_d - T_a). \tag{1.25a}$$

Parenthetically, it is noted that this is the same as

$$w_N = c_p(T_c - T_d) - c_p(T_b - T_a). \tag{1.25b}$$

The net work is represented in the figure by the area **abcda**, and the magnitude of that area is affected by the arbitrarily selected temperature ratio T_b/T_a. This is apparent when the two extreme situations shown by the dotted lines are considered: if compression is only to point $\boldsymbol{b_2}$, the heat supplied and heat rejected are practically the same and the net work is practically zero, the efficiency obviously being very low; if compression is to point $\boldsymbol{b_3}$ which is nearly at temperature T_c, the net work again is practically zero, although the efficiency has approached the Carnot value for the two extreme temperatures. Clearly, between these two extremes there must be an **optimum isentropic compression ratio** at which w_N is a maximum for the two limiting temperatures. A plot of w_N on T_b/T_a also is shown in Fig. 1.37. To find the optimum ratio, consider the differential of w_N with respect to T_b/T_a:

$$w_N = c_p T_a \left[\frac{T_c}{T_a} - \frac{T_b}{T_a} - \frac{T_d}{T_a} + 1 \right].$$

However, $T_d/T_c = T_a/T_b$ so that $T_d/T_a = T_c/T_b$. Thus

$$w_N = c_p T_a \left[\frac{T_c}{T_a} - \frac{T_b}{T_a} - \frac{T_c}{T_b} + 1 \right].$$

Now T_a, T_c, and c_p are fixed values. Therefore, at $d\,w_N/d\,T_b = 0$,

$$\frac{d(T_c/T_b)}{dT_b} = -\frac{d(T_b/T_a)}{dT_b}$$

$$-T_c T_b^{-2} = -T_a^{-1}$$

$$T_b/T_a = [T_c/T_a]^{1/2}, \tag{1.26a}$$

which is the optimum isentropic temperature ratio. The optimum isentropic pressure ratio would be

$$p_b/p_a = [T_c/T_a]^{\gamma/(2\gamma-2)}. \tag{1.26b}$$

1.3.6 *Variations in the Brayton Cycle*

It is apparent from Fig. 1.36 or Fig. 1.37 that the turbine exhaust gases at point $\boldsymbol{d}$ are at a temperature above that of the compressor delivery at point $\boldsymbol{b}$ for the usual compression ratios. Some of the rejected heat Q_L can therefore be utilized through a heat exchanger to reduce the amount of Q_H that must be obtained from outside the system of the working fluid. This internal heat exchange is called **regeneration** and was discussed in connection with vapor cycles.

The application of a regenerator to a Brayton cycle is illustrated in **Fig. 1.38**. The exhaust gases from the open-cycle gas turbine at $\boldsymbol{d}$ are cooled to $\boldsymbol{j}$ in the regenerator while the compressed gases at $\boldsymbol{b}$ are heated to $\boldsymbol{e}$ before any outside heat is added in the combustion chamber. In an ideal heat exchanger, the temperature of the compressed gas leaving the regenerator could be equal to the temperature of the exhaust gases entering. That is to say, $T_{e'} = T_d$. With constant specific

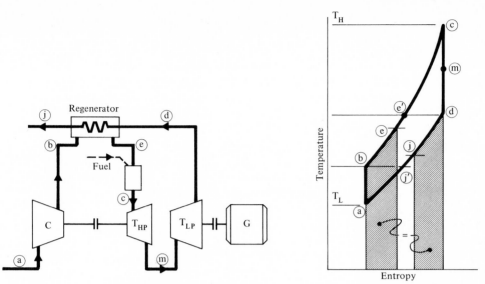

FIG. 1.38 An open-cycle gas turbine power plant with regeneration.

heat throughout the cycle, ideally $T_{j'}$ would equal T_b. This would indicate that the regenerator is 100% effective. Actually, effectiveness is defined in terms of change of enthalpy of the two streams of fluid, and is called **regenerator efficiency**:

$$\eta_R = \frac{h_e - h_b}{h_{e'} - h_b}. \tag{1.27a}$$

If the specific heat c_p were constant as assumed for the ideal gas cycles,

$$\eta_R = \frac{T_e - T_b}{T_{e'} - T_b}. \tag{1.27b}$$

In the ideal cycle shown, the work area is not affected at all by regeneration. The cycle efficiency is thus inversely proportional to the external heat addition. Without regeneration,

$$\eta_{th} = \frac{(T_c - T_b) - (T_d - T_a)}{(T_c - T_b)}.$$

With regeneration,

$$\eta_{th} = \frac{(T_c - T_b) - (T_d - T_a)}{(T_c - T_e)}.$$

In a real cycle where fluid friction must be overcome, the work of the compressor would be increased and the work of the turbine reduced because of pressure losses in both fluid streams within the regenerator. Note also that, even in the ideal cycle, regeneration is possible only where T_d is greater than T_b. As compression ratio and expansion ratio are increased, the possibility of regeneration disappears.

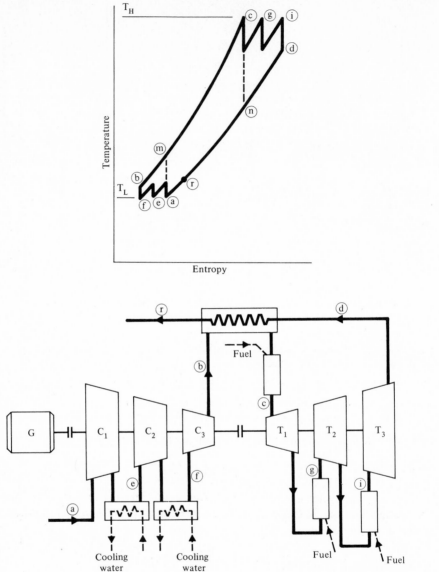

FIG. 1.39 A gas turbine power cycle with intercooling, reheat, and regeneration.

The gas turbine power cycle with **two stages of intercooling** during compression and **two stages of reheating** during expansion as well as **regeneration** is shown in **Fig. 1.39**. A simple cycle *amcna* between the same temperature limits and with the same pressure ratio is shown for comparison. The work area and the opportunity for regeneration have both been increased so that cycle efficiency is greatly improved. The equipment, on the other hand, has been made much more complex and probably will be justified only for very large stationary installations.

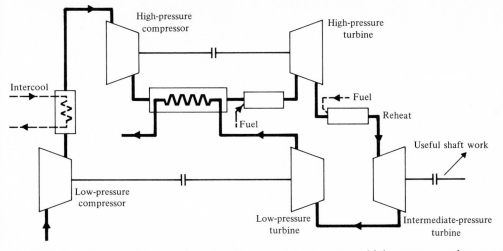

FIG. 1.40 A gas turbine power cycle with separate low-pressure, high-pressure, and power shafts.

An infinite number of compressor stages with intercooling, theoretically, would make the compression line **ab** isothermal. Similarly, an infinite number of reheat stages would make the expansion process **cd** isothermal. Perfect regeneration (100 % effective) from line **da** to line **bc** would then limit heat input to the isothermal line **cd**, and heat rejection to the isothermal line **ab**. The cycle illustrated therefore is an attempt to approach the Ericsson cycle of Fig. 1.31, which was demonstrated to have a cycle efficiency equal to the highest possible between T_L and T_H.

Intercooling and jacket cooling during the compression process not only reduce the work of compression but also decrease the specific volume of the gases and thus the size of the compression machine. Cooling may be justified for this reason alone. Reheating produces more work per pound of fluid, and thus it too reduces compressor size.

One of several possible gas turbine power-plant arrangements that have been used for automotive propulsion [A.36] is shown in **Fig. 1.40**. One intercooler and one reheat combustion chamber are used along with regeneration. A novel feature is the use of three separate shafts: one for the low-pressure compressor and turbine stages, one for the high-pressure compressor and turbine stages, and one for the useful shaft power to the wheels. Each shaft can run at a speed that is independent of the requirements for the other two shafts.

The top temperature in the internal-combustion gas turbine cycle is usually kept within allowable limits by large amounts of excess air supplied to the combustion chamber and mixed with the products of combustion. An alternative is the **injection of liquid water** into the combustion chamber as shown in **Fig. 1.41**. The gas temperature is reduced as the water rapidly vaporizes, and the resulting working fluid is a mixture of steam and gas [A.16]. Compared to operation with air, or air and products of combustion, the turbine work per pound of fluid is reduced slightly because of the lower combined γ-value. However, the compressor back work is greatly reduced relative to the turbine output because only the air must be compressed; the water is delivered by a pump with relatively small work absorption. The slightly higher turbine exhaust temperature (for the same number of expansion stages) allows greater regeneration with the water-injection cycle. Also, a water preheater, called an **economizer** (as in boiler practice), allows still greater heat recovery because of gas-to-liquid instead of gas-to-gas heat transfer.

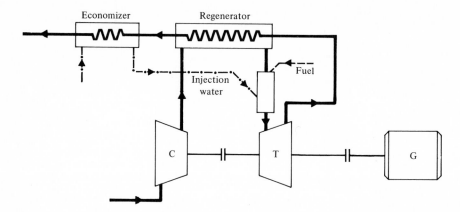

FIG. 1.41 A gas turbine power cycle with water injection, a regenerator, and an economizer.

The ideal cycle efficiencies plotted in Fig. 1.26 for a simple Brayton cycle can be improved upon by regeneration, intercooling during compression, and reheating during expansion. However, thermal efficiency for any of the cycles described will be reduced by compressor and turbine inefficiencies, the reasons for which will be discussed later. The ideal, simple, air-standard Brayton cycle with no losses is compared with four other cycles in **Fig. 1.42**. For these four cycles, a reasonably good value of 85 % is assumed for both compressor and turbine efficiency, and 70 % is assumed for regenerator efficiency. A top temperature of 1500 F and a compressor intake temperature of 60 F are assumed for all cycles. Constant values are still assumed for specific heats of the working fluid. In the cycle with water injection, $\frac{1}{2}$ lb water is injected per pound of air, and it is assumed that an economizer could cool the exhaust gases to 220 F.

First, it is noted that the efficiency of the simple cycle with the assumed machine losses (curve B) is much below the ideal values of curve A. The efficiency curve peaks and will actually go to zero as the compressor back work finally absorbs all the turbine output. At a lower top temperature, the curve would peak at a lower pressure ratio and would go to zero sooner.

When regeneration is added to the simple cycle of curve B, the efficiency is improved greatly at lower pressure ratios. However, curve C shows that the improve-

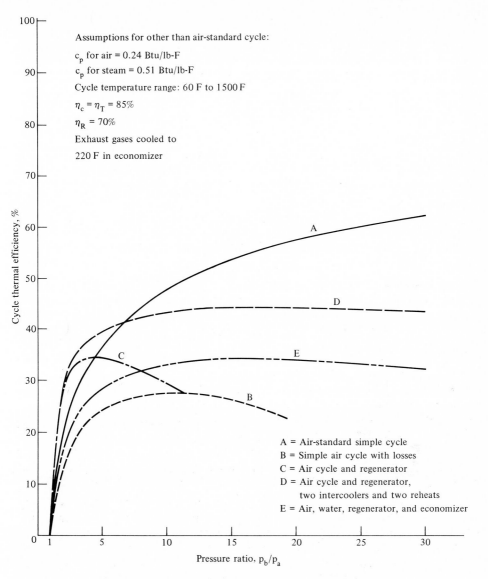

FIG. 1.42 A comparison of gas turbine cycle efficiencies.

ment becomes less as the pressure ratio rises, and finally the regenerator becomes useless at a pressure ratio of approximately 11 : 1, where the compressor delivery temperature eventually equals the exhaust temperature.

The complexities introduced by intercooling and reheating (curve *D*) result in a great improvement that does not disappear at higher pressure ratios. In fact, a nearly constant overall efficiency is found beyond about 10 : 1. Even without inter-

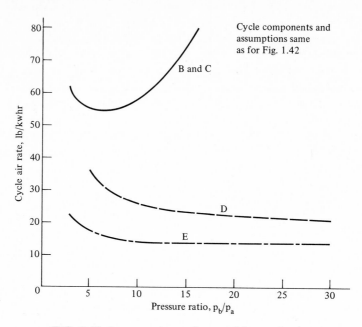

FIG. 1.43 A comparison of gas turbine cycle air rates.

cooling and reheating, the injection of water into the combustion chamber produces a considerable improvement over curve *B*. Curve *E* remains quite flat, since
the economizer can recover rejected heat even after the regenerator has become
useless. Reheat through further combustion is not possible because less air is present
than for the cycle of curve *D*, and so efficiencies are lower.

The mass rate of air flow through the compressor is a good indication of
machine size for a given power output. **Figure 1.43** compares values for cycles
B, C, D, and *E*, of the previous figure. The simple cycle (curves *B* and *C*) must have
the same specific rate of air flow with or without regeneration despite the great
differences in efficiency previously noted at lower pressure ratios. The intercooling
and reheating of cycle *D* allow much lower rates of flow as predicted from the
larger work area of the *T-s* diagram of Fig. 1.39. Not only is the air rate lowered,
but also it is less sensitive to pressure-ratio changes, and this provides more stable
operation with load variation. The very significant reduction in the amount of air
compressed when water is used to control the maximum temperature (and becomes
part of the working fluid) is shown by a comparison of curve *E* with either curve
D or the curve for *B* and *C*.

Whether a water-injection cycle is commercially important is less significant
to the present discussion than the introduction of the concept of evaporation to
control top temperature and the concept of mixed working fluids. Many combinations might be possible either for open cycles as described or for closed cycles with
thermal and mechanical separation of the fluids preceding the compressor inlet.

1.3.7 Deviations from Air-Standard Cycles

Actual engines operating with real gases must produce cycles that deviate from the ideal gas cycles that have been analyzed. Reciprocating engines and compressors will be described in Section 2 of Chapter 3, but the deviations from ideal operation can be dealt with at this point. The centrifugal and axial-flow compressors and turbines to be described in Chapter 3 have additional inherent deviations, and these too can be discussed beforehand.

The **actual working fluid** in a **reciprocating internal-combustion engine** is not pure air. During the compression stroke for both spark-ignition and compression-ignition engines, atmospheric air and water vapor are mixed with residual gases (clearance volume gases) from the previous cycle; in a carbureted or gaseous fuel spark-ignition engine, the fuel also is present during compression. On the expansion stroke, the working fluid is made up of products of complete or partial combustion and any excess air. For a hydrocarbon fuel and air, all the components aside from air have specific heat (c_p) values greater than 0.24 and ratios of specific heats (γ) less than 1.40. The expressions for thermal efficiency all indicate that the cycle efficiency is lowered as γ is reduced.

The **variation in specific heat** with temperature for the actual working gases always decreases the value of γ and thus the possible cycle efficiency. Since c_p and c_v increase with temperature above the values assumed for the air-standard cycle and the difference between the two is a constant, the ratio c_p/c_v is lower throughout the rest of the cycle than at the lowest temperature, generally assumed to be standard (approximately 520 R). To view the effect on the cycle through the change of specific heat rather than indirectly through the reflection of that change in γ, it is apparent that a higher specific heat will reduce the temperature rise for a given heat input thus lowering the average temperature at which heat is added to the cycle. Alternatively, it is apparent that to reach the same top temperature would require more heat, and at the same time, expansion through a given volume ratio would result in a higher final temperature and greater heat rejection.

Combustion is not a simple chemical reaction proceeding always in one direction. Molecular **dissociation** is encouraged at high temperatures and recombination will proceed at lower temperatures. Two simple oxidation equations can be written as examples:

$$CO + \tfrac{1}{2}O_2 \rightleftharpoons CO_2 + A \text{ Btu/mol} ; \qquad H_2 + \tfrac{1}{2}O_2 \rightleftharpoons H_2O + B \text{ Btu/mol}.$$

Association is exothermic, and dissociation endothermic. Top temperatures in internal combustion cycles are reduced by the reversal of the oxidation reaction. Though equal energy is returned to the working fluid as the chemical reaction goes to completion when temperatures fall during the expansion process, the available expansion ratio following addition of that part of the energy has been reduced. Thus thermal efficiency is lowered.

Processes are not truly adiabatic or isothermal as assumed. Events are not instantaneous. The real gases do have viscosity and therefore internal friction loss. All these reduce the net useful work output of the cycle.

Finally, there is a **pumping loss** in a naturally aspirated four-stroke cycle result-ing from the negative gauge pressure necessary on the induction stroke and the positive gauge pressure on the exhaust stroke so that gases can be forced into and out of the cylinder against the friction losses at surfaces, particularly in the relatively small valve openings. In a two-stroke cycle or a supercharged four-stroke cycle, pumping work external to the cylinder must be expended.

There is a great difference in the magnitude of fluid friction losses between a **gas turbine** cycle and an internal-combustion reciprocating engine cycle: processes are **nonflow** in the cylinder of a reciprocating engine, but **velocities are high** through the components of the gas turbine power plant. The **fluid friction** losses in a gas tur-bine plant require additional compressor work, result in an appreciable pressure drop in the combustion chamber or heat exchanger, and decrease work output of the turbine. In the **diffuser** or diverging duct after the compressor, it is not possible to convert all the kinetic energy of the stream to pressure head; whether this eventually constitutes a loss or not will be explored at greater length in Chapter 2. The **pressure drop in a combustion chamber with flow** is not wholly attribu-table to friction. It results in part from the pressure difference necessary to create the acceleration resulting from a large increase in specific volume in the ostensibly constant pressure combustion process. The **exhaust** gases must leave the turbine with some velocity. The kinetic energy can be appreciable and is certainly a loss in a stationary installation. In a jet propulsion system, this kinetic energy is not a loss if properly utilized.

One compensating factor exists for the gas turbine. The maximum cycle tem-peratures have to be limited to tolerable steady values at the turbine as mentioned previously. Dissociation and reassociation of products of combustion do not, therefore, usually take place in the turbine, and do not reduce the turbine work.

1.3.8 Thermodynamic Properties of Air and Other Gases

The working fluid dealt with so far in the discussion of gas cycles has been an idealized gas as described at the beginning of Section 1.3. It has been assumed to follow the perfect gas law of Eq. (1.14):

$$pv = RT$$

and to have constant values of specific heats as well as being frictionless. Its en-thalpy and internal energy have been taken as functions of temperature only. These assumptions will now be examined more closely.

Undoubtedly the most important gas for power cycles is air, a combination in very nearly constant proportions of oxygen and nitrogen primarily, plus argon, carbon dioxide, and traces of other gases. **Figure 1.44** indicates to what degree at least some of the ideal gas assumptions are reasonable for air. In the vicinity of the liquid-vapor region on the chart, enthalpy is seen to be a strong function of pressure as well as temperature; this will be dealt with further in a discussion of cryogenics and the production of liquid air in Chapter 5. At higher temperatures and moderate pressures, enthalpy lines are practically horizontal, which is to say that enthalpy is very nearly a function of temperature alone.

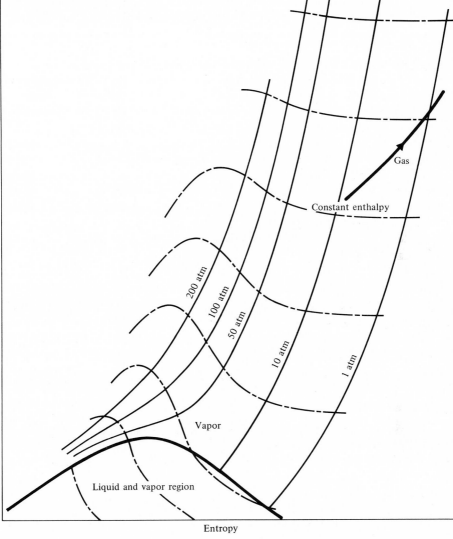

FIG. 1.44 An outline of a *T-s* diagram for air (see the detailed diagram in the appendix).

The assumption of constant specific heat is not valid for a real gas. **Figure 1.45** [A.9] is based on "zero pressure" values of c_p and on the Beattie-Bridgeman equation of state. It shows the strong dependence of constant pressure specific heat on temperature for various gases. A much weaker dependence on pressure can be noted also. Empirical functional relationships between c_p and T have been developed [A.39] and are often useful.

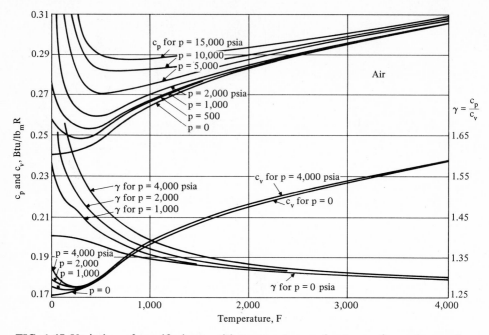

FIG. 1.45 Variation of specific heats with temperature and pressure for real gases (see Reference [A.10]). Above: air; opposite page: CO, H_2, CO_2, CH_4.

Over a range of pressures through which the specific heat is not appreciably affected by pressure and is therefore a function of temperature only, we can write

$$h - h_b = \int_{T_b}^{T} c_p \, dT, \qquad (1.28)$$

where h is the enthalpy above some arbitrary base temperature T_b. This implies that enthalpy is a function of temperature only, so far as that is true of the specific heat also. From the definition of enthalpy and the perfect gas relationship,

$$u = h - pv = h - RT. \qquad (1.29)$$

At reasonably low pressures and at temperatures well above the critical value, gases do conform to the perfect gas relationship very closely. Thus internal energy (u) is a function of temperature only. For a gas, we can write

$$ds = c_p \frac{dT}{T} - R \frac{dp}{p}$$

so that for an **isentropic process,**

$$c_p \frac{dT}{T} = R \frac{dp}{p},$$

$$\ln \frac{p'}{p_b} = \frac{1}{R} \int_{T_b}^{T} c_p \frac{dT}{T}. \qquad (1.30)$$

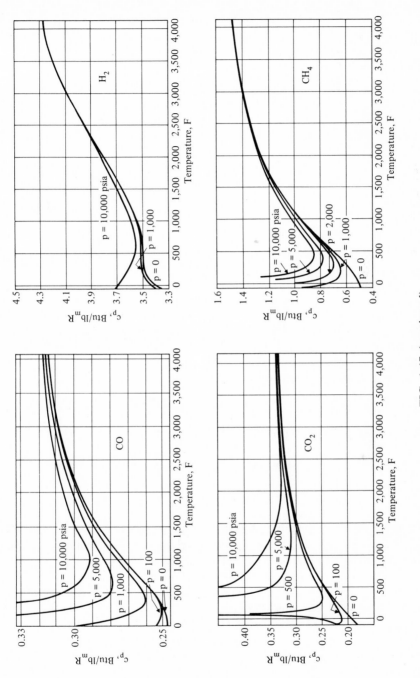

FIG. 1.45 (continued)

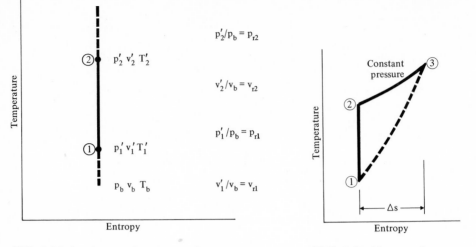

FIG. 1.46 An isentropic process between two points.

FIG. 1.47 Change of entropy.

The ratio p'/p_b which applies only to an **isentropic process** starting at p_b and T_b is defined as the **relative pressure** (p_r). It is apparent that the relative pressure is a function of temperature only, if the same is true for specific heat and thus for enthalpy.

An alternative expression for change of entropy in a gas is

$$ds = c_v \frac{dT}{T} + R \frac{dv}{v}.$$

Again, for an **isentropic process**,

$$\frac{dv}{v} = -\frac{1}{R} c_v \frac{dT}{T},$$

$$\ln \frac{v'}{v_b} = -\frac{1}{R} \int_{T_b}^{T} c_v \frac{dT}{T}. \tag{1.31}$$

The ratio v'/v_b is defined only for an isentropic process starting at v_b and T_b as the **relative volume** (v_r), and it is apparent that this term is a function of temperature only, so far as that is true for specific heat and enthalpy.

In the use of the properties relative pressure and relative volume, it is usually necessary to deal with ranges of pressure and specific volume not starting or ending at the arbitrary base point T_b, p_b, v_b. The uses of p_r and v_r are, however, not as restricted as it would seem. Between any two points *on an isentropic line* as shown in **Fig. 1.46**, pressure ratios and specific volume ratios are governed by the relative pressures and volumes since

$$\frac{p_1'}{p_2'} = \frac{p_1'/p_b}{p_2'/p_b} = \frac{p_{r1}}{p_{r2}} \tag{1.32a}$$

and

$$\frac{v'_1}{v'_2} = \frac{v'_1/v_b}{v'_2/v_b} = \frac{v_{r1}}{v_{r2}}. \tag{1.32b}$$

It cannot be emphasized too strongly that the terms "relative pressure" and "relative volume" have no significance except in isentropic processes.

Even for a perfect gas, entropy, unlike enthalpy, is a function of both pressure and temperature. As noted above, one expression for entropy is

$$ds = c_p \frac{dT}{T} - R \frac{dp}{p}.$$

Integrating, we find that

$$s - s_b = \int_{T_b}^{T} c_p \frac{dT}{T} - R \ln \frac{p}{p_b}.$$

The temperature-dependent component of this expression is usually designated as ϕ:

$$\phi = \int_{T_b}^{T} c_p \frac{dT}{T}. \tag{1.32c}$$

This property is a function of temperature only, so far as that is true for specific heat. It is a very useful function in the calculation of entropy change. In **Fig. 1.47** the entropy change from ① to ③ is

$$s_3 - s_1 = \phi_3 - \phi_1 - R \ln \frac{p_3}{p_1}. \tag{1.33a}$$

For the **isentropic process** from ① to ②, the change in ϕ is

$$\phi_2 - \phi_1 = R \ln \frac{p_2}{p_1}. \tag{1.33b}$$

For the **constant pressure process** from ② to ③, the change of entropy is simply

$$s_3 - s_2 = \phi_3 - \phi_2 = s_3 - s_1. \tag{1.33c}$$

Although the entropy change from ① to ③ can be calculated from Eq. (1.33a), it is often easier to calculate it through the intermediate point ② with Eq. (1.33c), as long as s_2 is equal to s_1.

It has now been shown that, where the perfect gas laws can be applied with sufficient accuracy and the specific heat is not appreciably affected by pressure, the following quantities are functions of temperature only: h, u, p_r, v_r, and ϕ. Numerical values of these quantities can therefore be presented at different values of temperature in tabular form.

For power engineering work, the most useful presentation of gas properties is found in *Gas Tables* by Keenan and Kay [B.10]. This book includes properties of air, products of combustion, and component gases, as well as compressible flow functions. It will be referred to simply as the Gas Tables in the following discussion. Besides the quantities mentioned in the previous paragraph, the tables present values for some or all the gases for specific heat at constant pressure (empirical), specific heat at constant volume ($c_v = c_p - R$), the ratio of specific heats ($\gamma = c_p/c_v$), the acoustic velocity ($a = \sqrt{\gamma g_c R T}$), the viscosity ($\mu$), the thermal conductivity (λ), and the Prandtl number ($c_p \mu/\lambda$), all of which are functions of temperature only, at least up to 200 psia. Also tabulated against temperature is the maximum flow per unit area per unit pressure resulting from isentropic expansion from zero velocity. This can be shown to be a function of initial temperature only. Various functional relationships that aid in calculations are presented in terms of the pressure ratio (r) and the exponent (n). Compressible flow functions including those from Shapiro, Hawthorne, and Edelmann [A.33, 34] and from Edmonson, Murnaghan, and Snow [A.8] are presented as well as other useful standard tables and conversion factors. To explain fully how all these tables are used would be too great a digression at this point. Instead, the use of the various terms in power-cycle calculations will be explained at appropriate places, and reference to the Gas Tables will occasionally be made.

The question of the practical application of the Gas Tables and their accuracy naturally arises since they are specified in general as "low-pressure" tables. In the discussion of Sources and Methods, pp. 199 to 213 of that publication, the question is answered by the authors. Calculated changes in enthalpy through an isentropic pressure drop from 200 psia show negligible error when comparison is made with the tables of Sage and Lacey [A.12] in which the pressure difference is taken into account. The temperature range is, of course, important, and these comparisons were made for temperatures above 30 F. The "low-pressure" tables are thus seen to be quite satisfactory for the usual Brayton and Diesel gas cycles in which pressures are usually below 200 and 300 psia respectively, and temperatures are usually above the normal ambient value. Even at much higher pressures and lower temperatures, these tables are useful for a first approximation since exact values of temperature and pressure are seldom known for either design or analysis calculations. On the other hand, dissociation is not taken into account, and the products of combustion are assumed to be products of complete combustion and stable; for the Otto cycle, which usually has high pressures and temperature, the Gas Tables should not be used.

The composition of dry air assumed by Keenan and Kaye is only slightly different from that adopted by the International Joint Committee on Psychrometric Data as described in Chapter 6 and listed in Table 6.1 of this book. Volume percentages used in the Gas Tables are 20.99 oxygen, 78.03 nitrogen, 0.98 argon, with other traces ignored. The combined value for the molecular weight of air is then 28.970 and the combined gas constant for air is 53.342 ft-lb/lb-R or 0.068549 Btu/lb-R [B.10, p. 202]. The products of combustion for which thermodynamic properties are given are those from the combination of air in various proportions with hydrocarbon fuels.

1.3.9 Stagnation Properties

The term **stagnation value** is an extremely useful concept in the analysis of the Brayton power cycle as well as in other aerodynamic work. Sometimes the stagnation value of a property truly represents a zero velocity situation; at other times the

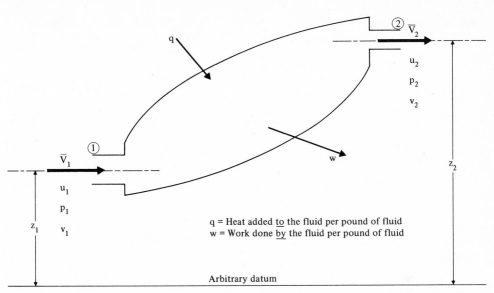

q = Heat added <u>to</u> the fluid per pound of fluid
w = Work done <u>by</u> the fluid per pound of fluid

Arbitrary datum

FIG. 1.48 A general steady-state, steady-flow, nonreactive system.

hypothetical stagnation condition is arrived at through a combination of property and energy terms to eliminate the kinetic energy as a separate item. The concept of stagnation enthalpy has already been referred to in Section 2.3 of this chapter following Eq. (1.9b). The definition is repeated here. It is best appreciated through a consideration of the first law of thermodynamics applied to a steady-flow non-reactive process. To repeat Eq. (1.9a) in a form applicable to **Fig. 1.48** between sections ① and ②,

$$[u_1 + p_1 v_1 + (\bar{V}_1^2/2g_c) + z_1] + q = [u_2 + p_2 v_2 + (\bar{V}_2^2/2g_c) + z_2] + w. \tag{1.34a}$$

It is assumed that $\bar{V}_1$ and $\bar{V}_2$ are constant velocities across sections ① and ②, that is, that the flow is one-dimensional. All units are either Btu per pound of fluid or foot-pounds per pound of fluid if Joule's equivalent ($J = 778$ ft-lb/Btu) is inserted as required. By the definition of enthalpy,

$$h = u + pv.$$

Therefore

$$[h_1 + (\bar{V}_1^2/2g_c) + z_1] + q = [h_2 + (\bar{V}_2^2/2g_c) + z_2] + w. \tag{1.34b}$$

A further grouping of terms is possible by the arbitrary **definition of stagnation enthalpy**:

$$h_o = h + (\bar{V}^2/2g_c).$$

The equation then becomes

$$h_{o1} + z_1 + q = h_{o2} + z_2 + w. \tag{1.34c}$$

For a power plant, the change of elevation is usually not significant (though it would be for a liquid pump). The equation is thereby further simplified to

$$h_{o1} + q = h_{o2} + w. \tag{1.34d}$$

This last equation indicates the usefulness of the concept of stagnation enthalpy in the analysis of a gas turbine power plant or any of its components. Between any two points in the cycle where there is either **heat added** or **work done**, the heat added to the fluid (q) or the work done by the fluid (w) is simply a change in stagnation enthalpy. Between two points where no heat is added or work done, the stagnation enthalpy remains constant. For many thermodynamic calculations, it is not at all necessary to know the value of enthalpy or the value of kinetic energy so long as the sum of the two is known.

The **stagnation temperature** (sometimes called **total** temperature) must be defined as the temperature corresponding to an enthalpy equal to the stagnation enthalpy. The stagnation enthalpy is greater than the actual enthalpy by the amount of the kinetic energy. For a fluid of constant specific heat,

$$T_o = T + \frac{\bar{V}^2}{2g_c c_p}, \tag{1.35a}$$

$$\frac{T_o}{T} = 1 + \frac{\bar{V}^2}{2g_c c_p T}. \tag{1.35b}$$

To put this in terms of the specific heat ratio and Mach number, the following relationships are used for a gas:

$$c_p = R\frac{\gamma}{\gamma - 1}, \qquad a = \sqrt{\gamma g_c RT}, \qquad M = \frac{\bar{V}}{a}.$$

Thus

$$\frac{T_o}{T} = 1 + \frac{\gamma - 1}{2}M^2. \tag{1.35c}$$

Except for extremely high velocities of flow, the ratio T_o/T is not great and the specific heat is very nearly constant. Equations (1.35a, b, and c) are therefore applicable as long as a value of c_p or γ is taken that applies to the temperature range T to T_o.

Up to this point, an actual stagnation process has not been specified. If a moving fluid could be brought to rest reversibly and adiabatically, which is to say isentropically, then an **isentropic stagnation pressure** would be achieved such that, for a gas,

$$\frac{p_o}{p} = \left(\frac{T_o}{T}\right)^{\gamma/(\gamma - 1)} \tag{1.36a}$$

$$= \left(1 + \frac{\gamma - 1}{2}M^2\right)^{\gamma/(\gamma - 1)}. \tag{1.36b}$$

Should the fluid actually be brought to rest, but with some loss accompanied by an increase of entropy, then the actual stagnation pressure would be less than the

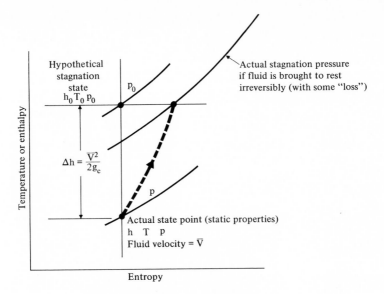

FIG. 1.49 Stagnation properties.

isentropic value. The concept of a hypothetical isentropic stagnation pressure, however, remains valid regardless of whether or not the fluid is brought to rest. In this book, the isentropic value will be intended unless otherwise specified. (As for temperature, the term *total* pressure is used by some authors.) **Figure 1.49** should aid in the understanding of stagnation properties and their relationship to "actual" or "static" properties.

1.3.10 Component Efficiencies for a Gas Turbine Power Plant

The rate of fluid flow through the compressor, diffuser, and turbine of a gas turbine power plant is high for the sizes of the components, and therefore the residence time is very low. Although heat gain or loss may be high per unit of time, the heat transferred to or from the working fluid per pound of fluid is usually negligible. Thus all processes are considered to be adiabatic except the combustion or purposeful heat transfer processes. While the kinetic energy of the fluid may be appreciable, it need not in general be evaluated separately and can be grouped with the property enthalpy. Thus Eq. (1.34d) can be applied between the inlet and outlet of each component. A temperature-entropy diagram can be useful for cycle and component analysis, but because all energy terms of interest will be found from enthalpy differences, enthalpy-entropy diagrams will be used to illustrate the following discussion.

The **compressor** and its process are shown in **Fig. 1.50**. In terms of the points shown, the **isentropic efficiency of compression** is usually defined as

$$\eta_{c_s} = \frac{\text{The isentropic work from } \boldsymbol{a} \text{ to } \boldsymbol{b'}}{\text{The actual work from } \boldsymbol{a} \text{ to } \boldsymbol{b}}.$$

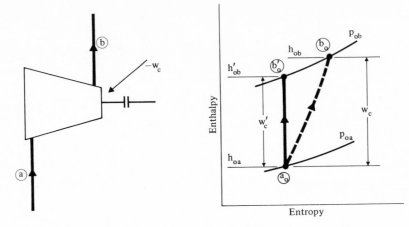

FIG. 1.50 Compression process.

From Eq. (1.34d), we see that this can be written in terms of stagnation enthalpy change, Btu/lb of fluid:

$$\eta_{C_s} = \frac{w'_C}{w_C} \tag{1.37a}$$

$$= \frac{h'_{ob} - h_{oa}}{h_{ob} - h_{oa}}. \tag{1.37b}$$

These equations consider only the effect on the fluid. There will also be a **mechanical friction loss** at the shaft bearings and seals, and that may be appreciable. If that mechanical loss per pound of working fluid is represented by F_C, then

$$\eta_{C_s} = \frac{h'_{ob} - h_{oa}}{h_{ob} - h_{oa} + F_C}. \tag{1.37c}$$

The pressure loss through the **burner and/or heat exchanger** will reduce the potential output of the turbine. This can be expressed either as the **burner** (and/or heat exchanger) **pressure coefficient** or as a percent pressure loss. The drop in pressure between the outlet of the compressor *b* and the inlet to the turbine *c* is shown in **Fig. 1.51**. Where there is no regenerator, points *e* and *b* are identical. The change in stagnation pressure is not likely to be exactly the same as the change in static pressure; the velocity almost always increases appreciably since the flow area is not usually increased in proportion to the change in specific volume from *b* to *c*. The pressure coefficient could be defined either in terms of the static pressures or in terms of the stagnation pressures. For consistency of approach in this discussion, stagnation pressures will be used. The use of static pressures would be as logical, but numerical values of the pressure coefficient would be different.

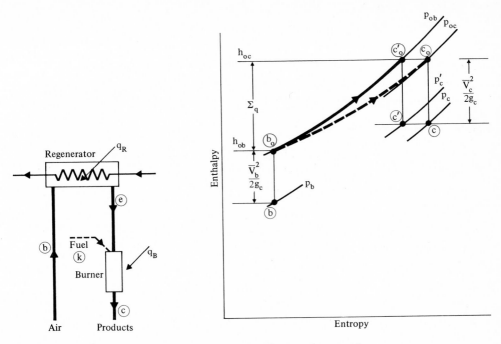

FIG. 1.51 Regenerator and burner pressure loss.

The **stagnation pressure coefficient** for this process will be defined as

$$E_{Ho} = \frac{p_{ob} - p_{oc}}{p_{ob}} \tag{1.38a}$$

$$= 1 - \frac{p_{oc}}{p_{ob}}. \tag{1.38b}$$

Because the coefficient is dimensionless, the units of pressure are not important, but absolute values must be used.

    Regenerator effectiveness (usually called efficiency) has already been defined by Eq. (1.27a). The same equation could be used for any heat exchanger. A "stagnation regenerator efficiency" could be defined in which the stagnation temperature of the outlet gases at 100 % efficiency would equal the stagnation temperature of the inlet heating medium, but it would really be meaningless because of the inclusion of the kinetic energy component.

    The **combustion efficiency of the burner** is quite different from the pressure coefficient. It indicates what fraction of the **heating value** of the fuel is actually transferred usefully to the working fluid. Either **higher** or **lower** heating value (HHV or LHV) could be used. The difference between the two is the energy carried away in the exhaust by water vapor that was formed in the combustion process when it is not condensed to liquid as it would be in a closed calorimeter determination of heating value. By definition (ASME Code),

$$\text{LHV} = \text{HHV} - 1030 \times 9 \times \text{H},$$

where H is the fraction of the fuel by mass that is hydrogen. The lower heating value of a fuel is commonly used as the criterion for ideal energy utilization in the combustion chamber of a gas turbine because the latent heat of condensation of any water formed in the reaction is considered unavailable for all practical purposes. Thus the combustion efficiency of the burner is defined as

$$\eta_B = \frac{\text{Heat actually transferred to the working fluid}}{\text{Lower heating value of the fuel used}}$$

$$= \frac{\text{Effective heating value}}{\text{Lower heating value}}. \tag{1.39a}$$

If the **mass of the fuel** can be neglected compared to the mass of the air, as it very often can be for an approximate calculation, the effective heating value is simply

$$\text{EHV} = q_B$$

$$\simeq h_{oc} - h_{oe} \text{ Btu/lb air}. \tag{1.39b}$$

An **approximate fuel/air ratio** (FA_a) can be calculated from Eq. (1.39b) :

$$FA_a = \frac{h_{oc} - h_{oe}}{\eta_B \text{LHV}} \text{ lb fuel/lb air (approx.)}. \tag{1.39c}$$

As the schematic diagram of Fig. 1.51 shows, air and fuel enter the burner but at different conditions, and the products of combustion leave at condition c . The **true fuel air ratio**, pounds of fuel per pound of air, will be designated by FA. The proper expression for the effective heating value is

$$q_B = (1 + FA)h_{oc(g)} - [h_{oe(\text{air})} + FA h_{k(\text{fuel})}] \text{ Btu/lb air}. \tag{1.39d}$$

For hydrocarbon fuels, the *Gas Tables* give enthalpy and other values for products of combustion, designated by "g" in Eq. (1.39d), for the fuel air/ratios resulting from 200 % theoretical air (50 % theoretical fuel) and for 400 % theoretical air (25 % theoretical fuel) as well as for zero fuel (i.e., the Air Tables in the appendix). Properties for mixtures between these ratios can be obtained by linear interpolation. Because all the tables except that for air alone give property values on a pound-mole basis, the correct molecular weight must be used to convert the property values to a pound basis. The combined molecular weight can be obtained by summation of the molecular weights of the various components, each multiplied by its volume fraction in the total. Alternatively, it can be obtained from Table 9 of the *Gas Tables* from the known ratio of the number of hydrogen atoms to the number of carbon atoms and the percent of theoretical fuel.

In any gas turbine power plant in which heat is added directly by combustion in the working fluid, the fuel/air ratio is always very lean (large amount of excess air) because temperatures at the entrance to the turbine must be limited. The

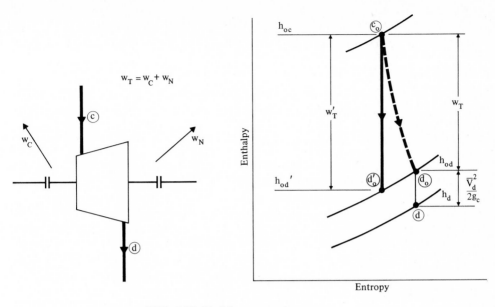

FIG. 1.52 Turbine expansion process.

chemically correct mixture strength for a wide range of hydrocarbon fuels with air is between 0.06 and 0.08 lb fuel per pound of air. With at least 100 % excess air (200 % theoretical air), the value of FA is quite small. Also, the enthalpy of the fuel, whether liquid or gaseous, is low compared to the enthalpy of combustion for the same amount of fuel. Thus the last term in Eq. (1.39d) can often be ignored. Where greater accuracy is required, the data of the National Bureau of Standards (referred to in the *Gas Tables*) can be used to find $h_{(fuel)}$; an example used by Keenan and Kaye (p. 220 of the *Gas Tables*) for liquid octane shows the value to be quite small:

$$h_{(fuel)} = 0.5T - 287 \text{ Btu/lb fuel,}$$

where T is expressed in degrees Rankine.

The efficiency of the **turbine** is calculated in a manner similar to that for the compressor. With reference to **Fig. 1.52**, the turbine efficiency is defined as

$$\eta_T = \frac{\text{The actual work from } \boldsymbol{c} \text{ to } \boldsymbol{d}}{\text{The isentropic work from } \boldsymbol{c} \text{ to } \boldsymbol{d'}}.$$

Again, Eq. (1.34d) can be applied:

$$\eta_T = \frac{w_T}{w_T'} \tag{1.40a}$$

$$= \frac{(h_{oc} - h_{od})_g}{(h_{oc} - h_{od}')_g}. \tag{1.40b}$$

These equations, as do the comparable compressor-efficiency equations, take into account only the effect on the working fluid. The mechanical friction loss for the turbine will be designated by F_T, and expressed in the same units as the enthalpy terms, usually Btu/lb fluid. Then

$$\eta_T = \frac{(h_{oc} - h_{od})_g - F_T}{(h_{oc} - h'_{od})_g}. \tag{1.40c}$$

The subscript "g" is used in Eqs. (1.40b) and (1.40c) because the gases are a combination of products of combustion and excess air when a burner rather than a heat exchanger is used, and the appropriate gas tables should be used for determination of enthalpy values. Even so, the error resulting from the use of air tables may be acceptably small when the percentage of excess air is high.

For many turbine/compressor arrangements, it is impossible to separate the mechanical friction loss of the turbine from that of the compressor. In such a situation, the total bearing and seal loss at the shaft can arbitrarily be divided equally between the two. The diagrams of Fig. 1.50 and 1.52 define the work of the compressor and the work of the turbine in terms of property changes in the working fluids. The net shaft output of the turbine/compressor power plant (w_N) is reduced by the shaft losses so that

$$w_N = (1 + FA)(w_T - F_T) - (w_C + F_C) \text{ Btu/lb air.} \tag{1.41}$$

For a gas turbine power plant used in jet propulsion, the **efficiency of the jet** will be examined in Chapter 2 [see Eq. (2.15)].

1.4 COMBINED CYCLES

Various vapor power cycles and gas power cycles have been discussed, and some of their advantages and limitations have been mentioned. For certain conditions, two or more different cycle types can be combined to achieve an operating economy that is not possible with one alone. Also, cycles are combined for reliability so that when one type cannot be used, the other can. It will be pointed out in Chapter 4 that where both power and either refrigeration or heat pumping are required, the heat rejection of one cycle might be advantageously employed in the other. Only a very limited number of examples of combined cycles will be given here.

1.4.1 Combinations of Thermally Independent Systems

A steam plant cannot be started up and put "on line" in a very short time; if it might be needed suddenly, it must be kept steaming. On the other hand, a diesel engine, a gas engine, or a gas turbine can be put on line from a cold start relatively quickly, and an installation incorporating several of these units can make possible additions to the plant capacity in small increments. In the sizes in which engine or gas turbine plants are built, their capital cost per kilowatt of rating is less than that for the same size steam plant. However, fuel costs may be higher. Large diesel engines and gas turbines have proved themselves to be reliable while producing "rated" power continuously over long periods of time. That rating is kept well below maximum possible power, which can be delivered only for periods of hours rather than days if large maintenance costs are to be avoided.

The load demand on a power plant is never absolutely steady. If only one unit, say a steam boiler and turbine combination, is installed with sufficient capacity to meet the maximum possible demand, the unit must operate continuously, most of the time at part load. For reliability and flexibility, more than one unit will usually be installed so that, at part load, one or the other can be shut down; where they are not all the same, the least economical will be shut down first. For these reasons, diesel engines or gas turbine power plants are sometimes used as **peaking** units. This means that the smaller units are operated only when it is necessary to carry the peak loads on the overall system. The existence of independent units of different kinds adds reliability to the systems and also allows operators to adjust the fuel type usage in accordance with varying availability and cost [A.5].

The addition of a different type of system may also be the most economical method of increasing the capacity of an existing power plant for which the load has increased through the years. In any steam power plant, the numerous internal auxiliary power requirements for feed water and cooling water pumps, lubrication pumps, forced draft and induced draft fans, compressed air, etc., might be provided by prime movers independent of the main system. However, their selection is usually based on reasons of convenience or operating simplicity rather than on thermodynamic considerations.

1.4.2 Thermally Interdependent Systems (Topping)

When the heat rejected from one power plant is used as the heat supply to another system, the arrangement is described as **topping**. The binary vapor cycles described in Section 1.2.8 are special examples of this. With the increased reliability of relatively large sized gas turbine power plants, they are being used more for central station installations, but their ratings are still far below the largest (and most economical) steam stations. In intermediate sizes, the economy of operation of a combined gas turbine/steam generation plant may be better than that for either type alone.

Where steam is needed at moderate temperature for process energy or for building heating, and where shaft power is needed at the same time for power generation, air compression, pump drive, or some other purpose, it is common practice to operate a "noncondensing" steam turbine. The exhaust from the turbine is delivered directly to the place at which heat is required, and after the latent heat has been transferred, the condensate is usually returned to the boiler. Alternatively, a gas turbine can be used as the prime mover, and its exhaust gases can be passed through a heat exchanger (which may be essentially a steam boiler) rather than through a regenerator. Steam or gas turbine efficiency will be reduced, but the overall plant economy may be greatly improved.

A fairly large power generation station is shown schematically in **Fig. 1.53**. This is the Horseshoe Lake Unit No. 7 of the Oklahoma Gas and Electric Company

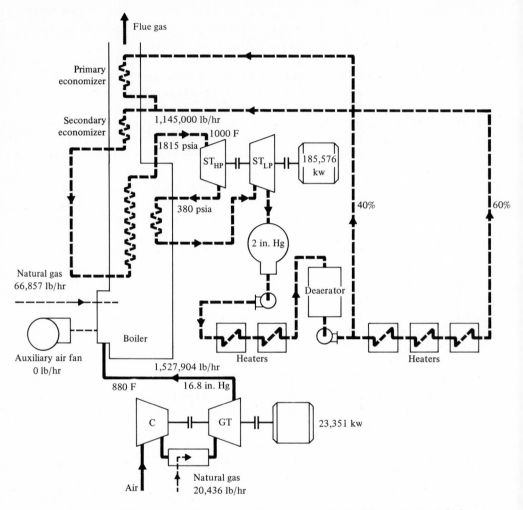

FIG. 1.53 A combined gas and steam turbine power plant. Schematic heat balance. Horseshoe Lake Unit No. 7 (Reference [A.38]).

[A.38]. About 30.6 % of the fuel consumption and 12.6 % of the power are accounted for in the gas turbine/generator unit. The exhaust from the gas turbine is supplied at a high temperature to the steam generator (boiler) where the combustion of additional fuel can be supported by the excess oxygen in the exhaust gases. The forced draft fans of the boiler are needed only for temporary peak loads or when the gas turbine is not in use. A bypass exhaust duct is provided for the gas turbine for periods when the steam cycle is not in use. Also, boiler air preheating is not required with the gas turbine and boiler combination. Flue gas loss is reduced by means of the primary economizer which allows the number of steam extraction

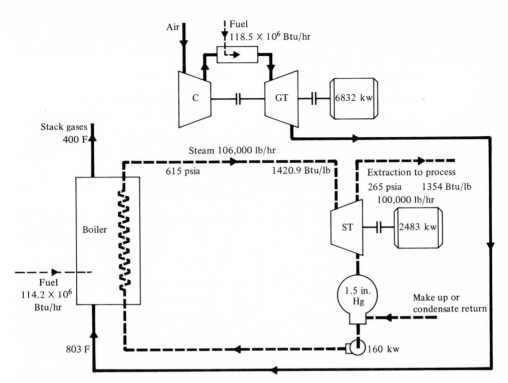

FIG. 1.54 A combined gas and steam turbine power cycle with steam extraction for process heat (Reference [A.23]).

feed-water heaters to be reduced. Full load economy is 9530 Btu per kwh compared with over 9900 Btu per kwh for a comparable steam cycle alone.

A system in which steam turbine extraction is used for process steam is shown in **Fig. 1.54**. Both a steam turbine and a gas turbine are used for shaft power, the latter topping the heat cycle of the former. Both the overall heat rate and the initial installation cost are reduced compared with a steam boiler and turbine used alone for shaft power with the same process steam extraction rate [A.23].

1.4.3 Reciprocating Engine and Gas Turbine Combinations

One of the most efficient heat cycles for a prime mover is realized by a combination of a reciprocating engine (with high but intermittent top temperature) and a gas turbine to allow complete expansion of the working fluid. This was mentioned in Section 3.4 of this chapter as an example of a practical Atkinson cycle. The **free-piston engine/compressor plus a turbine** illustrated schematically in **Fig. 1.55** is one possible arrangement [A.15]. Three identical reciprocating engine/compressor combinations are used here to achieve reasonably smooth output. These are sometimes called gassifiers or gas producers since they provide hot gas under pressure for the turbines.

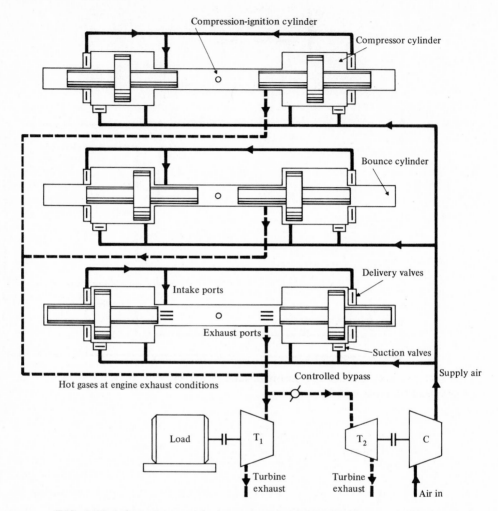

FIG. 1.55 A free-piston engine/compressor combined with a gas turbine.

The back work or work of compression of a conventional gas turbine Brayton cycle is a large part of the gross turbine work. In the combination of components shown, the work of compression is done mostly by the two-stroke, opposed piston, compression-ignition engines operating with the reciprocating compressors as free piston machines. No conversion to rotary motion is necessary for this internal work of the cycle. The high compression ratios and temperatures possible and the direct action of the power cycle forces on the compression cycle make this combination much more efficient than conventional reciprocating machines with rotating shafts. The net work output of the cycle is carried entirely by turbine No. 1 in the diagram. It operates at a speed dictated by the load, and its speed has no effect on the efficiency of the compression work. It is necessary to supply air to the free-piston compressors at a slight positive pressure for which purpose an auxiliary turbine is used to drive a low pressure-ratio compressor.

The method of operation of the combined plant is apparent from the figure. The inner faces of the smaller pistons form an opposed-piston compression-ignition engine, seen at its inner dead-center position in the second of the three assemblies and at its outer dead-center position in the lowest of the three. The outer faces of the smaller pistons operate in closed cylinders in which air is compressed, and reexpand in an efficient spring action to return the power pistons to the inner dead center. The compression ratio is not fixed, but varies to some extent with the load carried; also, the compression ratio and natural frequency of operation are affected by clearance in the bounce cylinders or by controlled bleed to or from these cylinders. Injection is triggered by the return of the pistons through a mechanism that is not shown. Only the outer faces of the larger pistons act as air compressors; the outward movement of the inner faces draws in air through open ports, and that air is delivered to the intake valves of the opposite ends of the same cylinder on the return stroke. The auxiliary compressor supply is thus utilized at reasonably constant rates.

It is not necessary that the reciprocating engine/compressor have the free-piston design. A piston/turbine arrangement with the piston motion linked to a crankshaft and the turbine geared to the same shaft has been developed. This has been referred to as the piston-turbine-compound (PTC) system [A.43].

A combination of a reciprocating engine with a gas turbine that has been in use for many years, both for aircraft spark-ignition engines and for heavier diesel engines, is **turbo-supercharging**. In this arrangement, the exhaust gases from the reciprocating engine are passed through a gas turbine connected by a free shaft to a supercharger, usually centrifugal. The back pressure at engine exhaust may decrease efficiency to some extent, but the main purpose of increasing output per cubic inch of cylinder displacement through supercharging is accomplished. The gear-train complexity of a supercharger driven directly from the engine shaft is avoided. The amount of supercharging is easily controlled by means of a bypass or "dump" valve to allow the exhaust gases to escape to atmosphere without going through the turbine. Turbo-supercharging has been carried a step further in one particular propeller driven aircraft. A two-stroke compression-ignition aircraft engine is supplied by a multistage air compressor and exhausts to a multistage turbine [A.31]. This is similar to a turbo-propeller aircraft propulsion system (Chapter 2), except that here, almost all net shaft work is taken from the reciprocating engine, while the gas turbine primarily drives the compressor. Some thrust is possible from the gas turbine exhaust discharged through a jet nozzle. High output diesel engines have been developed to use two-stage charging compressors with intercooling and after-cooling, the compressor stages being driven by two-stage exhaust-gas turbines [A.29].

PROBLEMS

1. It is true that work with loss implies a degradation of energy, i.e., an increase in entropy. When losses are involved in a continuously repeated working cycle, does the entropy of the working fluid undergo a net increase with each cycle?

2. If cycle A has a lower cycle efficiency than cycle B, does it necessarily have a lower relative efficiency?

3. Can liquid and vapor water be in equilibrium at 300 F and 20 psia?

4. Can water with a specific volume of 2000 cu ft/lb be at the triple point?

5. Dry saturated water vapor in a rigid container (constant volume) at 100 psia is cooled to 240 F. What percentage of the mass condenses?

6. Water at 150 psia and 2% wet ($x = 0.98$) is throttled to atmospheric pressure. Use a Mollier (h-s) chart to answer the following:

   a) Is the final state wet vapor, dry saturated vapor, saturated liquid, or superheated vapor?

   b) Did the temperature rise, fall, or remain constant?

7. Use an h-s diagram to find the work done by a steam turbine operating between 1000 and 100 psia with 85% efficiency, given that the throttle temperature is 800 F.

8. A single expansion turbine receives steam at the throttle conditions of 200 psia and 600 F. It normally exhausts to a condenser at atmospheric pressure. Assume a simple Rankine cycle and a turbine efficiency of 80%. By what percentage would the thermal efficiency of the cycle be increased if the condenser pressure were lowered to 1.0 psia?

9. The two-stage steam-jet air ejector shown in Fig. 1.12 uses the condensate from the main condenser as the cooling water for the two ejector condensers. Alternatively, cooling water from the same source as that used for the main condenser could have been used. What would be the prime advantage of using externally supplied cooling water rather than the condensate, and the prime disadvantage?

10. On the basis of (a) work per pound of steam and (b) cycle thermal efficiency, compare a simple ideal Rankine cycle at the throttle conditions of 1000 psia and 800 F with a similar cycle having reheat to 800 F at 100 psia. Assume that both have a final condenser pressure of 10 psia.

11. A reheat steam cycle has the first stage supply conditions of 1000 psia and 900 F. The reheat is at 40 psia and to the same temperature.

    a) Given that the efficiency of the first turbine is 85%, how much energy is added per pound of steam in the reheat coils?

    b) Assume that the same expansion efficiency exists in the second turbine. What is the cycle efficiency if the condenser pressure is 0.5 psia?

12. For the idealized vapor power cycle shown in Fig. 1.17, assume that the condenser pressure is 0.5 psia, the first extraction (i) is dry saturated steam at 600 psia, and the second extraction (j) is at 100 psia. What fraction of the total steam is extracted at these two points?

13. Using the values given in Fig. 1.19, find the overall thermal efficiency of the Ravenswood No. 3 Steam Power Plant.

14. a) In what respect is "piping friction loss" actually a loss in a power cycle?

    b) It is stated in the discussion of the mercury-water binary cycle that pressure loss between the mercury boiler and the turbine produces several degrees of superheat. Show this on a *T-s* diagram of the mercury cycle, not necessarily to scale.

15. A vapor power cycle inherently has two important advantages over a gas power cycle. What are they?

16. It is shown in the text that for a gas, both constant pressure and constant volume lines diverge vertically on a *T-s* diagram with linear scales. Prove that constant pressure lines are, however, parallel horizontally; that is, prove that one constant pressure line is the same shape as any other constant pressure line, but displaced horizontally, and that all isothermals are of equal length between two particular constant pressure lines. Prove that the same is true for constant volume lines.

17. In 1862, Beau de Rochas enunciated several desirable criteria for a reciprocating internal-combustion engine. Explain why he stated that the combustion chamber should have a small surface-to-volume ratio. What shape should the combustion chamber take? Can that shape be maintained on expansion?

18. By comparing Eq. (1.20c) with Eq. (1.21b), show without a doubt whether the Otto or the Diesel cycle would be more efficient for the same compression ratio.

19. a) Without reference to any other cycle, show that the thermal efficiency of an ideal Stirling power cycle with perfect regeneration is

$$\eta_{th} = 1 - (T_L/T_H).$$

b) The working fluid could accomplish the same cycle without regeneration. Show whether the thermal efficiency would then depend on the amount of work done in each cycle.

20. Although certain power cycles were devised long ago, they have become commercially important only quite recently. What were the **two principal** developments **in each case** that made practical
   a) the Stirling cycle?
   b) the Brayton cycle?

21. The efficiency formulas for the Carnot and the Brayton cycles have the same form, yet the Carnot is often referred to as an example of a cycle with the maximum possible thermal efficiency, while the Brayton is not. What is the distinction in this respect?

22. An air standard Brayton cycle must operate between an ambient air temperature of 500 R and a turbine inlet temperature of 1500 R. At what pressure ratio would
   a) the cycle efficiency be a maximum?
   b) the cycle efficiency be a minimum?
   c) the work per pound be a maximum?

23. A gas-turbine cycle with standard air at intake is to produce the maximum net useful work. For the ideal cycle, by what percentage could the cycle efficiency be improved if the allowable temperature at turbine inlet were increased from 2000 to 2500 F?

24. Why should we expect, for a given ratio of maximum to minimum temperature, that the efficiency of a simple gas-turbine cycle with regenerator will eventually decrease as the pressure ratio is increased?

25. Is it always useful to have a regenerator in a gas-turbine power cycle? Why?

26. When a gas turbine plant with a regenerator can reclaim all the energy lost as a result of turbine inefficiency, is the cycle's thermal efficiency then unaffected by turbine efficiency? Why?

27. Considering only the effect of variations in specific heat (i.e., ignoring friction losses and dissociation), would you expect the thermal efficiency of a practical gas turbine cycle to vary more or less than that of a practical spark-ignition reciprocating engine cycle from the value predicted for an air-standard cycle? Why?

28. Why is entropy not tabulated in the Gas Tables although it is tabulated in the Steam Tables?

29. What is the final temperature when air is compressed reversibly and adiabatically from 100 F and 5 atm to 20 atm? Compare the answer obtained from the Air Tables in the appendix with that obtained by calculation, assuming that $\gamma = 1.40$.

30. In defining "stagnation enthalpy," many books state that the value is reached when the fluid is "brought to rest reversibly and adiabatically." Discuss whether or not such a definition is needlessly limiting.

31. An air compressor on a test stand takes in standard air and compresses it through a total pressure ratio of 6:1. What compressor efficiency would produce an increase in entropy of 0.015 Btu/lb of air per degree R?

32. A compressor takes in air at 15 psia and 60 F and delivers it at a pressure of 90 psia. For a compressor efficiency of 80%, what is the increase in entropy, Btu per pound of air per degree R?

33. a) Use the Air Tables to calculate the work required per pound of air in a centrifugal compressor receiving air at 1 atm and 500 R and delivering it at 5 atm. Assume that the compressor is essentially adiabatic, and that its efficiency is 85%.
    b) What is the delivery temperature of the air?
    c) What would the exponent have to be in the polytropic compression equation ($pv^n = $ const) to achieve that temperature for the same pressure ratio?

34. Assuming there is no friction, would there be a pressure drop in a combustion chamber of constant diameter? Explain your answer briefly.

35. The inlet stagnation conditions for a gas turbine are 100 psia and 1500 F. The turbine exhaust stagnation pressure is 20 psia, and the efficiency is 80%. Assume that the gas has the same properties as air.
    a) What is the stagnation temperature of the exhaust gas?
    b) What is the entropy change of the gas?

36. The stagnation temperature of the gas entering a turbine is 1300 F, the expansion ratio is 4.5:1, and the turbine efficiency is 85%. Assume that the gas has the same properties as air. What mass rate of flow would produce a gross turbine shaft power equal to 50 horsepower?

37. Why does the combination of a free-piston engine/compressor and a gas turbine result in a higher thermal efficiency than a cycle using only a gas turbine or only a reciprocating engine?

38. Consider the free-piston engine/compressor and gas turbine system shown schematically in Fig. 1.55. Air is supplied at 520 R and 1 atm, and is compressed eventually to 30 atm before constant volume combustion in the compression-ignition cylinder. The top temperature of the cycle is 2400 F. Assume an ideal cycle with no

losses, and assume that the working fluid has properties as given in the "low pressure" Air Tables.

a) At what temperature and pressure would gas be available at the inlet to the gas turbine?

b) What is the value of the thermal efficiency for the ideal cycle?

REFERENCES

A. *Articles and Separate Publications*

1. AMANN, C. A., "Gasifier Can Drive Free Turbine Accessories," *SAE Jl.*, Aug., 1965, pp. 66–69.
2. ASME, *Properties of Steam at High Pressure*, American Society of Mechanical Engineers, New York, 1956, 5 pp.
3. BRODZELLER, L., "Jet Engine Provides Fast Startup, and Area Protection, at Low Cost," *Power Engineering*, Vol. 70, No. 10, Oct., 1966, pp.62–63.
4. "Chrysler's Auto Turbine Travels the Long Road Home," *SAE Jl.*, June, 1964, p. 30.
5. DELISTOVIC, J. A., "Peaking Plants," *Gas Turbine*, July/Aug., 1965, pp. 48–49.
6. DIERMAN, H. W., F. P. KUHL, and T. R. GALLOWAY, *General Design of the 1000-MW Unit at Ravenswood*, Consolidated Edison Co., Combustion Eng., Inc., Allis-Chalmers, 1963, 40 pp.
7. DOOLEY, J. L. and A. F. BELL, "McCullock Studies the Steam Car," *SAE Jl.*, June, 1962, p. 89.
8. EDMONSON, N., F. D. MURNAGHAN, and R. M. SNOW, *Bamblebee Report 26*, Dec., 1945.
9. ELLENWOOD, F. O., N. KULIK, and N. R. GRAY, "The Specific Heats of Certain Gases over Wide Ranges of Pressure and Temperature," *Cornell University Bulletin 30*, Oct. 1942.
10. ELSTON, C. E. and R. SHEPPARD, "First Commercial Supercritical Pressure Steam Turbine—Built for the Philo Plant," *ASME Paper* 55A–159.
11. GASPAROVIC, N., "Peak Load Coverage with Gas Turbines Today and Tomorrow," *Gas Turbine*, Vol. 6, No. 6, 1965, pp. 40–41.
12. GERHART, R. V., F. C. BRUNNER, H. S. MICKLEY, B. H. SAGE, and W. N. LACEY, "Thermodynamic Properties of Air," *Mechanical Engineering*, Vol. 64, 1942, pp. 270–274.
13. HACKETT, H. N., "Mercury for the Generation of Light, Heat, and Power," *Trans. ASME*, Vol. 64, No. 7, p. 647.
14. HODGSON, C. W., "Modern Thermal Station Design," *Consulting Engineer*, Sept., 1963, pp. 120–125.
15. HUBER, R., "Present State and Future Outlook of the Free-Piston Engine," *Trans ASME*, Vol. 80, 1958, p. 1779.
16. HUGHES, P. B. and B. D. WOOD, "Gas Turbines for Central Powerplants," Unpublished report for Ontario Hydro Electric Power Commission, 1948.
17. JENNINGS, B., "Process Refrigeration with Gas Turbines," *Proceedings National Engine Use Council*, 1964 meeting, pp. 91–95.

18. JONES, D. R., "The Empire District Electric Company Steam Power Modernization with Gas Turbine Addition," *Westinghouse*, Apr. 1965, 14 pp., figures, and tables.

19. KELLER, C., "The Nuclear Gas Turbine," *Gas Turbine*, July–Aug., 1965, pp. 42–45.

20. KOLFLAT, A., *Industrial Power Plants*, a Series of Project Studies, *Consulting Engineer*, Apr., 1958.

21. MANGAN, J. L., and R. C. PETTIT, "Combined Cycle with Unified Boiler has High Efficiency," *Power Engineering*, Sept., 1963, pp. 47–49.

22. McANNENY, A. W., "Binary Cycle Gas Fired Turbine for Gas Compression Service," *ASME paper*, 62-GTP-17.

23. McCONNEL, J. E., and K. C. WEIN, "Combined Gas-Steam Cycles Offer Savings to Industry," *Power*, Vol. 107, No. 5, May, 1963, pp. 68–71.

24. McVAY, C. G., and A. J. FIEHN, "Supercritical Mine-Mouth Plant Nears Completion—Part 1." *Power Engineering*, Vol. 70, No. 10, Oct., 1966, pp. 54–57.

25. MILLER, E. H. and B. M. CAIN, "Large Steam-Turbine Generators for the 1960's," *ASME Paper*, 59 APWR 3, 1959.

26. NELSON, L. M., "Gas Turbine Fuel is Sewage Treatment By-product," *Power Engineering*, Vol. 70, No. 10, Oct., 1966, pp. 58–59.

27. PENNY, N., "The Development of the Glass Ceramic Regenerator for the Rover 2S/150 R Engine," *SAE Paper*, 660361, June, 1966.

28. PERCIVAL, W. H., "Fluorochemical Vapor Machine," *SAE Paper*, 931B, Oct., 1964.

29. ROBINSON, R. R. and J. E. MITCHELL, "Development of a 300 psi (21.1 kg/cm²) Bmep Continuous Duty Diesel Engine," *CIMAC* International Congress, 1965.

30. ROWLEY, L. N. and B. G. A. SKROTZKI, "Gas Turbines" *Power*, Oct., 1946, pp. 667–686.

31. SAMMONS, H. and E. CHATTERTON, "The Napier Nomad Aircraft Diesel Engine," *SAE Trans*, Vol. 63, 1955, p. 107.

32. SCANLAN, J. A., and JENNINGS, B. H., "Bibliography—Free-Piston Engines and Compressors," *Mechanical Engineering*, Apr., 1957, p. 339.

33. SHAPIRO, A. H., and G. M. EDELMAN, *Jl. Applied Mechanics*, Vol. 14, 1947, A154–162.

34. SHAPIRO, A. H. and W. R. HAWTHORNE, *Jl. Applied Mechanics*, Vol. 14, 1947, A317–337.

35. SNOKE, D. R. and G. L. MRAVA, "Only Whisper from New Portable Powerplant," *SAE Jl.*, Nov., 1964, p. 72.

36. STARKMAN, E. S., "What Engine Will Power the Car of Tomorrow?" *SAE Jl.*, Vol. 72, No. 4, 1964, pp. 34–39.

37. STEWART, W. L., A. J. GLASSMAN, and R. P. KREBS, *The Brayton Cycle for Space Power*, SAE paper 741A, Sept., 1963, 12 pp.

38. STOUT, J. B., J. J. WALSH, and A. G. MELLOR, "First Large Combined-Cycle Unit Justified by Increased Efficiency," *Electric Light and Power*, Vol. 41, No. 5, May, 1963, pp. 32–35.

39. SWEIGERT, R. L. and M. W. BEARDSLEY, "Empirical Specific Heat Equations Based upon Spectroscopic Data," *Georgia School of Technology Bulletin*, Vol. 1, No. 3, June, 1938.

40. TABOR, H. and L. BRONICKI, "Establishing Criteria for Fluids for Small Vapor Turbines," *SAE Paper*, 931C, National Transportation, Powerplant and Fuel Lubricants Meeting, Oct., 1964.

41. WALSH, J. J., "Design Criteria for Supercritical Units," *Consulting Engineer*, July, 1966, pp. 100–103.

42. WILSON, W. B. and T. G. HINIKER, "Gas Turbines Reduce Fuel and Power Bills in the Process Industries," *Power Engineering*, Vol. 69, No. 12, Dec., 1965, pp. 54–57.

43. WITZKY, J. E., "Forecast of Future Power Plant Developments," *ASME Paper* 64-OGP-8, 1964.

B. *Books*

1. ASME, *Thermodynamic and Transport Properties of Gases, Liquids, and Solids*. New York: McGraw-Hill, 1959.

2. DILLIO, C. C. and E. P. NYE, *Thermal Engineering*. Scranton: International Textbook, 1963.

3. DURHAM, F. P., *Aircraft Jet Power Plants*. Englewood Cliffs: Prentice-Hall, 1951.

4. ENNIS, W. D., *Applied Thermodynamics for Engineers* (third edition). New York: D. Van Nostrand, 1913.

5. ENNIS, W. D., *Vapors for Heat Engines*. New York: D. Van Nostrand, 1912.

6. GAFFERT, G. A., *Steam Power Stations* (third edition). New York: McGraw-Hill, 1946.

7. FAIRES, V. M., *Thermodynamics of Heat Power*. New York: Macmillan, 1958.

8. HAYWOOD, R. W., *Analysis of Engineering Cycles*. Oxford: Pergamon Press, 1967.

9. HILSENRATH, J., et al., *Tables of Thermal Properties of Gases*, NBS Circular 564, 1955.

10. KEENAN, J. H. and J. KAYE, *Gas Tables*. New York: Wiley, 1945.

11. KEENAN, J. H. and F. G. KEYES, *Thermodynamic Properties of Steam*. New York: Wiley, 1936.

12. LEWIS, A. D., *Gas Power Dynamics*. Princeton: D. Van Nostrand, 1962.

13. MACKAY, D. B. *Design of Space Powerplants*. Englewood Cliffs: Prentice-Hall, 1963.

14. POTTER, P. J., *Power Plant Theory and Design* (second edition). New York: Ronald Press, 1959.

15. PREDVODITELEV, A., et al., *Tables of Thermodynamic Functions of Air*. Glenridge, N. J.: Association Technical Services, 1962.

16. ROSSINI, F. D., D. D. WAGMAN, W. H. EVANS, S. LEVINE, and I. JAFFE, *Selected Values of Chemical Thermodynamic Properties*, Circular of the NBS 500, Feb., 1952.

17. SCHMIDT, E., *Thermodynamics, Principles and Applications to Engineering*. New York: Dover, 1966.

18. SKROTZKI, B. G. A., and W. A. VOPAT, *Power Station Engineering and Economy*. New York: McGraw-Hill, 1960.

19. SKROTZKI, B. G. A., *Basic Thermodynamics—Elements of Energy Systems*. New York: McGraw-Hill, 1963.

20. SOLBERG, H. L., O. C. CROMER, and A. R. SPALDING, *Thermal Engineering*. New York: Wiley, 1960.

21. ZERBAN, A. H. and E. P. NYE, *Steam Power Plants*. Scranton: International Textbook, 1952.

22. ASME, *Thermodynamic and Transport Properties of Steam*. New York: American Society of Mechanical Engineers, 1967.

23. *Steam Tables: Properties of Saturated and Superheated Steam*. New York: Combustion Engineering, Inc., 1940.

CHAPTER 2

AIRCRAFT AND MISSILE PROPULSION

2.1 INTRODUCTION

Some of the power cycles examined in Chapter 1 are applicable to aircraft and missile propulsion. The thermodynamic considerations in adapting particularly the Brayton cycle to aircraft propulsion and some thermodynamic aspects of rocket propulsion will be examined in this chapter. The machines which utilize or form components of power cycles will be described, and their thermodynamic features will be examined in more detail in Chapter 3.

The reciprocating internal-combustion engine was adapted to aircraft propulsion from the earliest days of powered flight, the beginning of this century. It was the power plant with the lowest weight-to-horsepower ratio. Today, it is still the most widely used power plant for relatively small propeller-driven aircraft. There is no need to examine this application of the Otto cycle separately here. The net power output of the engine is utilized as shaft power to drive the propeller. Engine details are different for aircraft applications, but the thermodynamic cycle remains unchanged.

Jet propulsion utilizing a gas turbine power plant was first successful in aircraft during World War II. The turbojet operates on an open Brayton cycle in which part of the useful energy is derived from the kinetic energy of the leaving gases. Also, the ram effect of incoming air can be utilized to decrease the compressor work. For these reasons, the thermodynamic cycle is different from the Brayton cycle for stationary or automotive power. Variations on the so-called turbo-propeller aircraft propulsion system utilize both shaft power and working fluid kinetic energy; they will be discussed in this chapter. It was mentioned in Chapter 1 that the open Lenoir cycle finds an application in the pulse-jet propulsion engine described briefly in this chapter.

In the discussion of rocket propulsion in this chapter, many of the terms used are applicable to currently developing technology in plasma and electrostatic ionic propulsion as well as to chemical rocket systems. There will be no discussion of energy sources other than liquid and solid fuels until magnetohydrodynamics and electrogasdynamics are introduced in Chapter 7 as an example of direct energy conversion.

This chapter is concerned with the propulsion of a body by the acceleration of a fluid relative to that body. It is assumed that the basic laws of fluid mechanics are

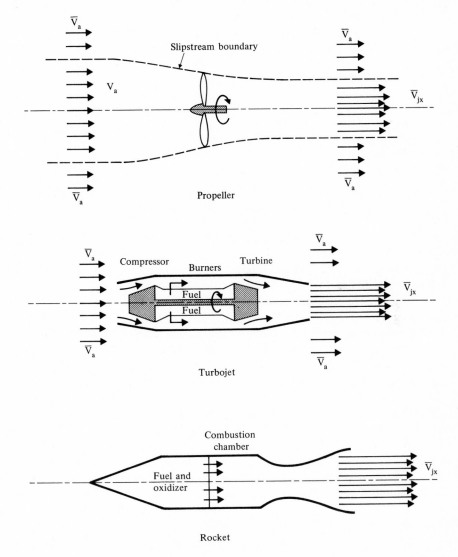

FIG. 2.1 Basic propulsion systems.

understood by the reader, but a brief recapitulation will precede a comparison of propulsion systems. Where it is necessary to consider the ideal rate of flow through a nozzle or the nozzle discharge velocity, the formulas will be used without derivation; it is assumed that nozzle theory has been or will be covered in fluid mechanics or aerodynamics courses.

2.2 BASIC PROPULSION SYSTEMS

The creation of acceleration in a fluid in order to produce a propulsive thrust force on a body is illustrated by the three basic propulsion systems shown schematically in **Fig. 2.1**. The **propeller** accelerates a part of the fluid in which it is operating from the relative velocity $\bar{V}_a$ between the body and the undisturbed stream (usually the speed of flight for an aircraft) to a greater relative velocity. In the idealized diagram, an imaginary slipstream boundary is shown. Only the axial component of jet velocity is useful for forward propulsion, and that is called $\bar{V}_{jx}$. The propulsion fluid does not pass through the power plant used to rotate the propeller shaft.

The **turbojet** propulsion system is different in that the air brought in at the front of the power plant is mixed with fuel in the burners, and the acceleration results primarily from the increase in temperature and specific volume of the jet fluid compared to the inducted air. Two different accelerations are involved: the air is accelerated from the relative velocity of $\bar{V}_a$ (usually the speed of flight) to the relative velocity $\bar{V}_{jx}$ in the jet; the fuel initially has zero velocity relative to the power plant and is accelerated to $\bar{V}_{jx}$. The whole shaft power of the gas turbine is utilized in driving the compressor. Turbo-propeller and ducted turbo-fan systems combining the two just described will be discussed in Section 2.6. In these systems, additional turbine shaft power is required for the propeller.

The basic difference between the propeller or the air-breathing turbojet and the **rocket** is that the rocket utilizes no fluid from its surroundings for propulsion. Both the fuel and the oxidizer are carried with the power plant and are accelerated from zero relative velocity to a high value at the nozzle outlet. Again, only the axial component of jet velocities is useful for propulsion, and that is represented by $\bar{V}_{jx}$. A combination of the rocket and turbojet systems will be described in Section 2.8.

2.3 FORCES RESULTING FROM FLUID ACCELERATION

A general open system through which fluid flows is shown in **Fig. 2.2**. This two-dimensional example shows a positive acceleration of fluid in both the x- and the y-directions. A component of acceleration could be added in the z-direction as well, where x, y, and z are mutually perpendicular. Fluid will enter the system at ① with a uniform velocity $\bar{V}_1$ across A_1, at a rate $\dot{m}_1$ pounds per second, and with a pressure p_1. Additional fluid at a rate $\dot{m}_3$ will be assumed to join $\dot{m}_1$ with a velocity $\bar{V}_3$, but through negligible flow area. Both fluids will leave the system at ② with uniform velocity $\bar{V}_2$ across A_2 and with a pressure p_2. The ambient pressure p_a acts on the boundaries of the system. The external drag forces are not included in an evaluation

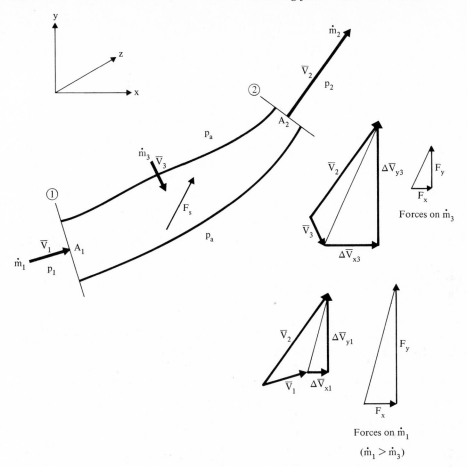

FIG. 2.2 Forces required to accelerate fluid in an open system.

of a propulsion system since the power plant can be mounted or enclosed in various ways.

The force required to accelerate a particle of mass m is

$$F = \frac{m}{g_c} \frac{d\bar{V}}{dt}.$$

Therefore a force must act on the fluid passing through in such a way that, in the x-direction,

$$F_x = \frac{\dot{m}_1}{g_c} (\bar{V}_2 - \bar{V}_1)_x + \frac{\dot{m}_3}{g_c} (\bar{V}_2 - \bar{V}_3)_x, \qquad (2.1a)$$

where $(\bar{V}_2 - \bar{V}_1)_x$ and $(\bar{V}_2 - \bar{V}_3)_x$ are the components of ΔV_1 and ΔV_3 in the x-direction as shown on the diagram. Identical expressions can be written for F_y and F_z in the y- and z-directions respectively.

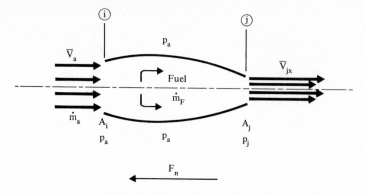

FIG. 2.3 A general propulsion system.

The forces to cause this acceleration come from the difference in the pressure forces at sections ① and ② and from a force exerted **by the system boundaries and internal components** on the fluid. This last force is designated as F_S in the diagram. Its component in the x-direction must be

$$F_{Sx} = F_x - (p_1 A_{1x} - p_2 A_{2x}). \tag{2.1b}$$

An equal and opposite force in the x-direction must be exerted by the fluid on the system. However, that is not the only force on the system. The ambient pressure p_a acts on the projected areas of the system in the x-, y-, and z-directions. The component of the sum of all forces on the system in the x-direction will be called the **net thrust**, F_n, acting toward the left:

$$\begin{aligned} F_n &= F_{Sx} + [p_a(A_x - A_{2x}) - p_a(A_x - A_{1x})] \\ &= F_{Sx} + p_a(A_{1x} - A_{2x}), \end{aligned} \tag{2.1c}$$

where A_x, A_{1x} and A_{2x} are total, inlet, and outlet areas projected in the x-direction. Combining Eqs. (2.1a, b, and c), we find that

$$\begin{aligned} F_n &= F_x - [(p_1 A_{1x} - p_2 A_{2x})] + [p_a(A_{1x} - A_{2x})] \\ &= \frac{\dot{m}_1}{g_c}(\bar{V}_2 - \bar{V}_1)_x + \frac{\dot{m}_3}{g_c}(\bar{V}_2 - \bar{V}_3)_x - A_{1x}(p_1 - p_a) + A_{2x}(p_2 - p_a). \end{aligned} \tag{2.1d}$$

For the general propulsion system, illustrated in **Fig. 2.3**, the thrust producing force is axial and positive in the direction of flight. The relative velocity between the intake air at i and the system is equal in magnitude to the speed of flight $\bar{V}_a$. The pressure at i is the ambient pressure. The fuel has zero velocity relative to the aircraft. The axial velocity $\bar{V}_{jx}$, by definition, is in the direction of flight. For such a system, the net thrust is

$$F_n = \frac{\dot{m}_a}{g_c}(\bar{V}_{jx} - \bar{V}_a) + \frac{\dot{m}_F}{g_c}(\bar{V}_{jx}) + A_j(p_j - p_a). \tag{2.2a}$$

This expression applies to the **turbojet** propulsion system.

For a propeller, only one fluid is involved, and it is accelerated from $\bar{V}_a$ to $\bar{V}_{jx}$. Pressures ahead of and behind the propeller must equalize to p_a. The expression for net thrust for a **propeller** reduces to

$$F_n = \frac{\dot{m}}{g_c}(\bar{V}_{jx} - \bar{V}_a). \tag{2.2b}$$

This expression provides a first approach to the determination of net thrust for a turbojet, because the mass of fuel per second is usually small compared to the mass of air, and the jet pressure is very close to atmospheric. Therefore, Eq. (2.2b) can be used for a **turbojet** with **fuel neglected** and p_j **assumed equal to** p_a.

For a rocket, the expression for net thrust is simplified because both the fuel and the oxidizer are carried with the system and have no initial velocity relative to it. The net thrust (or simply the thrust) of a **rocket** therefore is

$$F = \frac{\dot{m}_P}{g_c}(\bar{V}_{jx}) + A_j(p_j - p_a), \tag{2.2c}$$

where $\dot{m}_p$ is the mass rate of flow of all propellant material through the nozzle.

The **gross thrust** for a propulsive system is a term used primarily for turbojets. It is a hypothetical value used to eliminate the effect of intake conditions on the value of thrust produced, and it is particularly useful in comparing the data for different speeds of flight, including stationary test-bed data. Gross thrust is defined by

$$F_g = F_n + \frac{\dot{m}_a}{g_c}(\bar{V}_a). \tag{2.3}$$

There is no significance to the terms "net" and "gross" applied to the thrust of a rocket.

It will be useful to speak of the thrust produced per unit mass rate of flow of propulsive fluid. For a turbojet with a fuel-to-air ratio designated by FA, the **specific net thrust** is

$$F_{ns} = \frac{(1 + FA)}{g_c}(\bar{V}_{jx}) - \frac{\bar{V}_a}{g_c} + \frac{A_j}{\dot{m}_a}(p_j - p_a) \text{ lb thrust/lb air per sec.}$$

The corresponding **specific gross thrust** is

$$F_{gs} = \frac{(1 + FA)}{g_c}(\bar{V}_{jx}) + \frac{A_j}{\dot{m}_a}(p_j - p_a) \text{ lb thrust/lb air per sec.}$$

It would be useful, of course, to remove the one remaining reference to the geometry of the power plant, A_j, and to deal as far as possible with fluid properties. For the temperatures and pressures of gases at the jet outlet, it is reasonable to assume that the perfect gas laws apply. From the continuity equation,

$$A_j\bar{V}_{jx}\rho_j = \dot{m}_a + \dot{m}_F; \qquad \frac{A_j}{\dot{m}_a} = \frac{(1 + FA)}{\bar{V}_{jx}\rho_j} = \frac{(1 + FA)}{\bar{V}_{jx}}\frac{R_jT_j}{p_j}.$$

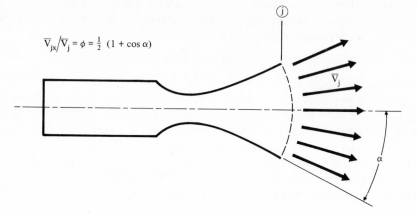

FIG. 2.4 A diverging nozzle with uniform velocity distribution.

The specific net thrust equation can be written

$$F_{ns} = (1 + FA) \left\{ \frac{\bar{V}_{jx}}{g_c} + \frac{R_j T_j}{\bar{V}_{jx}} \left[1 - \left(\frac{p_a}{p_j} \right) \right] \right\} - \frac{\bar{V}_a}{g_c}. \tag{2.4}$$

For specific gross thrust, F_{gs}, the final term drops off.

The term **equivalent jet velocity** is used in an attempt to express the gross thrust as a product of the mass rate of flow and velocity. It is the **hypothetical** jet velocity that would give a particular **gross thrust** if the jet did expand exactly to a pressure equal to the atmospheric pressure ($p_j = p_a$). Thus

$$F_g = (\dot{m}_a + \dot{m}_F) \frac{\bar{V}_{je}}{g_c},$$

$$F_{gs} = (1 + FA) \frac{\bar{V}_{je}}{g_c}. \tag{2.5a}$$

From this definition, it is seen that the equivalent jet velocity must be

$$\bar{V}_{je} = \bar{V}_{jx} + \frac{g_c A_j}{\dot{m}_a + \dot{m}_F} (p_j - p_a)$$

$$= V_{jx} + g_c \frac{R_j T_j}{\bar{V}_{jx}} \left[1 - \left(\frac{p_a}{p_j} \right) \right]. \tag{2.5b}$$

For a **rocket**, only one mass rate of flow is involved, and

$$F = \frac{\dot{m}_P}{g_c} \bar{V}_{je}. \tag{2.6}$$

Most rockets have diverging nozzles to achieve supersonic velocities. It can be shown [A.2] that the axial component of a conically diverging nozzle is related to the actual jet velocity as follows:

$$V_{jx} = \phi V_j, \tag{2.7a}$$

where

$$\phi = \tfrac{1}{2}(1 + \cos \alpha), \tag{2.7b}$$

the angle α being half the cone angle. See **Fig. 2.4.** Thus the relationship between effective jet velocity and actual jet velocity for a rocket can be written

$$
\begin{aligned}
V_{je} &= \phi \bar{V}_j + \frac{g_c R_j T_j}{\bar{V}_j}\left[1 - \left(\frac{p_a}{p_j}\right)\right] \\
&= V_j\left\{\phi + \frac{1}{M_j^2 \gamma}\left[1 - \left(\frac{p_a}{p_j}\right)\right]\right\},
\end{aligned} \tag{2.7c}
$$

where M_j is the Mach number at the jet exit. (Mach number is the ratio of the local fluid velocity to the speed of sound in that fluid. For a gas, $M = \bar{V}/\sqrt{\gamma g_c RT}$.)

2.4 PROPULSIVE EFFICIENCY

Regardless of the thrust force in any propulsion system, it must be realized that no useful work is being done unless there is forward propulsion. The **thrust power** is therefore defined as

$$TP = F_n \bar{V}_a,$$

where $\bar{V}_a$ is the forward velocity of the system.

It is now possible to define **propulsive efficiency** and to examine the difference between a propeller or a simplified turbojet (fuel ignored) on the one hand and a rocket on the other. Propulsive efficiency is the ratio between the useful power output (which is the thrust power) and the sum of this output and the loss (which are more directly obtained than the input). The lost energy is the kinetic energy of the jet, dissipated as the jet stream comes to rest relative to the surroundings; there must be a loss wherever the effective jet velocity is not equal in magnitude to the speed of flight through the surroundings.

Useful output

  a) Propeller or simplified turbojet:

$$
\begin{aligned}
TP &= F_n \bar{V}_a \\
&= (\dot{m}/g_c)(\bar{V}_{je} - \bar{V}_a)\bar{V}_a.
\end{aligned} \tag{2.8}
$$

  b) Rocket:

$$
\begin{aligned}
TP &= F\bar{V}_a \\
&= (\dot{m}/g_c)(\bar{V}_{je})\bar{V}_a.
\end{aligned} \tag{2.9}
$$

Lost power

  Propeller, simplified turbojet, and rocket:

$$\text{K.E. Loss/sec} = (\dot{m}/2g_c)(\bar{V}_{je} - \bar{V}_a)^2. \tag{2.10}$$

Efficiency of propulsion

a)   Propeller or simplified turbojet:

$$\eta_P = \frac{\text{Useful output}}{\text{Output} + \text{Loss}}$$

$$= \frac{(\bar{V}_{je} - \bar{V}_a)\bar{V}_a}{(\bar{V}_{je} - \bar{V}_a)\bar{V}_a + \frac{1}{2}(\bar{V}_{je} - \bar{V}_a)^2}$$

$$= \frac{2\bar{V}_a}{2\bar{V}_a + (\bar{V}_{je} - \bar{V}_a)}$$

$$= \frac{1}{1 + (\Delta\bar{V}/2\bar{V}_a)} \tag{2.11a}$$

$$= \frac{2}{1 + (\bar{V}_{je}/\bar{V}_a)}. \tag{2.11b}$$

b)   Rocket:

$$\eta_P = \frac{\text{Useful output}}{\text{Output} + \text{Loss}}$$

$$= \frac{\bar{V}_{je}\bar{V}_a}{\bar{V}_{je}\bar{V}_a + \frac{1}{2}(\bar{V}_{je} - \bar{V}_a)^2}$$

$$= \frac{2(\bar{V}_{je}\bar{V}_a)}{\bar{V}_{je}^2 + \bar{V}_a^2}$$

$$= \frac{2(\bar{V}_{je}/\bar{V}_a)}{1 + (\bar{V}_{je}/\bar{V}_a)^2} \tag{2.12a}$$

$$= \frac{2(\bar{V}_a/\bar{V}_{je})}{1 + (\bar{V}_a/\bar{V}_{je})^2}. \tag{2.12b}$$

For the two basically different systems, it can be seen that as $(\bar{V}_{je}/\bar{V}_a)$ approaches unity, the propulsive efficiency approaches 100 %. For the propeller and the turbojet, the flight velocity cannot exceed the jet velocity if the thrust is to be positive, and $\bar{V}_a$ approaches $\bar{V}_{je}$ only as a limit at which the thrust would equal zero. On the other hand, the jet velocity of a rocket is independent of forward motion, and $\bar{V}_{je}$ can be less than, equal to, or greater than $\bar{V}_a$.

The ideal values of propulsive efficiency from Eqs. (2.11) and (2.12) are plotted in **Fig. 2.5.** Actual overall efficiencies of flight are much lower.

2.5 *THE TURBOJET PROPULSION SYSTEM*

The gas turbine power plant (Brayton cycle) and its components, except the intake diffuser and the jet nozzle, have been discussed at length in Section 1.3.5 and Section 1.3.10. The application of the cycle to aircraft propulsion and these two additional components will be dealt with here.

Figure 2.6 shows the ideal cycle as well as a cycle with losses in each component. The **ram effect** resulting from the forward velocity of the aircraft $(\bar{V}_a)$ is the useful kinetic energy of this relative velocity. Whether or not there are internal friction

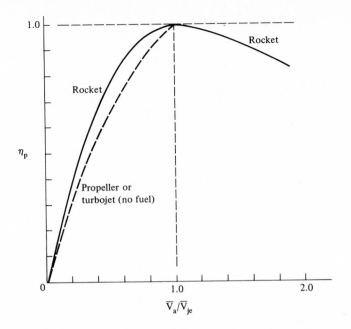

FIG. 2.5 Ideal propulsive efficiency.

losses in the inlet diffuser, the ram process can be considered adiabatic as were other turbine cycle processes. The stagnation enthalpy, with or without friction, at the diffuser outlet (compressor inlet) is

$$h_{oa} = h_a + \frac{\bar{V}_a^2}{2g_c J}.$$

The **ram efficiency** (η_R) is defined in terms of pressure drop because energy is constant. Point **n** is the compressor inlet.

$$\eta_R = \frac{\text{Actual pressure rise}}{\text{Ideal pressure rise}}$$

$$= \frac{p_{on} - p_a}{p_{oa} - p_a}$$

$$= \frac{p_{on}/p_a - 1}{p_{oa}/p_a - 1}. \tag{2.13}$$

This definition of ram efficiency has a serious defect. At zero forward velocity the denominator is zero and the numerator is equal to zero (no loss) or negative. Even with a finite value of $\bar{V}_a$, the denominator could be negative if p_{on} dropped below the static ambient pressure. Such situations create meaningless values of ram efficiency.

A more useful gauge of ram effect is given by the percent of stagnation pressure that is lost. Compare this with Eqs. (1.38a) and (1.38b) for a burner. The **stagnation**

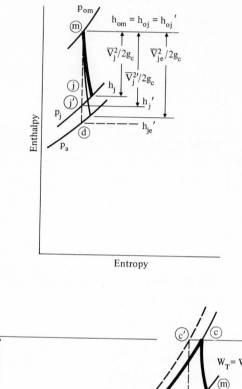

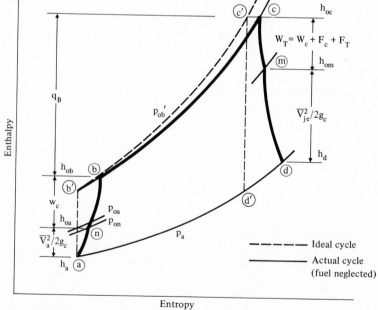

FIG. 2.6 An actual turbojet cycle (compare Fig. 1.36).

pressure coefficient is defined as

$$E_R = \frac{p'_{on} - p_{on}}{p'_{on}} = \frac{p_{oa} - p_{on}}{p_{oa}} = 1 - \frac{p_{on}}{p_{oa}}. \tag{2.14}$$

By definition, effective jet velocity ($\bar{V}_{je}$) assumes expansion to atmospheric pressure. An accompanying diagram in Fig. 2.6 shows the real and ideal jet expansion process for p_j greater than p_a. It is often convenient to use the effective jet velocity ($\bar{V}_{je}$) in a discussion of the cycle and propulsive efficiency. However, it must be remembered that it was defined in terms of gross thrust and hypothetical expansion to a jet pressure equal to ambient pressure. It has no significance in connection with actual jet efficiency.

Because the purpose of a jet, like that of a nozzle, is to create a high fluid velocity, its efficiency is defined in terms of velocity rather than energy. **Jet efficiency** is the velocity existing at the end of expansion compared to the velocity that would exist after isentropic expansion to the same static pressure. With the symbols of Fig. 2.6,

$$\eta_J = \frac{\bar{V}_j}{\bar{V}'_j} = \left[\frac{h_{om} - h_j}{h_{om} - h'_j} \right]^{1/2}. \tag{2.15}$$

If the same definition is applied to the **effective jet velocity**, an expression can be found in terms of specific gross thrust that is useful particularly where thrust can be measured as on a test stand:

$$\eta_{Je} = \frac{\bar{V}_{je}}{\bar{V}'_{je}}$$

$$= \frac{1}{1 + FA} \frac{F_{gs}}{[2J/g_c(h_{om} - h'_{je})]^{1/2}}. \tag{2.16a}$$

Alternatively, it may be desirable to express the effective jet efficiency entirely in terms of fluid properties. From Eq. (2.5b),

$$\bar{V}_{je} = \bar{V}_j + \frac{g_c R_j T_j}{\bar{V}_j} [1 - (p_a/p_j)].$$

From the diagram,

$$\bar{V}_j = [2g_c J(h_{om} - h_j)]^{1/2}.$$

Combining these, we have

$$\eta_{Je} = \frac{(h_{om} - h_j) + (R_j T_j/2J)[1 - (p_a/p_j)]}{[(h_{om} - h'_{je})(h_{om} - h_j)]^{1/2}}. \tag{2.16b}$$

The actual jet efficiency and the effective jet efficiency are not necessarily numerically equal.

If the weight of fuel is neglected for the cycle of Fig. 2.6, the cycle efficiency, propulsive efficiency, and overall thermal efficiency are very easily correlated. The **overall thermal efficiency** (fuel neglected) is

$$\eta_o = \frac{\text{Useful output}}{\text{Heat input}} = \frac{F_{ns}\bar{V}_a}{J(h_{oc} - h_{ob})} = \frac{(\bar{V}_{je} - \bar{V}_a)\bar{V}_a}{g_c J(h_{oc} - h_{ob})}. \tag{2.17}$$

Inspection of the diagram reveals the net useful energy, as for Eqs. (1.25a) and (1.25b). The **thermal efficiency** of the cycle is

$$\eta_{th} = \frac{\text{Heat input} - \text{Heat rejected}}{\text{Heat input}}$$

$$= \frac{\text{Net useful energy}}{\text{Heat input}} = \frac{(\bar{V}_{je}^2 - \bar{V}_a^2)}{2g_c J(h_{oc} - h_{ob})}. \tag{2.18}$$

The **propulsive efficiency,** fuel neglected, was shown in the development of Eq. (2.11) to be

$$\eta_P = \frac{(\bar{V}_{je} - \bar{V}_a)\bar{V}_a}{(\bar{V}_{je} - \bar{V}_a)\bar{V}_a + \frac{1}{2}(\bar{V}_{je} - \bar{V}_a)^2}$$

$$= \frac{2(\bar{V}_{je} - \bar{V}_a)\bar{V}_a}{(\bar{V}_{je}^2 - \bar{V}_a^2)}. \tag{2.19}$$

The **product** of the thermal efficiency (η_{th}) and the propulsive efficiency (η_P) is:

$$\eta_{th}\eta_P = \frac{(\bar{V}_{je} - \bar{V}_a)\bar{V}_a}{g_c J(h_{oc} - h_{ob})} = \eta_{OA}.$$

If an overall efficiency were based on the heating value of the fuel, the combustion efficiency of the burner would be taken into account also. This **fuel-based overall efficiency** would be

$$\eta_F = \frac{\text{Useful output}}{(h_{oc} - h_{ob})/\eta_B}$$

$$= \eta_{th}\eta_P\eta_B. \tag{2.20}$$

The overall efficiency based on fuel consumption is perhaps more usefully expressed as the specific fuel consumption. Unlike the output for other power plants, the output of a jet propulsion system may be expressed as thrust alone (if desired). The **thrust specific fuel consumption** is defined as

$$\text{tsfc} = \frac{\text{Lb fuel per hour}}{\text{Net thrust}} = \frac{3600\,\dot{m}_F}{F_n}$$

$$= \frac{3600\,\text{FA}}{F_{ns}} \text{ lb fuel/lb thrust-hour.} \tag{2.21}$$

Although horsepower may seem to have little relevance in jet propulsion, it is a useful and familiar unit of power. Thus the specific fuel consumption is sometimes calculated as **thrust horsepower specific fuel consumption:**

$$\text{thpsfc} = \frac{3600\,\dot{m}_F}{THP}$$

$$= \frac{3600\,\text{FA}}{F_{ns}V_a/550} \text{ lb fuel/hp-hr.} \tag{2.22}$$

Note that for a particular fuel heating value, this is inversely proportional to the fuel-based overall efficiency (η_F).

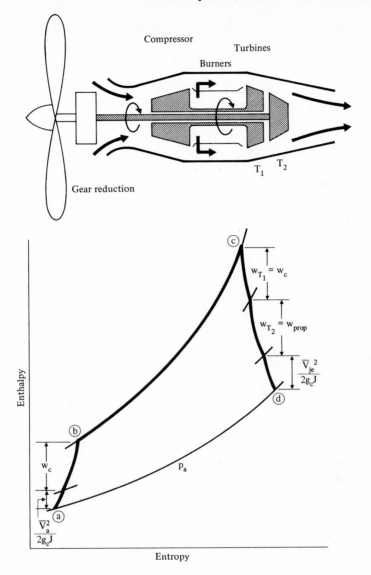

FIG. 2.7 The turbo-propeller.

2.6 TURBO-PROPELLERS, TURBO-FANS, AND THRUST AUGMENTATION

The three topics to be discussed in this section, turbo-propellers, turbo-fans, and thrust augmentation devices, are grouped together because the first two could be considered special examples of the last. In all three systems, the thrust produced by a simple turbojet propulsion system is supplemented or augmented by some addition to the basic turbojet.

The turbo-propeller system illustrated in **Fig. 2.7** is an attempt to utilize the advantages of both the propeller (normally driven by a reciprocating piston engine) and the turbojet in the area of operation in which neither alone is completely satisfactory.

The conventional propeller achieves good thrust at low speed because very large air quantities are accelerated, but to moderate "jet" velocities. At relatively high speeds of flight, the conventional propeller becomes very inefficient because of high relative velocities between the propeller and the air. The tip speed is limited to a total relative velocity (forward plus tangential) of something less than Mach 1.0 to avoid shock losses. On the other hand, the simple turbojet is inefficient at low flight speeds because large jet velocities are necessary to achieve appreciable thrust, and high $\bar{V}_j/\bar{V}_a$ ratios are inefficient. In intermediate ranges, the combination turbo-propeller is economical especially for long flights at moderate speeds. Even with a separate shaft from the later turbine stages as shown, a gear reduction is necessary to reduce propeller speed below compressor/turbine speed.

The *h-s* diagram of Fig. 2.7 indicates the increased power requirements of the turbine. In addition to the turbine work (w_{T1}) equal to the compressor work plus losses, there must be shaft work available (w_{T2}) equal to the propeller work (w_{prop}) plus losses. The jet velocity is thus reduced. Although any proportion of the useful work could go to the propeller, it has been usual to have this well over 50 % of the available energy.

The cycle efficiency of the turbo-propeller combination is therefore

$$\eta_{\mathrm{th}} = \frac{\text{Net useful work}}{\text{Heat added}}$$

$$= \frac{(w_{T2}) + (V_{je}^2 - V_a^2)/2g_cJ}{(h_{oc} - h_{ob})}. \tag{2.23}$$

The **turbo-fan** and **ducted-fan** arrangements shown in **Fig. 2.8** are essentially the same from a cycle point of view as the turbo-propeller system. Additional shaft work is required from the turbine or from a separate turbine to drive what might be called fans or blowers or low-pressure-ratio compressors. The difference between the first of these and the turbo-propeller is that a cowling surrounds the fan tips to provide more efficient operation at higher speeds of flight with smaller diameter, and therefore smaller tip speed, for a given shaft speed. Also, part of the discharge of the fan goes to the main compressor so that this could be considered the first compressor stage. The air that passes outside the compressor intake provides forward thrust in exactly the same way that air from a propeller does. It does not join the products of combustion to pass through the main propulsion jet.

The second arrangement illustrated does duct the air accelerated by the fan to the rear of the power plant, where it is mixed with the turbine exhaust to augment the mass rate of flow through the propulsion jet.

The third arrangement utilizes extensions of the blades of a free turbine, the blades being twisted, of course, to provide thrust rather than torque. All the gases from the compressor-drive turbine pass through this last turbine stage, which can

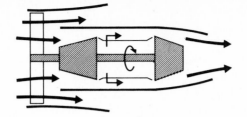

Turbo-fan

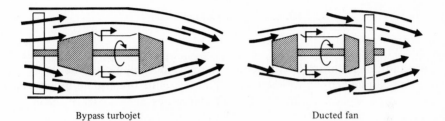

Bypass turbojet

Ducted fan

FIG. 2.8 Turbo-fan and ducted-fan arrangements.

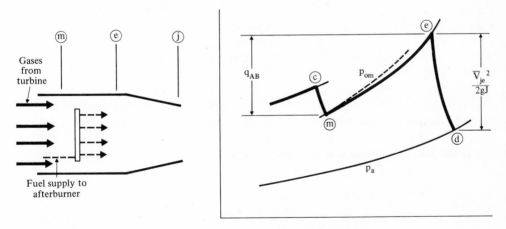

FIG. 2.9 Afterburner thrust augmentation.

operate at a slower speed to accommodate the greater radius of the ducted fan. The two streams of air and gas proceed to the thrust nozzle together.

Afterburning is the most effective means of thrust augmentation, but it is uneconomical and thus is used only as required rather than continuously as are the turbo-propeller and turbo-fan arrangements. Because a turbojet must operate with an excess of air in order to limit the gas temperature at the entrance to the turbine, oxygen is present in the turbine exhaust to support further combustion. The introduction of additional fuel at the afterburner, as illustrated in **Fig. 2.9**

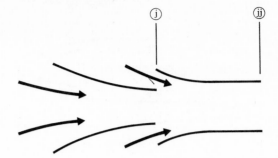

FIG. 2.10 Static jet augmenter.

provides additional thrust by increasing the specific volume of the gases. The temperature can go well above the turbine-limited value because no moving parts are encountered beyond the afterburner. The efficiency of this part of the cycle is, however, lower than that of the basic cycle because the available expansion ratio (P_{oe}/P_a) is less than the original (P_{oc}/P_a). The fraction of the energy added from m to e that can be utilized is potentially less than the useful fraction of the heat added at the higher pressure before point c. Note that the temperature at which heat is rejected has been raised by afterburning. Recall Fig. 1.37.

Under favorable conditions, the thrust of the basic turbojet can be increased with no addition to the system except the **static jet augmenter** shown in **Fig. 2.10**. The high velocity at the jet outlet j produces a low pressure which, along with a viscous entrainment of the surrounding air, induces a flow of atmospheric air through the collar of the augmenter duct. The mass rate of flow at section jj is greater than that through the turbojet system, but the velocity will be reduced because of the entrainment of the augmenting air. If the product $\dot{m}\bar{V}$ has been increased, the gross thrust will have been increased proportionately.

2.7 RAMJET AND PULSE-JET ENGINES

Two very simple air-breathing propulsion devices operate without a turbine/compressor power plant to supply air; instead, they depend on their forward velocity to provide sufficient "ram" pressure. Consequently, they are not self launching. Both are shown in **Fig. 2.11**.

The **ramjet** is the simpler of the two, being little more than an open duct with controlled inlet diffusion for maximum pressure rise and a fuel supply that burns continuously. It is sometimes referred to as an aero-thermodynamic-duct (abbreviated athodyd), and in Europe as the Lorin power plant. It has been used at the rotor tips of helicopters to produce torque since the tangential velocity can be high even at low flight speeds or hovering. The ramjet is useful from about Mach 0.7 to about Mach 3.0. The scramjet, which is a ramjet with supersonic combustion, may raise this limit much higher [A. 6].

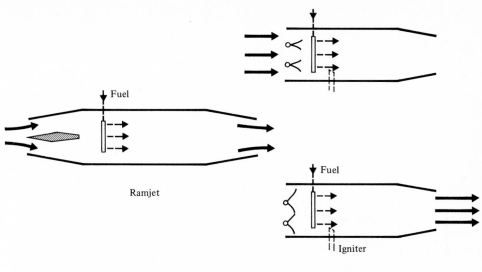

FIG. 2.11 Ramjet and pulse-jet engines.

In the same figure, a **pulse-jet** is shown at two different times in its operating cycle. At the air intake, long thin vanes are pivoted to open when the stagnation pressure created at the front is greater than the duct pressure. As the intake vanes return to their closed position, the fuel-air mixture is fired. The products escaping from the rear provide the forward thrust. Ignition is intermittent, but timed only by the valve motion which assumes a natural frequency of operation for the whole system. In the best known example of this device, the German V-1 winged missile or buzz-bomb of the second World War, the frequency was about 40 cps. The frontal drag of this system limits its operation to subsonic flight. A reference to this device as an open Lenoir cycle was made in Section 3.4 of Chapter 1.

2.8 ROCKETS

The effective jet velocity, propulsive efficiency, and other basic considerations for a rocket have already been discussed in the earlier sections of this chapter where propellers, turbojets, and rockets were described briefly and compared. Certain aspects of the rocket are peculiar to it alone, and a group of terms that relate primarily to rocket science and technology has been developed. These terms will be dealt with here. Many of them apply to any type of rocket, but the discussion in this chapter will be concerned primarily with liquid fuel and solid fuel "chemical" rockets.

Jet velocities are much higher for a rocket than for a turbojet, and diverging nozzles are necessary to achieve these supersonic values. Nozzle geometry is established by the pressure ratio. However, this cannot be constant for a rocket

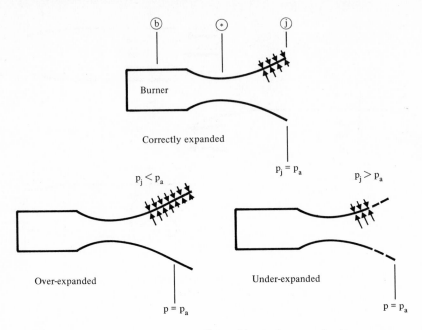

FIG. 2.12 Propulsion and drag effect of jet and atmospheric pressures.

climbing vertically through the atmosphere. A compromise in design or a variable shape ("rubber" nozzle) is therefore necessary. A **correctly expanded** nozzle is one in which the jet pressure is exactly equal to the ambient pressure. Inspection of **Fig. 2.12** shows that an **over-expanded** nozzle in which p_j is less than p_a (diverging section too long for conditions of operation) will have an additional drag force because of the atmospheric pressure on the extension of the nozzle beyond the section at which $p_j = p_a$. On the other hand, an **under-expanded** nozzle in which jet pressure does not get as low as ambient pressure does not fulfill its maximum possible forward thrust potential in that it fails to provide additional surface on which jet pressure can act positively, and also it fails to achieve maximum possible jet velocity.

Probably the most significant single criterion in the evaluation of rockets is the **specific impulse.** It is the ratio of the thrust force to the rate of propellant expenditure and is often given only in its net units, seconds:

$$I_s = \frac{F}{\dot{m}_P} \text{ lb thrust/lb propellant per sec.}$$

The units of thrust must be pounds-force, while those of mass of propellant per second are pounds-mass. The thrust, of course, is the product of the mass rate of flow and the effective jet velocity, so that

$$I_s = \dot{m}_P \bar{V}_{je}/g_c \dot{m}_P = \bar{V}_{je}/g_c \text{ sec.}$$

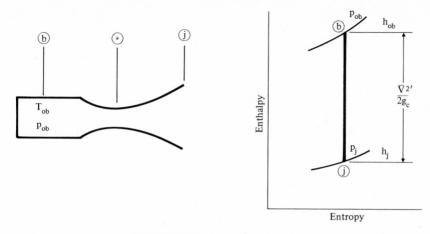

FIG. 2.13 Ideal nozzle expansion.

During the time that the thrust remains constant, the product of thrust and time is the **total impulse.** [Where thrust varies with time, the total impulse would be the integral of $(F\,dt)$ or the area under a curve of F on a base of time.]

$$I_T = Ft = I_s \dot{m}_p t$$
$$= I_s \cdot \text{(Total propellant consumed) lb-sec.}$$

Another term, the **specific propellant consumption,** evaluates the same factors as does the specific impulse, but inversely. It could be defined as the mass in pounds of propellant to produce a total impulse of one pound-second:

$$W_s = \dot{m}_p/F = 1/I_s \text{ lb/lb-sec.}$$

It should be noted here that the specific propellant consumption is extremely important in the overall evaluation of a rocket propulsion system when the weight of the propellant (as in a chemical rocket) is a large fraction of the total weight of the whole system. It is not quite so important by itself if the propulsion system involves other large weights as in a magnetohydrodynamic rocket.

Further evaluation of a rocket involves the effectiveness of the expansion nozzle. For an ideal gas in the burner with stagnation temperature T_{ob} and stagnation pressure p_{ob}, the ideal velocity at the jet is that attained by isentropic expansion to p_j as shown on the *h-s* diagram of **Fig. 2.13.** With a constant specific heat assumed,

$$\bar{V}_j' = \sqrt{2g_c c_p (T_{ob} - T_j)}$$
$$= \sqrt{2g_c c_p T_{ob} [1 - (p_j/p_{ob})^{\gamma-1/\gamma}]}$$
$$= \sqrt{\frac{2g_c \gamma R T_{ob}}{(\gamma - 1)} [1 - (p_j/p_{ob})^{\gamma-1/\gamma}]}$$
$$= \sqrt{\frac{2g_c \gamma R_U T_{ob}}{(\gamma - 1)M_w} [1 - (p_j/p_{ob})^{\gamma-1/\gamma}]}. \tag{2.24}$$

In the last form of this equation, R_U is the universal gas constant (approximately 1544 ft lb/lb mole-R) and M_w is the molecular weight. The significance of the molecular weight of the gas or gases is hidden in the previous forms of the equation, but is displayed in the last. It is interesting to note also that the term in the square brackets is the ideal gas-cycle efficiency for a Brayton cycle with an expansion process from p_{ob} to p_j. For this reason, the symbol η is sometimes used for that part of the expression for ideal jet velocity.

The **nozzle velocity coefficient** (ϕ) is the ratio between the ideal and the actual jet velocity:

$$\bar{V}_j = \phi \bar{V}_j'.$$

The **ideal mass rate of flow** of propellant depends on the throat area (A^*) at which section the velocity must be the acoustic velocity (Mach number unity) if the pressure ratio is greater than the critical value. The stagnation temperature and stagnation pressure with no loss would be the same at the throat and in the burner. Equation (1.35c) relates static and stagnation temperatures as a function of Mach number. Thus the throat velocity is

$$\bar{V}^* = \sqrt{g_c \gamma R T_*} = \sqrt{\frac{2 g_c \gamma R T_{ob}}{\gamma + 1}} \text{ ft/sec.}$$

From Eq. (1.36b), the density, where M is unity, is given by

$$\rho^* = \frac{p^*}{R T^*} = \frac{p_{ob}}{R T_{ob}} \left(\frac{2}{\gamma + 1} \right)^{1/\gamma - 1} \text{ lb/ft}^3.$$

From the continuity equation,

$$\dot{m}_P' = A^* \rho^* \bar{V}^*$$

$$= A^* p_{ob} g_c \gamma \left(\frac{\gamma + 1}{2} \right)^{(\gamma+1)/(2-2\gamma)} \sqrt{\frac{M_w}{\gamma g_c R_U T_{ob}}}. \tag{2.25}$$

The actual mass rate of flow may be less than this, equal to it, or even greater than the ideal. The ideal rate assumes a **frozen flow**, that is a gas in which no chemical reactions are taking place. For a chemical rocket, this is not likely to be so. Continued chemical reaction in the nozzle might mean a higher molecular weight than that assumed for completed reactions. Heat loss to the surroundings will produce a cooler and therefore a more dense gas. The pressure and temperature in the burner will depend on the degree to which equilibrium is reached. At lower temperature and for uncombined gases, the value of γ is likely to be higher. The **nozzle discharge coefficient** (ϵ) is the ratio of the ideal to the actual mass rate of flow and may be equal to unity or less or more than unity:

$$\dot{m}_P = \epsilon \dot{m}_P'.$$

The comparison of rocket nozzles of different sizes is facilitated through the use of a **propellant flow parameter** (or synonymously, a **weight flow coefficient**) for which the symbol C_W will be used:

$$C_W = \frac{\dot{m}_P}{A^* p_{ob}} \text{ sec}^{-1}.$$

Note that this is not a dimensionless coefficient. The **ideal** weight flow parameter (C'_W) can be calculated from the expression for the ideal mass rate of propellant flow (Eq. 2.25) for various values of T_{ob}/M_w and of γ.

The magnitude of the thrust force is strongly dependent on the throat area and the stagnation pressure in the burner chamber. A dimensionless **thrust coefficient** (C_T) is defined by

$$F = C_T A^* p_{ob}.$$

The ratio of the thrust coefficient to the propellant flow parameter is identical to the specific impulse.

Another term commonly used to evaluate a rocket nozzle is the **characteristic velocity** ($\bar{V}_c$). It is the ratio of the effective velocity to the thrust coefficient and is immediately seen to be the inverse of the propellant flow parameter on a slug rather than a pound-mass basis:

$$\bar{V}_c = \bar{V}_{je}/C_T \text{ ft/sec}$$
$$= (Fg_c/W_P)(A^*P_{ob}/F) = g_c/C_W.$$

The interdependence of several of these terms indicates some redundancy. However, each is useful to illustrate characteristics in different ways.

The significance of the characteristic velocity is indicated partly by the fact that with A^* equal to unity and W_P/g_c equal to unity, $\bar{V}_c$ would be numerically equal to the combustion chamber stagnation pressure. It could be described as the pressure (psi) necessary to force a unit mass rate of flow (1 slug/sec) through a unit throat area (1 sq in.). Examination of the expression for the ideal rate of propellant flow shows that $\bar{V}_c$ has the same order of magnitude as the sonic velocity at the combustion chamber temperature. Because the characteristic velocity ($\bar{V}_c$) is the ratio of effective velocity to the thrust coefficient and therefore the ratio of $A^* p_{ob}$ to W_P/g_c, it combines chamber, nozzle, and propellant characteristics and is an indication of the effectiveness of use of the gas generator or combustion process. It therefore is referred to as a "figure of merit" for the combination of propellant and combustion chamber design.

Combustion chambers are found in many forms. To indicate size by a single linear dimension would be difficult. Because of the significance of the nozzle throat area in determining mass rate of propellant escape, the **characteristic length** (L_c) of the combustion chamber is defined as

$$L_c = \frac{\text{Combustion volume}}{A^*}.$$

The dimensions are usually inches. It gives a relative indication of the residence time for the combustion process.

In the definitions of all the preceding terms, the relationship between the mass of the rocket alone and the mass of the propellant carried has been ignored. This important characteristic is included in the criterion called **propulsion vacuum velocity** ($\bar{V}_{P\text{ vac}}$). It is determined as the burn-out velocity (the velocity achieved when all propellant has been expended) when both drag and gravity are ignored and when the initial velocity is zero. Such a situation would exist for an orbital launch in a direction at right angles to a radius from the center of the earth and at an altitude where drag is negligible. Even for operation under other conditions, the removal of gravity forces and drag forces from any term that attempts to evaluate a propulsion system alone is justifiable since those forces will depend on the vehicle configuration and the flight program and not on the propulsion system itself. It will be obvious that, if launch velocity is not zero, the "propulsion vacuum velocity" will be the change of velocity over the time period of propellant consumption. Consider the acceleration at any instant:

$$\text{Mass}\cdot\text{Acceleration} = \text{Net force,}$$

$$\frac{m}{g_c}\cdot\frac{d^2x}{dt^2} = F = -I_s\left(\frac{dm}{dt}\right).$$

Here m is the instantaneous value of the mass of the rocket and its propellant in pounds and dm/dt is the rate of propellant consumption. The negative sign indicates that the mass of the body is decreasing. The change of velocity is the integral of the acceleration over the time period. At t_1 the velocity is zero (or $\bar{V}_1$) and the mass is the sum of the masses of the propellant (m_P) and the empty rocket (m_{MT}). At burnout time t_2, the velocity is the "propulsion vacuum velocity" (or this plus V_1) and the mass is that of the empty rocket (m_{MT}):

$$\frac{d^2x}{dt^2} = -g_c\frac{I_s}{m}\frac{dm}{dt}; \qquad \bar{V}_{P\,vac} = -g_c\int_{t_1}^{t_2}\frac{I_s}{m}\,dm.$$

It is reasonable to assume for the purpose of the definition of propulsion vacuum velocity that the specific impulse is constant. This does not necessarily mean that the thrust force is constant:

$$\bar{V}_{P\,vac} = g_c I_s \ln\frac{m_1}{m_2}$$

$$= g_c I_s \ln\left(1 + \frac{m_P}{m_{MT}}\right)$$

$$= \bar{V}_{je} \ln\left(1 + \frac{m_P}{m_{MT}}\right). \tag{2.26}$$

In Chapter 14 of Reference [B.11], Siefert summarizes typical values of the important performance parameters for both liquid and solid propellant rockets shown in Table 2.1.

Table 2.1

Parameter	*Symbol*	*Units*	*Typical Values*
Effective exhaust velocity	$\bar{V}_{je}$	ft/sec	3000–12,000
Specific impulse	I_s	sec	100–400
Thrust coefficient	C_T	—	1.1–2.0
Characteristic velocity	$\bar{V}_c$	ft/sec	2000–6000
Characteristic length	L_c	in.	10–50

The significance of the **molecular weight** of the propellant fluid was mentioned in the discussion of the ideal jet velocity, Eq. (2.24). That equation indicates that a high burner stagnation temperature and a low molecular weight are desirable in the attainment of high jet velocity. Even without that equation, the conclusions of the following paragraphs can be drawn [B.11, Chapter 14].

From the principle of equipartition of energy, we know that individual molecules of all gases at the same temperature will have the same translational kinetic energy which is proportional to the product of mass and the square of velocity. If the gases are allowed to expand to a vacuum, the velocity will be proportional to the inverse of the square root of the molecular weight. To compare two propellant gases of molecular weights M_{w1} and M_{w2}, we see that

$$\frac{\bar{V}_1}{\bar{V}_2} = \sqrt{\frac{M_{w2}}{M_{w1}}}.$$

The propulsive effect is the result of change of momentum, which is the product of mass and velocity, so that the **momentum per molecule** will be in proportion to

$$\frac{\text{Momentum}_1}{\text{Momentum}_2} = \frac{M_{w1}}{M_{w2}} \sqrt{\frac{M_{w2}}{M_{w1}}} = \sqrt{\frac{M_{w1}}{M_{w2}}}.$$

However, for the **same mass rate of flow** or rate of expenditure of propellant, the number of molecules must be inversely proportional to the molecular weight. The total momentum will vary according to

$$\frac{\sum \text{Momentum}_1}{\sum \text{Momentum}_2} = \frac{M_{w2}}{M_{w1}} \sqrt{\frac{M_{w1}}{M_{w2}}} = \sqrt{\frac{M_{w2}}{M_{w1}}}.$$

The specific impulse is the ratio of the thrust force to the propellant mass rate of expenditure. It is therefore dependent on molecular weight in the same way that the total change of momentum is for a given mass rate of flow. Thus

$$\frac{I_{s1}}{I_{s2}} = \sqrt{\frac{M_{w2}}{M_{w1}}}.$$

Consequently, the specific impulse would be greatest for hydrogen alone, the gas with the lowest molecular weight.

The primary classification for rocket motor types is made according to the fuel used which may be **liquid**, **solid**, or sometimes a combination of the two. While the **oxidizer** is usually oxygen itself or some chemical source of oxygen, the term is applied to anything that will provide a suitable exothermic chemical reaction with the **fuel**. A **hybrid** rocket combines a solid fuel with a liquid oxidizer.

Highly simplified diagrams of liquid- and solid-fuel rockets are shown in **Fig. 2.14**. The former are more complicated, requiring a pressurization system to supply the fuel and the oxidizer separately at high pressure to the burner nozzles. The storage tanks for fuel and oxidizer can be of relatively light-weight construction because they are not subjected to the high burner pressure. The pressurization system may embody a gas turbine driving the two supply pumps or compressors.

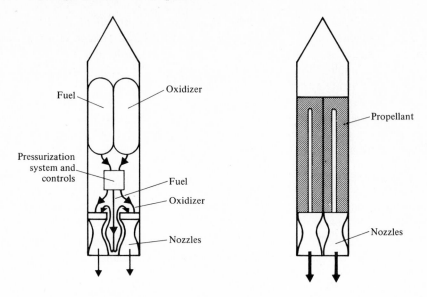

FIG. 2.14 A bipropellant liquid fuel rocket and a monopropellant solid fuel rocket.

The turbine itself requires a separate gassifier or combustion chamber as a source of hot gas. The nozzles can be cooled by fuel on its way to the burner passing over the outer surfaces from the exit end forward as shown in the diagram. Perhaps the greatest advantage of the liquid-fuel type rocket is that its fuel supply and therefore its thrust can be programmed, even including shutdown and restart, more easily and reliably than can the burning rate of a solid-fuel rocket.

The most commonly used liquid fuel has been kerosene used with oxygen as **bipropellants**. Both have been stored together cryogenically to form a **monopropellant**, which means simply that both fuel and oxidizer are stored and supplied as a premixed single substance. Other monopropellants are liquid oxygen and hydrazine (N_2H_4), nitric acid (HNO_3) and unsymmetrical dimethyl hydrazine (referred to as UDMH), and hydrogen peroxide with a hydrocarbon fuel. Single-component monopropellants, in the presence of a catalyst or at higher temperatures, decompose and release energy with no "oxidizer." Examples are hydrazine (N_2H_4), ethylene oxide (C_2H_4O), and nitromethane (CH_3NO_2). Liquid hydrogen used with liquid oxygen as a bipropellant could be classed as a *high-energy* fuel. Other examples are various combinations of fluorine (F_2) and hydrogen, or fluorine and hydrazine. These can improve specific impulse and reduce weight by 10 to 20 % compared to kerosene and oxygen. Because of the effect of temperature on dissociation and the effect of molecular weight already discussed, maximum specific impulse is not usually obtained at the chemically correct fuel-oxidizer ratio.

The whole fuel storage volume in a solid-fuel rocket is integral with the burner space and must therefore be built to withstand maximum burner pressure. It is possible to program thrust output by means of the internal configuration of the solid charge. This is shown in **Fig. 2.15**. Shutdown and restart have been made possible by interruptions in the solid fuel structure, but the degree of control possible is not the same as with liquid fuel systems.

A solid-fuel rocket must use a monopropellant. This may be homogeneous as in the "double-base" propellant, nitroglycerin-nitrocellulose: $C_3H_5(NO_3)_3 - C_6H_7O_2(NO_3)_3$. Alternatively, it may be a composite-type solid fuel with discrete particles of oxidizer distributed throughout a fuel matrix.

For a solid fuel, the **burning rate** is usually defined as the rate at which the burning surface recedes in a direction normal to itself. This is sometimes called the linear burning rate, and is expressed in inches per second. Its value for a given fuel structure is strongly dependent on the chamber pressure and to some extent on the temperature of the fuel. In addition to this normal burning, there can be an erosion of the fuel surface caused by high gas velocities with subsequent chemical action in the particle stream. This is referred to as erosive burning, and is largely independent of pressure and temperature.

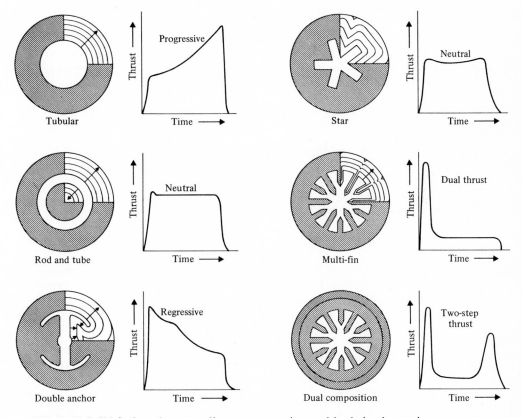

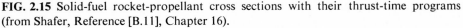

FIG. 2.15 Solid-fuel rocket-propellant cross sections with their thrust-time programs (from Shafer, Reference [B.11], Chapter 16).

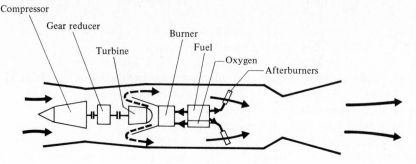

FIG. 2.16 A turbo-rocket.

A serious limitation to the thrust duration of a solid-fuel rocket is that the nozzle cannot be cooled regeneratively as can that of the liquid-fuel rocket.

When a rocket operates in the atmosphere rather than in space, several different means are available to improve the specific impulse. Air augmentation of the jet gases [A.8] is possible in a **ducted rocket** in which there can be afterburning of fuel in the ducted air stream, and also in a **shrouded rocket** in which the air augmentation occurs through a jet ejection action as described in conjunction with Fig. 2.10. A third possibility, where atmospheric air is available, is a combination of turbojet and rocket principles to provide a **turbo-rocket** [A.3]. This is shown schematically in **Fig. 2.16**. Part of the oxygen supply is carried in the craft along with the fuel supply. Some fuel and oxygen are supplied in an over-rich mixture (which limits the temperature) to a turbine which drives the intake air compressor. The fuel-rich turbine exhaust goes with the fresh air to afterburners, which are supplied with additional fuel. Just as the turbo-propeller propulsion system spanned the operating region between the limits where either propeller propulsion or turbojet propulsion was clearly superior, the turbo-rocket can operate best in between the regions of turbojet and rocket superiority.

PROBLEMS

1. In connection with the ideal propulsive efficiencies plotted in Fig. 2.5, the following statement was made: "Actual overall efficiencies are much lower." What two major assumptions were made that account for this discrepancy?
2. In Fig. 2.5, why does the propulsive efficiency curve for a turbojet terminate at a certain point while that for a rocket does not?
3. A simplified turbojet and a propeller have the same expression for propulsive efficiency. Why is a turbojet system much less efficient than a propeller drive at about 200 mph?
4. Why is the overall efficiency of a turbojet aircraft improved in general
   a) at higher altitudes?
   b) at greater speeds of flight?

5. Which is more indicative of economy for an aircraft: thrust specific fuel consumption or thrust horsepower specific fuel consumption?

6. What is the propulsive efficiency of a turbojet aircraft that develops a specific gross thrust of 50 lb per pound of air per second with a fuel-to-air ratio of 0.013 when flying at 650 mph? Assume that the jet pressure is practically equal to the atmospheric pressure.

7. What would the specific gross thrust have to be to produce a propulsive efficiency of 52% for a turbojet aircraft flying at 550 mph? Assume that the fuel-to-air ratio is 0.015, and the jet pressure is practically equal to the atmospheric pressure.

8. Does the ram efficiency affect the stagnation enthalpy at the entrance to the compressor of a turbojet aircraft?

9. Why is the ram efficiency defined in terms of pressure, and the nozzle efficiency in terms of velocity, while the compressor and turbine efficiencies are defined in terms of enthalpy?

10. A turbojet aircraft flies at 600 mph (880 fps) at an elevation of 15,000 ft where the atmospheric pressure is 8.29 psia and the temperature is +5 F. The stagnation pressure coefficient of the inlet diffuser is 70%. The compressor is 80% efficient. How much work is done by the compressor per pound of air to deliver it to the combustion chamber at a stagnation pressure of 40 psia?

11. A turbojet engine in an aircraft flying in standard air at an altitude of 25,000 ft ($p_a = 5.45$ psia) has a turbine exit stagnation temperature and stagnation pressure of 1000 F and 20 psia, respectively, and the jet pressure is 10 psia. (Neglect fuel.) For a jet nozzle efficiency of 90%, what is
a) the specific gross thrust?
b) the Mach number of the gas with respect to the exit nozzle? (Take $\gamma_j = 1.36$)
c) the increase in entropy of the gas (assumed the same as air) as it passes through the jet nozzle?

12. Assume that a turbojet could fly at 800 mph at standard sea level with an ideal Brayton cycle (ignoring fuel) under conditions that would give the maximum useful work with a limiting temperature of 2250 R at the turbine inlet. (See Chapter 1.)
a) What would the theoretical propulsive efficiency be?
b) With no other losses, by what percentage would the specific net thrust change if the burner pressure coefficient were changed from 0 to 0.05?

13. Consider a turbojet powerplant with component efficiencies at flight conditions as follows:

    Ram stagnation pressure coefficient, 0.96.
    Compressor isentropic efficiency, 0.88.
    Burner stagnation pressure coefficient, 0.95.
    Combustion efficiency, 0.97.
    Turbine isentropic efficiency, 0.85.
    Jet effective efficiency, 0.92.

Given that the speed of flight is 600 mph at an elevation of 30,000 ft, where the atmospheric pressure is 4.36 psia and the temperature is 412 R, find the following for a mass rate of air consumption of 65 lb/sec and a fuel-to-air ratio of 0.015.

The lower heating value of the fuel is 19,500 Btu/lb. The compressor pressure ratio is 4.5:1.

a) Maximum stagnation temperature in the system.
b) Net thrust.
c) Fuel-based overall thermal efficiency.
d) Thrust specific fuel consumption.

Assume that all gases have the properties of air as given in the Air Tables. Sketch a temperature-entropy diagram approximately to scale for the whole "cycle."

14. A turbopropeller aircraft flying at 400 mph can deliver 5000 hp to the propeller shaft and requires 8000 hp for the compressor with an air consumption in the power plant of 300,000 lb of air per hour. The limiting stagnation temperature at the turbine inlet is 1400 F, and the stagnation pressure there is 78 psia. The jet can expand to 13 psia. Assume that the turbine efficiency and the jet efficiency both are 88%, while the propeller efficiency is 78%. What total drag force can be overcome in steady level flight?

15. A turbojet is to fly at standard sea level at 500 mph. Assume that the stagnation pressure at the burner inlet and outlet will be 5 atm, and that the stagnation temperature at burner outlet will be 1240 F.

a) For an ideal cycle (no losses), by what percentage could the gross thrust be increased by the use of an afterburner with a stagnation temperature limit of 1840 F?

b) What would the ideal cycle efficiencies be with and without the burner in operation?

16. a) Draw an *h-s* diagram for an ideal ramjet "cycle" in which compression and expansion are both isentropic, jet pressure equals ambient pressure, and there is no loss of stagnation pressure throughout.

b) Calculate the ideal burner stagnation pressure for a flight speed of Mach 3.0. [Recall Eq. 1.36(b). Take ambient pressure as 1 atm.]

17. For an ideal ramjet as described in the previous problem, prove (a) that the ratio of the jet velocity to the speed of flight is equal to the ratio of the acoustic velocity at jet conditions to the acoustic velocity at ambient conditions and (b) that this ratio is the square root of $T_{oj} : T_{oa}$.

18. Why does a tubular design in a solid propellant rocket fuel chamber give a progressive thrust-time program?

19. In a rocket, any divergence of the jet stream reduces the effective thrust for a given jet velocity. Why is a diverging nozzle used?

20. A rocket motor usually uses a "fuel-rich" mixture while a turbojet engine always uses excess air. Explain these two facts.

21. A rocket motor combustion chamber has a stagnation temperature of 5000 F. The propellant fluid has an effective molecular weight of 25, and its value of γ is 1.25. Assuming an ideal mass rate of flow through the nozzle, calculate the characteristic velocity.

22. Compare the possible specific impulses of a rocket using (a) liquid oxygen and hydrogen, and (b) liquid oxygen and pentane (C_5H_{12}) as fuel, other things remaining equal. Assume for this purpose complete combustion and no dissociation.

REFERENCES

A. *Articles*

1. BEAM, P. E., JR., R. E. CUTLER, J. E. BROCK, and I. J. GERSHON, "Regenerative Turboprops Promise Endurance, Payload Hikes," *SAE Jl.*, Apr., 1964, pp. 62–65.
2. MALINA, J. F., "Characteristics of a Rocket Motor Unit Based on the Theory of Perfect Gases," *Jl. Franklin Institute*, Vol. 230, No. 4, 1940.
3. LOMBARD, A. A. and J. G. KEENAN, "The Turborocket," *Flight International*, Oct. 29, 1964, pp. 752–754.
4. SEIFERT, H. S., "Chemical Rocket Fundamentals," Chapter 14 in Reference B.11.
5. SHAFER, J. I., "Solid Rocket Propulsion," Chapter 16 in Reference B.11.
6. STULL, F. D., "Scramjet Combustion Prospects," *Astronautics and Aeronautics*, Vol. 3, No. 12, 1965, pp. 48–52.
7. SUTHERLAND, G. S., "Solid and Liquid Rockets—a Comparison," *SAE Paper*, 42A. Apr., 1958.
8. YAFFEE, M. L., "Air Augmented Missile Studies Increasing," *Aviation Week and Space Technology*, Nov. 22, 1965, pp. 49–61.

B. *Books and Separate Publications*

1. BARRERE, M., A. JAUMOTTE, B. F. DE VEUBEKE, and J. VANDENKERCKHOVE, *Rocket Propulsion*. New York: Elsevier, 1960.
2. BLASINGAME, B. P., *Astronautics*. New York: McGraw-Hill, 1964.
3. BUSSARD, R. W. and R. D. DeLAUER, *Fundamentals of Nuclear Flight*. New York: McGraw-Hill, 1965.
4. CORLISS, W. R., *Propulsion Systems for Space Flight*. New York: McGraw-Hill, 1960.
5. DOW, R. B., *Fundamentals of Advanced Missiles*. New York: Wiley, 1958.
6. DURHAM, F. P., *Aircraft Jet Power Plants*. Englewood Cliffs: Prentice-Hall, 1951.
7. HESSE, W. J., *Jet Propulsion*. New York: Pitman, 1958.
8. HILL, P. G. and C. R. PETERSON, *Mechanics and Thermodynamics of Propulsion*. Reading, Mass.: Addison-Wesley, 1965.
9. MORGAN, H. E., *Turbojet Fundamentals*. New York: McGraw-Hill, 1958.
10. PURSER, P. E., M. A. FAGET, and N. F. SMITH (Eds), *Manned Spacecraft: Engineering Design and Operation*. New York: Fairchild, 1964.
11. SEIFERT, H. S. (Ed), *Space Technology*. New York: John Wiley, 1959.
12. SMITH, C. W., *Aircraft Gas Turbines*. New York: Wiley, 1956.
13. SMITH, G. G., *Gas Turbines and Jet Propulsion for Aircraft*. New York: Aerosphere, 1944.
14. SUTTON, G. P., *Rocket Propulsion Elements*. New York: Wiley, 1963.
15. TURCOTTE, D. L., *Space Propulsion* (first edition). New York: Blaisdell, 1965.
16. WIECH, R. E., JR., and R. F. STRAUSS, *Fundamentals of Rocket Propulsion*. New York: Reinhold, 1960.
17. ZUCROW, M. J., *Aircraft and Missile Propulsion*, Vols. I and II, New York: Wiley, 1958.

CHAPTER 3

ENGINES, COMPRESSORS, AND TURBINES

3.1 INTRODUCTION

In the previous two chapters, "Power Cycles" and "Aircraft and Missile Propulsion," compression and expansion processes were discussed as part of thermodynamic cycles, but the configuration and general nature of the machines necessary were not described in any detail. Occasionally a reciprocating, a centrifugal, or some other type of machine was mentioned, but only in passing. This will also be true in general for the chapters that follow.

In this chapter, some of the details and characteristics of various types of engines and compressors will be presented. The discussion will still be limited to thermodynamic aspects of the machines. Terminology relating to physical detail will be introduced only insofar as it is an aid to the understanding of the thermodynamic processes. Some remarks on mechanical characteristics will be included to help the reader appreciate certain advantages and disadvantages.

Both compression and expansion machines will be discussed; where these machines are **prime movers,** the purpose is the production of power. The usual practice of calling reciprocating and rotary machines **engines**, and centrifugal and axial-flow machines **turbines** will be followed. It is common practice to call all three classes of machines **compressors** where the only object is to raise the pressure of the working fluid. Pumps that handle liquids, and blowers and fans that create fluid flow, but against relatively small pressures, will not be considered in any detail since their problems are primarily fluid-mechanic rather than thermodynamic in nature.

3.2 RECIPROCATING COMPRESSORS AND ENGINES

The term **reciprocating** is applied to a machine when a piston moves back and forth within a cylinder. This motion may be linked to rotary motion of a shaft or may act directly in a reciprocating manner as in a free-piston machine such as that

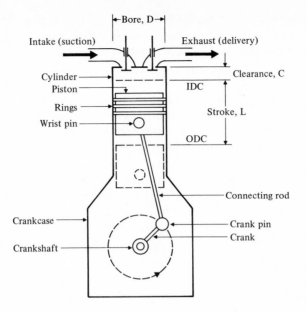

Single-acting vertical engine or compressor

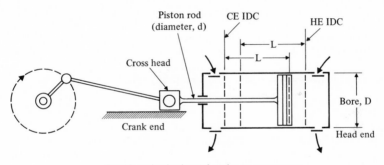

Double-acting horizontal engine or compressor

FIG. 3.1 Reciprocating engine and compressor nomenclature.

described in Chapter 1, Section 4.3. Linkage to a rotating shaft is much more common. Reciprocating compressors and engines are one type of **positive displacement** machine. That is to say, a surface (usually solid) sweeps through a measurable volume in each cycle. Rotary machines are also usually this type. Steady-flow machines such as centrifugal and axial-flow compressors and turbines are not positive displacement types.

3.2.1 General Description

Some reciprocating machine terminology is defined in **Fig. 3.1**. Most readers of this book will be familiar with the majority of the terms to be employed, and the diagrams will serve simply as a reminder. To prevent confusion later, four terms will be explained here. A **double-acting** machine uses both the head-end and crank-end faces of the piston in separate working cycles, and almost invariably accomplishes a cycle at each end in each crankshaft revolution. A **single-acting** machine produces or absorbs work through the action of the fluid on the head-end side of the piston only; although some two-stroke cycle engines do draw the fresh charge into the crankcase by the upward piston motion and then compress it slightly by the downward motion of the piston for subsequent delivery to the head-end side, these are still essentially single-acting machines. The term **inner dead center** will be used to denote the position of the piston producing minimum fluid volume, and **outer dead center** to denote maximum fluid volume. The alternative terms head-end and crank-end dead center could be confusing in a double-acting machine, and the synonymous terms top and bottom dead center, though the most commonly used in industry, are ambiguous for any cylinder orientation other than upright. It should be emphasized that the diagrams are schematic only and particularly that the valves take many different positions in the almost innumerable engine and compressor designs.

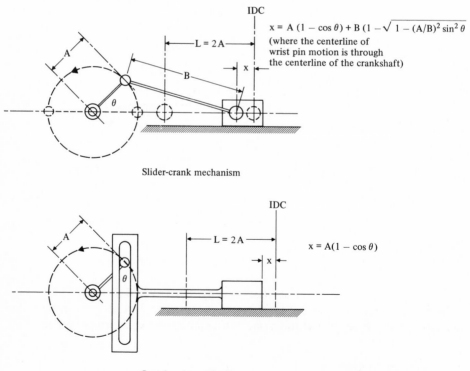

IDC

$$x = A\,(1 - \cos\theta) + B\,(1 - \sqrt{1 - (A/B)^2 \sin^2\theta}\,)$$

(where the centerline of wrist pin motion is through the centerline of the crankshaft)

Slider-crank mechanism

IDC

$$x = A(1 - \cos\theta)$$

Scotch yoke mechanism

FIG. 3.2 Reciprocating mechanisms.

Two methods of converting reciprocating motion to rotary motion or vice versa are shown in **Fig. 3.2**. The ordinary slider-crank mechanism is the most common method. The Scotch-yoke mechanism provides a simple harmonic motion of the piston for constant angular velocity of the crank, and is sometimes desirable. The velocities and accelerations of the pistons can be determined by differentiation of the displacement expressions with respect to time (the second differential for acceleration) or by various techniques of kinematics, but this text will not be directly concerned with engine and compressor dynamics.

Some of the possible arrangements for cylinders are illustrated in **Fig. 3.3**. The upright and V arrangements are the most common for both compressors and

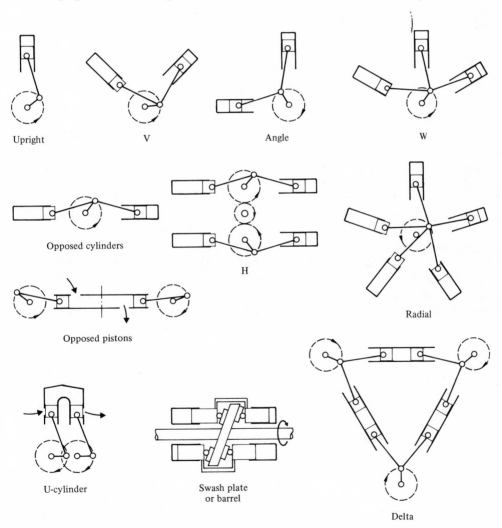

FIG. 3.3 Various cylinder arrangements.

engines, and are familiar to most readers from automotive practice where multi-cylinder engines usually have six cylinders in line for the upright arrangement or four in line on each side of the V for an eight-cylinder engine. A 90-degree V with one cylinder horizontal and the other vertical is called an angle arrangement and is sometimes used for two-stage compressors or for engine/compressor combinations. The upright arrangement can be inverted as in some small aircraft engines where visibility above the propeller shaft is important. The opposed-cylinder and opposed-piston arrangements are inherently well balanced, the former having the advantage of a single crankshaft, and the latter the advantage of requiring no cylinder head. In addition, the relative piston velocity (rate of change of volume) is doubled for a given crank and piston speed in the opposed-piston design. As shown, this arrangement lends itself to cylinder porting and straight flow-through of gases for scavenging, where the openings of the inlet and exhaust ports are controlled by the position of the pistons. The U cylinder is a variation on the opposed-piston arrangement. The W, H, delta, and radial configurations are variations on those already mentioned. On all multishaft arrangements, the various shafts must retain fixed relative rotations through gearing or other linkages. The radial engine is most familiar in conventional, air-cooled, aircraft engines where five, seven, or nine cylinders are possible in one bank, and two or even three banks have been used. Vertical-shaft radial engines are used in large stationary power plants with vertical-shaft generators mounted below. An odd number of cylinders per bank is necessary with alternate cylinders firing in successive revolutions for four-stroke cycle radial engines, but any number of cylinders can be used for two-stroke engines. The swash-plate design has been tried for engines and is common for small compressors and pumps. For each pair of opposed cylinders on a compressor, the shaft and swash-plate rotation forces one piston toward inner dead center while the other piston is pulled toward outer dead center. Slipper blocks and swivel bearings transmit thrust between the inclined plate and the pistons. For an engine, the motion of each piston away from inner dead center produces shaft rotation through the relative motion of the slipper block and the plate. Several pairs of cylinders can be evenly spaced around the shaft for a nearly steady compressor delivery or engine power output.

Two stroke and **four-stroke** cycles have been mentioned in Section 3.3 of Chapter 1, and the ideal pressure-volume diagrams for the Otto cyle with two or four strokes were shown in Fig. 1.25. Reciprocating compressors almost invariably operate on a two-stroke cycle (one crankshaft revolution). The fluid is compressed and delivered on the stroke from outer to inner dead center, and on the return stroke, after expansion of clearance volume fluid, a new charge is brought into the cylinder as the piston moves toward outer dead center. Internal combustion engines operate on either a two-stroke or a four-stroke cycle. In the latter, a complete crankshaft revolution between working cycles is utilized to drive exhaust gases out on one stroke and draw in a new charge on the return stroke so that a working cycle occurs only once in two crankshaft revolutions. The two-stroke cycle, therefore, has potentially greater power output per cubic inch of cylinder volume for a given crankshaft speed of rotation.

3.2.2 Volumetric Efficiency

Reciprocating and rotary compressors and engines are often referred to as **positive displacement** machines because a piston or other moving element sweeps through a definite volume on each cycle. The **displacement volume** is that volume (usually cubic feet or cubic inches) through which the piston or other element moves during the full induction stroke. For the **single-acting** engine illustrated in Fig. 3.1, the **piston displacement** (PD) would be

$$PD = \left(\frac{\pi}{4} D^2 L\right) n, \tag{3.1a}$$

where n is the number of cylinders. For one cylinder of the **double-acting** engine in Fig. 3.1, the piston displacement would be

$$PD = \frac{\pi}{4}(2D^2 - d^2)L. \tag{3.1b}$$

The theoretical maximum rate of charge induction for either an engine or a compressor on a **two-stroke cycle** would be

$$V_{th} = (PD)N \text{ cfm}, \tag{3.2a}$$

where N is in revolutions per minute and dimensions are in feet. For a **four-stroke cycle**,

$$V_{th} = (PD)\frac{N}{2} \text{ cfm}. \tag{3.2b}$$

In Eqs. (3.2a) and (3.2b), N and $N/2$, respectively, are cycles per minute (cpm). Actually any engine or compressor takes in less than this volume of working fluid when the volume is measured or calculated at ambient temperature and pressure for a normally aspirated machine, or at supercharger or blower discharge conditions for a machine with a forced air supply. For consistency, the volumetric efficiency (η_v) of an engine or compressor is best defined in terms of the mass rate of flow or the mass of fluid per cycle. Thus the volumetric efficiency is

$$\eta_v = \frac{\text{Actual mass of fluid per cycle}}{\text{Theoretical mass of fluid per cycle}}$$

$$= \frac{\dot{m}/\text{cpm}}{PD/v_a}, \tag{3.3a}$$

where $\dot{m}$ is the actual mass rate of flow, lb/min, and v_a is the specific volume of the fluid at ambient (or blower discharge) conditions, cu ft/lb. This becomes a volume ratio when the actual rate of flow is expressed as a volume at ambient conditions, and divided by the theoretical piston displacement:

$$\eta_v = \frac{\dot{m}v_a}{(PD)(\text{cpm})}, \tag{3.3b}$$

which is the same as Eq. (3.3a).

For a compressor, $\dot{m}$ should be the mass rate of flow discharged from the machine. Because of leakage, this may be less than the mass rate of flow induced. The point of measurement should be specified. For an internal combustion engine,

$\dot{m}$ is usually the air supplied and not the exhaust gas discharged, because it is assumed that the air flow is what limits the potential output per cubic foot of displacement and the fuel could be supplied separately.

For a reciprocating machine, the **clearance volume** is simply the minimum cylinder volume, which is the cylinder volume when the piston is at the inner dead-center position. The term **isentropic compression ratio** was used in the discussion of ideal gas cycles and defined following Eq. (1.18) as the volume ratio from the beginning to the end of the isentropic compression process. For a four-stroke engine operating on the Otto cycle, this is consistent with

$$CR = \frac{\text{Clearance volume} + \text{Piston displacement}}{\text{Clearance volume}}.$$

This last definition is used for reciprocating machines, as a general rule. However, for the ideal cycle of a two-stroke engine, as illustrated in Fig. 1.25, we assume that inlet and exhaust ports or valves are open during an ineffective part of the piston stroke, i.e., when the pressure is constant near outer dead center. For consistency with the discussion of ideal cycles, that part of the piston displacement should be disregarded in the calculation of compression ratio. Since its usage is not uniform, care should be exercised in the use of the term "compression ratio."

For compressors, the term **clearance factor** (CF) is sometimes used and defined as

$$CF = \frac{\text{Clearance volume}}{\text{Piston displacement}}.$$

This is *not* the inverse of compression ratio. The effect of clearance factor on the volumetric efficiency of a reciprocating compressor will be examined in Section 2.4 of this chapter. See Eqs. (3.12a and c).

3.2.3 Indicated Horsepower, Brake Horsepower, and Mechanical Efficiency

A distinctive feature of reciprocating compressors and engines is that instruments known as **indicators** can be applied to them to record or display pressure-volume or pressure-time diagrams. Special provisions may have to be made to connect the pressure transducer into the cylinder head. The indicators are actuated by the pressure of the working fluid itself, by the motion of the piston or crosshead, or by the crank. Such instrumentation [B12, Chapter 5] will not be described in detail in this text. As a general rule, purely mechanical devices (in which the pressure piston is acted on by the working fluid and moves a stylus, while a reducing mechanism from the slider-crank of the machine moves the recording chart to represent the change of volume) can be used up to shaft speeds of several hundred rpm. Beyond that, but still limited, are the mechanical devices in which a drum is rotated to provide a time base, while a piston is moved by cylinder pressure changes. At speeds approaching 1000 rpm, electronic instrumentation is much more satisfactory, and this can be used at speeds of several thousand rpm, now common in reciprocating engines and compressors. The electronic pressure transducer in the cylinder head may be of the piezoelectric, strain gauge, variable inductance, or

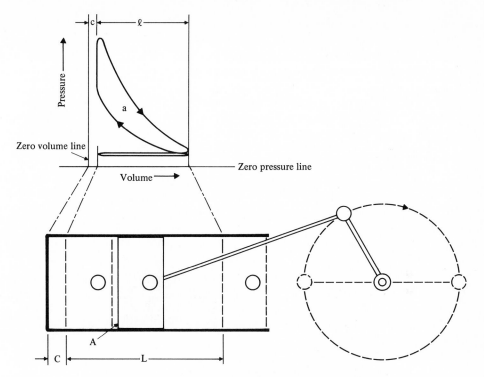

FIG. 3.4 A pressure-volume diagram for an engine.

variable capacitance type. The diagram is usually displayed on a cathode-ray oscilloscope with a constant speed time sweep triggered by either the crank position or the pressure rise. A pressure-time diagram can be converted to a pressure-volume diagram if that is desirable. Also, a *p-t* diagram may be converted electronically to a *dp/dt* diagram, which is useful in combustion studies.

It will be shown that the enclosed area of a pressure-volume diagram of a cycle represents the work done in that cycle, and the product of that work and the number of cycles per minute (or per second) represents the power of the fluid cycle. The horsepower calculated from the *p-v* diagram of an indicator is called the **indicated horsepower** (ihp). **Figure 3.4** shows a pressure-volume diagram produced by the working cycle of an engine. The stroke L has been reduced to a distance l on the diagram by the ratio l/L inches per foot. The vertical pressure scale of the diagram is S psi/inch. The zero-volume and zero-pressure lines are not located automatically; they can be placed only if the clearance volume and the absolute pressure at some point in the cycle are known. It will be seen, however, that these datum lines are not necessary for the calculation of indicated horsepower.

The major part of the *p-v* diagram of Fig. 3.4 is reproduced in the first two diagrams of **Fig. 3.5**. The exhaust and suction lines, which represent very little net work, are ignored until the final diagram of the figure. For a nonflow process,

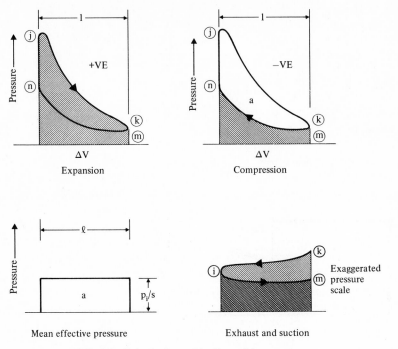

FIG. 3.5 Calculation of indicated horsepower.

such as the expansion from j to k, or the compression from m to n, the work
is

$$W = \int p \, dV. \tag{3.4a}$$

For the expansion from j to k, the work done on the piston is the area under the
line jk, and for the compression from m to n, the work done by the piston on the
fluid is represented by the area under mn. (There is no volume change and no work
done from n to j.) The net work of the cycle, therefore, is represented by the
enclosed area, a in². To convert this to foot-pounds of work, we might imagine an
indicated mean effective pressure ($P_i =$ imep) acting throughout the stroke length
L in the direction in which the work is positive. On the diagram, this would be
represented by the mean height of the area, so that the indicated mean effective
pressure would be

$$P_i = \frac{a}{l} S \text{ psi}. \tag{3.4b}$$

In one cycle, the work done would be the force from P_i acting on the net piston
area A (square inches) through stroke L (feet):

$$W = P_i A L. \tag{3.4c}$$

For N cycles per minute (N = rpm for a two-stroke cycle and rpm/2 for a four-stroke cycle), the rate of doing work, which is power, would be

$$\text{ihp} = \frac{P_i A L N}{33,000}, \qquad (3.4\text{d})$$

where one horsepower is defined as 33,000 ft-lb/min.

The final diagram of Fig. 3.5 illustrates the exhaust and suction strokes of the four-stroke cycle, but with an exaggerated pressure scale. The net work area *kimk* is opposite in sign to the work area of the major part of the cycle, and if appreciable, must be subtracted from it. The clockwise loop for such a diagram produces a positive area and the counterclockwise loop a negative area (the opposite being true for mirror-image diagrams). If areas are obtained by means of a **planimeter**, and if the lines are traced in their natural sequence, the negative loops will automatically be subtracted from the positive loops, and the net work areas will be recorded.

The preceding discussion of Figs. 3.4 and 3.5, though based on the example of an engine, is applicable as well to a compressor diagram. The only difference is that work by the piston on the fluid might be considered positive for a compressor, though this would be contrary to our adopted convention.

The **brake horsepower** of a reciprocating machine derives its name from the fact that the power output of an engine can be measured by means of a braking arrangement or dynamometer [B12, Chapter 2] to apply a torque to resist the turning of the shaft. The term is applied also to compressors to denote the shaft power input or to any machine to denote that the power in question is that transmitted through a shaft. The synonymous term **shaft horsepower** is decidedly preferable, but brake horsepower is much more commonly used. Both terms will be used in this book.

To calculate shaft or brake horsepower, it is necessary to know the **torque** transmitted by the shaft and the speed of rotation. **Figure 3.6** is applicable to a compressor, an engine, or any other machine. For power input, the torque input to the shaft would be in the direction of rotation; for power output, the resisting torque

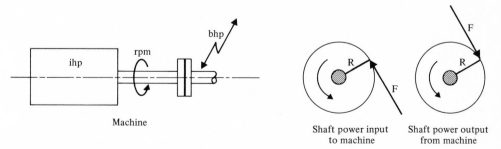

FIG. 3.6 Shaft or brake horsepower.

would be in the opposite direction. The torque T could be thought of as a force F (pounds) acting at a radius R (feet). Then

$$T = FR \text{ lb-ft.} \tag{3.5a}$$

The work done per revolution can be visualized best by considering the relative motion of the shaft and the force acting at the radius R. If the shaft is imagined to be stationary, then the force F acts through a distance $2\pi R$ in one revolution. At N rpm (ω rad/sec), the brake or the shaft power is

$$\text{Power} = (2\pi RF)N \text{ ft-lb/min} \tag{3.5b}$$

$$\text{bhp} = 2\pi TN/33{,}000 \tag{3.5c}$$

$$= T\omega/550. \tag{3.5d}$$

Between the shaft power (bhp) and the power of the fluid cycle (ihp), there is always some friction loss, so that the useful power is less than the input. The **friction horsepower** is the difference between the two:

$$\text{bhp} = \text{ihp} \pm \text{fhp,} \tag{3.6a}$$

where the sign is positive for a compressor and negative for an engine. The **mechanical efficiency** (η_m) takes only this friction loss into account, and is not concerned with other cycle losses. For a **compressor,**

$$\eta_m = \text{ihp/bhp.} \tag{3.6b}$$

For an *engine,*

$$\eta_m = \text{bhp/ihp.} \tag{3.6c}$$

The indicated mean effective pressure is a measure of the energy expenditure per cycle per unit of cylinder displacement volume. It is often difficult to measure the indicated horsepower, but easier (and sometimes more significant) to measure the brake horsepower. The term **brake mean effective pressure** ($P_b = $ bmep) is therefore common in describing the specific output of an engine or (less frequently) the input to a compressor. It could be defined simply as bearing the same relationship to brake horsepower that imep bears to indicated horsepower:

$$P_b/P_i = \text{bmep/imep} = \text{bhp/ihp.} \tag{3.7}$$

For an engine, this ratio is the mechanical efficiency.

The overall thermal efficiency of an engine is often expressed inversely as the **brake specific fuel consumption** (bsfc). When the fuel rate is appropriately expressed in pounds per hour, the specific fuel consumption is

$$\text{bsfc} = \frac{\text{lb fuel/hr}}{\text{bhp}} = \text{lb/bhp-hr.}$$

Although it may seem inappropriate for a turbine/generator power plant, the use of the term **indicated heat rate**, mentioned in Section 2.3 of Chapter 1, is understandable now that the terms shaft horsepower and indicated horsepower have been explained.

3.2.4 Reciprocating Compressors

A distinction must be made between the nonflow compression process (illustrated as part of the cycle in Fig. 3.5) for which the work is $\int p \, dv$, and a combination of induction, compression, and delivery. This combination must require the same energy as steady flow from one pressure to another does. It has already been shown in the development of Eq. (1.34d) that, when the elevation change is negligible, the work of induction, compression, and delivery is the change in stagnation enthalpy minus the heat added. When that equation is rearranged, we find that

$$-w = (h_{o2} - h_{o1}) - q \text{ Btu/lb.}$$

This equation applies to reversible or irreversible and to adiabatic or nonadiabatic operation. Conventionally, the work of compression is negative.

This can be restated in terms of specific volume and absolute pressure if a **reversible process** is assumed. Any change in the kinetic energy will also be assumed negligible, so that h can be used in place of h_o. The heat transferred in a reversible process is, from the definition of entropy,

$$đq_{\text{rev}} = T \, ds.$$

From this relationship, we find that

$$T \, ds = dh - v \, dp,$$

and the work of compression becomes

$$-w_C = (h_2 - h_1) - \int_1^2 T \, ds$$
$$= \int_1^2 v \, dp \text{ Btu/lb.} \tag{3.8a}$$

If the functional relationship between p and v is known, this expression can be calculated. For a **gas**, it can be assumed that the compression and expansion processes will follow the general polytropic:

$$pv^n = \text{constant.}$$

Thus

$$-w_C = \int_1^2 v_1 (p_1/p)^{1/n} \, dp$$
$$= p_1^{\,n} v_1 \frac{[p_2^{1-(1/n)} - p_1^{1-(1/n)}]}{1 - (1/n)}$$
$$= \frac{n}{n-1} p_1 v_1 [(p_2/p_1)^{(n-1)/n} - 1]. \tag{3.8b}$$

For the special case of a **reversible adiabatic** compressor,

$$-w_C' = \frac{\gamma}{\gamma - 1} p_1 v_1 [(p_2/p_1)^{(\gamma-1)/\gamma} - 1] \tag{3.9a}$$
$$= c_p T_i(J)[(p_2/p_1)^{(\gamma-1)/\gamma} - 1]. \tag{3.9b}$$

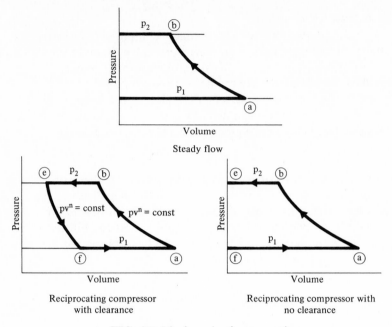

FIG. 3.7 Ideal work of compression.

For a **reversible isothermal** compressor, Eq. (3.8a) becomes

$$-w_{Ct} = p_1 v_1 \int_1^2 \frac{dp}{p}$$

$$= p_1 v_1 \ln (p_2/p_1) \tag{3.10a}$$

$$= p_1 v_1 \, 1_n (v_1/v_2.) \tag{3.10b}$$

It is interesting that Eq. (3.8b) for polytropic compression with constant pressure suction and delivery could be developed directly from the idealized pressure-volume diagrams of **Fig. 3.7**. We will consider the general case (with clearance volume) assuming that during the reexpansion of the clearance volume gas, the value of the exponent (n) is the same as the value during compression. The work of the cycle will be the sum of the work values for the individual processes. For each process

$$W = \int p \, dV.$$

Therefore

$$W = p_a V_a^n \int_a^b \frac{dV}{V^n} + p_2(V_e - V_b) + p_e V_e^n \int_e^f \frac{dV}{V^n} + p_1(V_a - V_f)$$

$$= \frac{p_b V_b - p_a V_a}{1-n} + p_2(V_e - V_b) + \frac{p_f V_f - p_e V_e}{1-n} + p_1(V_a - V_f).$$

For the ideal cycle,

$$p_a = p_f = p_1; \quad T_a = T_f = T_1;$$
$$p_b = p_e = p_2; \quad T_b = T_e = T_2.$$

Also, the amount of gas drawn in per cycle from f to a at T_1 and p_1 equals the amount of gas discharged from b to e at T_2 and p_2. This will be called m_g pounds. Therefore

$$(V_a - V_f) = m_g V_1; \quad (V_b - V_e) = m_g V_2.$$

Also,

$$W = \frac{p_2(V_b - V_e)}{1 - n} - p_2(V_b - V_e) - \frac{p_1(V_a - V_f)}{1 - n} + p_1(V_a - V_i)$$

$$= \frac{n}{1 - n}[p_2(V_b - V_e) - p_1(V_a - V_f)],$$

so that

$$-W_C = m_g \frac{n}{n - 1}[p_2 v_2 - p_1 v_1].$$

The work per pound of gas compressed and discharged is

$$-w_C = \frac{n}{n - 1} p_1 v_1 \left[\frac{p_2 v_2}{p_1 v_1} - 1\right]$$

$$= \frac{n}{n - 1} p_1 v_1 \left[\left(\frac{p_2}{p_1}\right)^{(n-1)/n} - 1\right],$$

which is exactly the same as Eq. (3.8b). What is more, it is easily shown that the whole reasoning could have been applied to the diagram of the ideal compressor without a clearance volume, and the result would have been the same. Therefore the **work per pound of gas** is not affected by either the clearance volume or the size

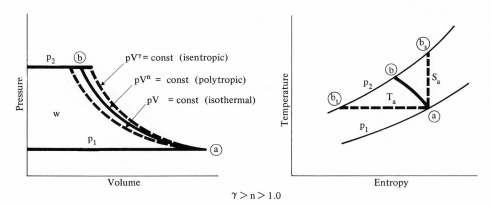

$$\gamma > n > 1.0$$

FIG. 3.8 A comparison of compression processes.

of the compressor. We shall see, however, that the amount of gas compressed and delivered per cycle between p_1 and p_2 is affected by both.

The work of compression for a given gas between two pressures clearly depends on the value of the exponent in the relationship $pv^n = $ const. **Figure 3.8** compares the work areas on the p-v diagram and the process lines on the T-s diagram for isentropic compression ($n = \gamma$), and isothermal compression ($n = 1.0$). The isothermal process can be approached only by continuous cooling. While reversible

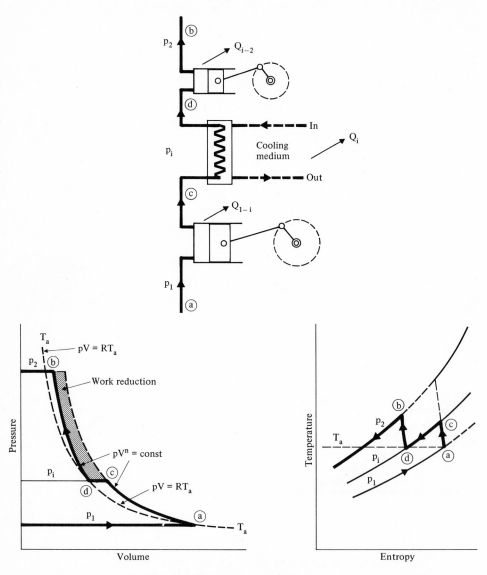

FIG. 3.9 Two-stage gas compression.

adiabatic (isentropic) compression appears to require the maximum work, poly-tropic compression in a real compressor with some cooling but with mechanical and fluid friction loss will usually require more work input than the ideal adiabatic work. The definition of compression efficiency as given in connection with the gas turbine power cycle, preceding Eq. (1.37a), is seen now to be only one of several different possible definitions. That definition was called the **isentropic compression efficiency** (η_{Cs}), and is justified there because of the requirements of the Brayton cycle. To repeat it in terms of pressure,

$$\eta_{Cs} = \frac{\text{The isentropic work of compression from } p_1 \text{ to } p_2}{\text{The actual work of compression from } p_1 \text{ to } p_2}.$$

When real work per pound can be measured, but enthalpies cannot be determined easily, the ideal work is best determined from Eq. (3.9a).

For a compressor in which an attempt is made to approach constant tem-perature compression, the **isothermal compression efficiency** (η_{Ct}) could justifiably be used as a measure of approach to this more demanding criterion. Its definition would be

$$\eta_{Ct} = \frac{\text{The isothermal work of compression from } p_1 \text{ to } p_2}{\text{The actual work of compression from } p_1 \text{ to } p_2}.$$

The ideal work for a gas would be calculated from Eq. (3.10a). In general, enthalpy differences cannot be used for nonadiabatic compression unless heat transfer quantities are known also.

Intercooling provides a means of approach to isothermal compression and is usually justified for overall pressure ratios greater than five or six to one. **Figure 3.9** illustrates the saving in work for two-stage compression (one intercooling heat exchanger), the assumption being made that intercooling can return the gas to its original temperature, an idealization usually not realized. Of course, some work input reduction is achieved by cylinder jacket and head cooling, and that saving is represented by the difference in area enclosed by the adiabatic process, $pv^\gamma = $ constant, and the polytropic process, $pv^n = $ constant; only the latter is shown.

The amount of work saved by intercooling depends on the intercooling pres-sure or intermediate pressure, p_i. The **optimum value of intermediate pressure** is that for which the total work of compression from p_1 to p_2 is a minimum. From Eq. (3.8b), with $p_a v_a = p_i v_d = p_1 v_1$,

$$-\sum w_c = \frac{n}{n-1}\left\{ p_a v_a \left[\left(\frac{p_i}{p_a}\right)^{(n-1)/n} - 1 \right] + p_i v_d \left[\left(\frac{p_b}{p_i}\right)^{(n-1)/n} - 1 \right] \right\}$$

$$= \frac{n}{n-1} p_1 v_1 \left[\left(\frac{p_i}{p_1}\right)^{(n-1)/n} + \left(\frac{p_2}{p_i}\right)^{(n-1)/n} - 2 \right]. \qquad (3.11a)$$

For fixed values of p_1, p_2, v_1, and n, the expression for the work of the two stages can be differentiated with respect to p_i, which is the variable. When the first derivative is equated to zero, the optimum value of intermediate pressure is found to be

$$p_i = \sqrt{p_1 p_2}. \tag{3.11b}$$

For three-stage compression, the two optimum intermediate pressures would be found to be $\sqrt[3]{p_1^2 p_2}$ and $\sqrt[3]{p_1 p_2^2}$.

Although the above derivations apply to perfect gases undergoing processes of the type $pv^n = \text{constant}$, it is noted in Chapter 4 that the intermediate pressure of Eq. (3.11b) gives values very close to the optimum for two-stage vapor compression in refrigeration cycles.

The effect of the **clearance factor** and of the **delivery pressure** on the volumetric efficiency in the ideal cycle of a reciprocating compressor are shown in **Fig. 3.10**. Because the clearance volume gas at point e reexpands to f as the piston moves from the inner dead-center position, the effective suction stroke at the piston is reduced; no new gas is drawn into the cylinder until the cylinder pressure drops to the suction pressure at f. The volumetric efficiency was defined in Eq. (3.3a) as

$$\eta_v = \frac{\text{Actual mass of fluid per cycle}}{\text{Theoretical mass of fluid per cycle}}.$$

If it is assumed that the whole suction stroke could have been utilized to bring in fluid at $p_1 = p_a$, and that the fluid temperature could remain constant throughout that stroke, then

$$\eta_v = \frac{(V_a - V_f)}{(V_a - V_e)}$$

$$= \frac{(V_a - V_e) - (V_f - V_e)}{(V_a - V_e)}$$

$$= 1 + \text{CF} - \frac{V_f}{(V_a - V_e)}$$

$$= 1 + \text{CF} - \text{CF}\left(\frac{V_f}{V_e}\right). \tag{3.12a}$$

The clearance factor (CF) was defined at the end of Section 2.2 of this chapter as follows:

$$\text{CF} = \frac{\text{Clearance volume}}{\text{Piston displacement}}$$

$$= \frac{V_e}{(V_a - V_e)}.$$

It is governed entirely by the physical dimensions of the compressor. The volume ratio (V_f/V_e), on the other hand, depends on the pressure ratio (p_2/p_1) and on the pressure-volume relationship for the reexpansion line ef. For a gas following the polytropic relationship $pV^n = \text{constant}$,

$$\frac{V_f}{V_e} = \left(\frac{p_2}{p_1}\right)^{1/n}$$

$$\eta_v = 1 + \text{CF}\left[1 - \left(\frac{p_2}{p_1}\right)^{1/n}\right]. \tag{3.12b}$$

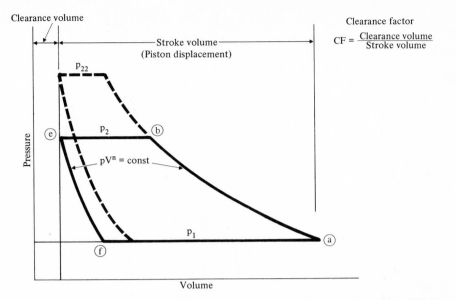

FIG. 3.10 The effect of clearance factor and delivery pressure on volumetric efficiency.

This expression was developed for an idealized compressor cycle. Actually, the specific volume (v_a) of the gas at *a* is not likely to be that found at the intake of the machine, in terms of which volumetric efficiency is sometimes expressed as it was in Eq. (3.3b). The pressure in the cylinder during the suction stroke must be less than the intake line pressure, and the temperature is likely to be higher because of heat transfer from the cylinder walls. Therefore, the expression for volumetric efficiency for an idealized gas compression cycle should be revised to

$$\eta_v = \left\{ 1 + CF \left[1 - \left(\frac{p_b}{p_a} \right)^{1/n} \right] \right\} \frac{v_1}{v_a},\tag{3.12c}$$

where v_1 is the specific volume at compressor intake conditions, usually ambient conditions.

One method of control for reciprocating gas compressors is the automatic variation of the clearance factor by the opening and closing of clearance volume pockets so that the volumetric efficiency is changed to obtain different rates of discharge over a fixed pressure range. It was shown earlier that the work required per pound of fluid delivered is not affected by the clearance volume.

In the calculation of gas properties in compressor cycles, the "low pressure" Gas Tables (see Chapter 1, Section 3.8) can usually be used without great error except where very high pressure ranges and very low temperatures are encountered. The temperatures and pressures in the compressors of conventional power and refrigeration cycles are usually such that dissociation and variation of enthalpy with temperature need not be considered.

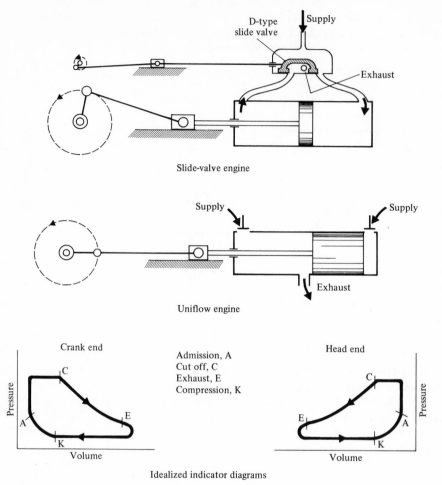

D-type
slide valve

Supply

Exhaust

Slide-valve engine

Supply

Supply

Exhaust

Uniflow engine

Crank end

Admission, A
Cut off, C
Exhaust, E
Compression, K

Head end

Idealized indicator diagrams

FIG. 3.11 Reciprocating vapor engines.

3.2.5 *Reciprocating Vapor Engines*

The steam engine, developed through the eighteenth century, was the most important prime mover throughout the nineteenth and, in some applications, into the twentieth century. Although the internal combustion engine, the steam turbine, and the gas turbine have replaced the reciprocating steam engine in most applications, its simplicity and reliability make it worthy of consideration in special applications even today [A.8]. Other vapors besides steam can be used.

Two designs for reciprocating vapor engines are shown schematically in **Fig. 3.11**. Also shown are typical pressure-volume indicator diagrams defining some of the common terms. The simple **D slide valve** shown is driven by an eccentric from the main crankshaft and is always over 90° ahead of the crank. The same cylinder ports are used for both admission and exhaust at each end. In the **uniflow** engine shown, admission of vapor to each end is accomplished through separately actuated and independently timed valves. The exhaust is through one set of ports for the two ends, timed only by the motion of the piston which uncovers the ports near each outer dead-center position. Cooling of the two ends of the cylinder and therefore initial condensation of incoming vapor are reduced.

The idealized indicator diagrams shown could not be achieved by either of the two engines illustrated, but could be approximated by an engine with separately timed admission and exhaust valves. Four key points are shown on each diagram: A indicates the opening of the vapor admission valve, C is the point at which that valve closes, E is the beginning of exhaust, and K is the beginning of compression of the trapped vapor when the exhaust valve closes. The pressure rise from K to A reduces throttling loss at admission, and cushions the deceleration of the piston toward inner dead center. The work per cycle is represented by the sum of the two indicator diagrams. It should be noted that the effective piston area is less for the crank end than for the head end in these double-acting engines.

3.2.6 Internal Combustion Engines

In this section and in Section 3.2.7, some aspects of internal combustion engines that do not seem to be specifically thermodynamic in nature will be discussed. These aspects, notably the phenomena of combustion and the problems of fuel supply, are extremely important in the determination of allowable compression ratio for a reciprocating internal combustion engine. The significance of compression ratio in the determination of ideal and actual thermal efficiencies has been dealt with. It is important to appreciate related restrictions imposed on engine design. The treatment will be descriptive.

Combustion chambers of reciprocating internal combustion engines exist in many shapes and arrangements. The advantages and disadvantages of these shapes relate primarily to the optimization of the combustion process. The severest limitation on compression ratio and therefore on potential thermal efficiency in a **spark-ignition** engine is the problem of **end-gas detonation**, commonly called combustion "knock." It will be described very briefly here.

The fuel supply to a spark-ignition engine is usually carbureted so that a mixture of fuel and air is compressed after the induction stroke or period. Because of an **ignition-delay** period during which preflame reactions take place, the electric spark for ignition must be passed well before the inner dead-center piston position is reached. Although all combustion energy is presumed to be added at constant volume and therefore instantaneously in the ideal Otto cycle, the combustion actually requires finite time for completion in a real engine. With reference to **Fig. 3.12**, this passage of time is not so apparent on the pressure-volume diagram as on the pressure-time diagram because the ignition delay and most of the combustion take place in what might be called the ineffective crank angle near the inner dead-center position while the piston is practically stationary. Experiments usually show that maximum efficiency is achieved when the maximum pressure in the cycle occurs

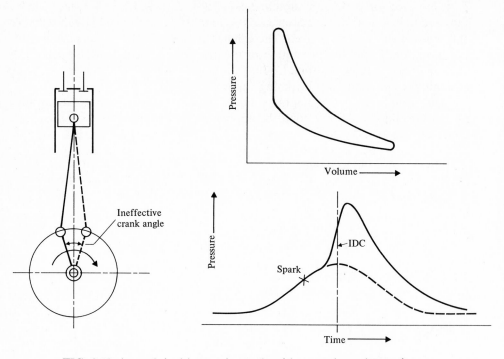

FIG. 3.12 A spark-ignition engine cycle without end-gas detonation.

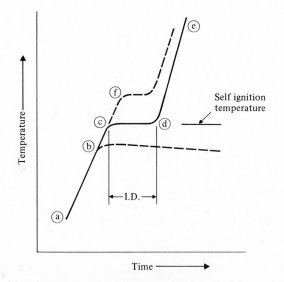

FIG. 3.13 Self-ignition temperature and ignition-delay period for a combustible mixture.

about 10 or 12° after inner dead center. (One can readily appreciate the order of magnitude of the time period being discussed by calculating the time required for say 10° of crank rotation at 1000 or 5000 rpm.)

In a spark-ignition engine, the combustible mixture is not normally raised to such a high temperature on the compression stroke that **auto-ignition** or **self ignition,** of the charge will take place. The energy release and local heating of the electric spark are necessary to initiate combustion. However, combustible mixtures will ignite spontaneously if conditions are suitable. The temperature-time diagram of **Fig. 3.13** indicates that the self-ignition temperature for a particular mixture is not a unique value, but depends on the rate of temperature rise. (It also depends on pressure.) If a combustible mixture is compressed only from *a* to *b*, self ignition or spontaneous ignition does not occur. If the same mixture is compressed from *a* to *c*, it will be found that the temperature will rise further because of combustion from *d* to *e*, but only after an ignition-delay period indicated by the time from *c* to *d*. Had the mixture been rapidly compressed initially to *f*, the self-ignition temperature would have been higher, and the ignition-delay period would have been shorter.

In special research engines with transluscent heads, the progress of the visible flame front across the combustion chamber has been recorded by high-speed photography techniques [A.24]. The gases ahead of that flame front at any instant are called **end gases,** as noted in **Fig. 3.14**. Combustion is not completed in the flame front itself, but is continued in the luminescent region behind it. A combustible mixture in the end-gas zone was compressed along with the rest of the charge on the compression stroke, and is further compressed as combustion in the region behind the flame front increases pressure throughout the whole combustion space. Also, it is heated by energy radiated from the flame to solid surfaces, and transferred by conduction to the end gases.

End-gas detonation occurs when the end gases are raised to their self-ignition temperature and the ignition-delay period passes before the flame front sweeps

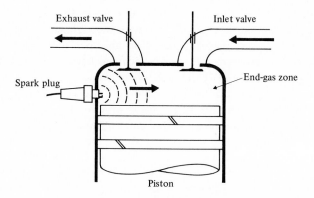

FIG. 3.14 Combustion space in a spark-ignition engine.

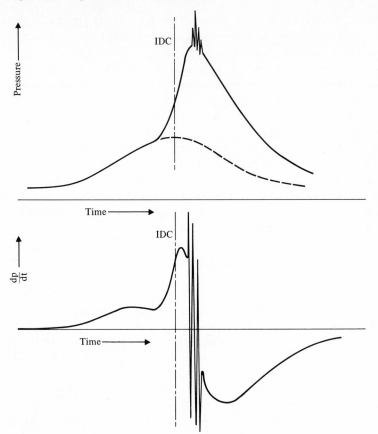

FIG. 3.15 Spark-ignition engine *p-t* and *dp/dt* diagrams with end-gas detonation.

the whole combustion space. The resulting combustion is not gradual as was that initiated by the orderly propagation of the flame front. Instead, it can be so rapid that it is classed as detonation, which is sometimes defined as combustion rapid enough to force a pressure-wave propagation at super-acoustic velocity. Such a pressure wave striking the opposite wall is reflected back and forth producing the characteristic knocking sound. Combustion with end-gas detonation is evident on the pressure vs. time and on the rate of pressure rise vs. time diagrams shown in **Fig. 3.15**. The maximum pressure of the cycle may not be greatly increased. It is the extremely high rate of pressure rise that can physically damage an engine either directly or through imperfect lubrication resulting from the scouring effect of the detonation at surfaces normally protected by surface films.

The tendency toward knocking combustion is aggravated by the necessity for long flame-front travel, poor turbulence and mixing of the charge (except in strati-fied charge engines, which will be mentioned later), hot surfaces remote from

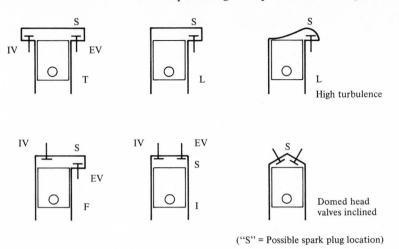

("S" = Possible spark plug location)

FIG. 3.16 Some cylinder head types for spark-ignition engines.

the spark plug, as well as low self-ignition temperature of the mixture, high charge density, high inlet temperature, high cylinder wall temperature, solids in the combustion chamber, and high compression ratio. The self-ignition temperature is raised, and knocking therefore suppressed, by dilutents in the mixture including atmospheric moisture. A mixture ratio, which is near chemically correct (stoichiometric), is not easily ignited, so that either a lean or a rich mixture allows other conditions to be more severe. The cylinder wall temperatures are affected by the rate of heat release per unit volume, so that high brake mean effective pressure which occurs under unthrottled or supercharged running conditions promotes detonation.

Cylinder-head geometry is very important as already indicated. It is clear that if the inlet and exhaust valves of Fig. 3.14 were interchanged, knocking tendency would be aggravated because the exhaust valve is the hottest spot in the engine aside from the spark plug. When the six different cylinder head types illustrated in **Fig. 3.16** for *s-i* engines are compared, the obsolescence of the *T*-head is understandable. Even with the spark plug located centrally rather than over the hot exhaust valve, the flame travel distance to the farthest surface is too great. Also, such a high ratio of surface-to-volume would result in high heat loss. In addition, a double camshaft for valve operation is necessary. The simple *L*-head brings the spark plug close to the exhaust valve, and the design is greatly improved by the streamlining of the head, as in the high-turbulence shape shown. This design makes possible a compact combustion space with very little gas at the far side of the cylinder and good cooling to avoid auto-ignition in that end-gas zone. (Only one valve appears in the *L* arrangement and some others, because one valve is directly behind the other.)

The *F*-head requires a more complicated valve-gear arrangement, but allows room for a large inlet valve in the cooler region with a compact combustion space to one side, above the exhaust valve. A streamlined version of the *F*-head is usually referred to as the high-turbulence Ricardo head after its developer. The simple valve-in-head, or *I*-head arrangement produces too flat a combustion space at high compression ratios. It is improved by the hemispherical shape of the inclined-valve, domed-head arrangement. In this, the piston face can be dished to produce a nearly spherical combustion chamber (theoretically the best because of its minimum surface-to-volume ratio during combustion) or it can be domed to produce a minimum clearance volume. The advantage achieved by minimum-flame travel length in suppressing end-gas detonation may be offset by excessively high rates of pressure rise and rough operation.

Combustion knock or detonation in a **compression-ignition** engine is quite a different problem. Air alone is brought in during the induction stroke, or period, and is compressed along with residual gases, essentially inert, from the previous cycle. Fuel is injected directly into the combustion chamber toward the end of the compression stroke. There is no spark to initiate combustion. The timing of the fuel injection affects the time for the start of combustion, and the duration of injection can control the rate of pressure rise.

The compression-ignition engine depends on auto-ignition for its combustion. If a fuel with a low self-ignition temperature and a short ignition-delay period is used, combustion will start soon after the start of injection, as long as a high enough temperature has been reached on the compression stroke. If the fuel has too high a self-ignition temperature, and its ignition-delay period is too long, a relatively large amount of fuel will be in the combustion space when ignition occurs, and the rate of pressure rise can be great enough to cause combustion knock. The effect is similar to that of end-gas detonation in a spark-ignition engine. It is apparent, then, that most of the factors that suppress knock in a spark-ignition engine will encourage knock in a compression-ignition engine. Thus high self-ignition temperature, long ignition delay, low compression ratio, low brake mean effective pressure, low cylinder wall temperature, and very lean mixture ratios will all aggravate knock and rough running in a compression-ignition engine. On the other hand, turbulence which promotes fine atomization and good distribution of the fuel will suppress knock as it did in the spark-ignition engine.

In the reciprocating compression-ignition engine, the greatest difficulty is to achieve an intimate mixing of the air with the fuel, which is injected separately toward the end of the compression stroke and, at higher loads, on into the expansion stroke. The open-type combustion chambers illustrated in **Fig. 3.17** are found mostly in the larger slow-speed engines. Turbulence is encouraged by the clearance volume shape in some designs, by tangential rather than radial porting in two-stroke engines, and also by the location and direction of spray of the injector. To achieve rapid enough combustion for high-speed operation, additional encouragement of fuel and air mixing is required. In the swirl- (or turbulence) chamber design, the fuel is usually sprayed into the air against the direction of air motion to improve spray atomization. In the precombustion-chamber design, the chamber is the major portion of the clearance volume. The rapid pressure rise in the volume

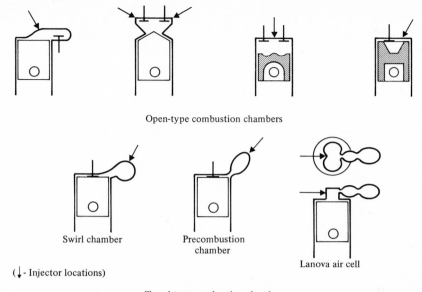

Open-type combustion chambers

Swirl chamber   Precombustion
                          chamber

($\downarrow$ - Injector locations)   Lanova air cell

Closed-type combustion chambers

FIG. 3.17 Combustion chamber types for compression-ignition engines.

is contained by the walls of the chamber and not immediately transmitted to the piston, so that running is smoother. As the piston moves from an inner dead-center position, the rush of burning fuel and air from the restricted opening of the precombustion chamber creates the high turbulence needed for continued even mixing and complete combustion.

The air-cell design provides a chamber into which air without fuel is compressed, so that its subsequent escape through an orifice will promote mixing and remove burning gases from the region of the injector to avoid carbon deposit there. The Lanova arrangement illustrated is unique in that fuel is injected into a chamber in front of the air cell and the air in the cell is further compressed by the initial burning. As the piston moves from inner dead center, the expansion of the air assists the rapid movement of the burning mixture into the swirl-producing shape of the remainder of the combustion chamber. (Valves are not shown in this illustration.)

Large engines of either the spark-ignition or the compression-ignition type may have more than one inlet valve and more than one exhaust valve. While large and smooth passageways are important for the reduction of pumping loss in the induction of the charge and removal of the exhaust gases, the improvement of volumetric efficiency is even more important in the achievement of maximum output per cubic inch of piston displacement.

Various attempts have been made to enable spark-ignition engines to utilize less expensive fuels, and still avoid the possibility of combustion knock by prevent-

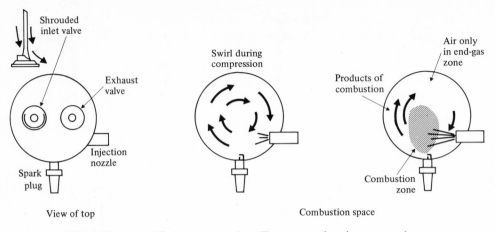

FIG. 3.18 A stratified charge engine (Texaco combustion process).

ing the presence of a combustible mixture in the end-gas zone. One method of accomplishing this is the example of a **stratified-charge** engine illustrated in **Fig. 3.18**. This system is referred to as the Texaco Combustion Process [A.3]. In this system, air is drawn in alone during the suction stroke, and fuel is injected late in the compression stroke. Compression ratios are not high enough to initiate combustion, and a spark plug is used. Combustion is limited to a zone near the spark plug. To ensure stratification of the charge, an organized swirl of air is started during the suction stroke as air is drawn in through a **shrouded** inlet valve, which directs the air to one side. This swirl persists during compression and combustion. Thermal efficiency is potentially high compared with that for other spark-ignition engines because relatively high compression ratios can be used without knock. Supercharging to improve specific output is possible. The main advantage achieved is economy through the use of a wide range of inexpensive fuels, and poor burning efficiency is offset by this saving. Compression ratios and thus pressures need not be as high and injection timing need not be as precise as in compression-ignition engines. Other engines can be classed as stratified charge engines; for instance, the Hesselman design burns inexpensive low-volatile fuels, which produce limited regions of combustible mixture simply because a large fraction of the fuel is not vaporized at the time combustion starts in the region of the spark plug.

High compression ratios without end-gas detonation are also possible in **dual-fuel** reciprocating internal-combustion engines. The major part of the combustion energy is supplied by a gaseous fuel, the amount of which is limited to produce a relatively lean mixture with the air. A pilot charge of liquid fuel is injected separately to produce a locally rich mixture and initiate combustion. This is similar to the stratified charge concept; end-gas detonation is avoided simply because the mixture in the region remote from initial burning is too lean for auto-ignition. In a four-stroke dual-fuel engine, the gaseous fuel and air can be drawn into the cylinder together. In a two-stroke engine, the gas may be admitted at a slight pressure through its own valve early in the compression stroke; alternatively, it can be supplied at higher pressure with the pilot charge of liquid fuel late in the compression stroke to aid atomization and distribution of the liquid. The dual-fuel engine is best adapted to large stationary installations, where efficiency is more important than power per cubic inch, although the latter can be raised with lean mixtures by supercharging, which would be prohibited at high-compression ratio by end-gas detonation in a conventional spark-ignition engine, particularly in large cylinder sizes.

3.2.7 Carburetion and Fuel Injection for Internal-Combustion Engines

In reciprocating **spark-ignition** engines, it is a relatively simple matter to supply a gaseous fuel in proper proportions with air as the load changes. However, the metering, atomization, and distribution of liquid fuel, particularly in a multi-cylinder engine, is more difficult. Only in a limited number of applications is an engine subjected to a steady load demand that can be met by fixed rates of fuel and air supply. In most stationary installations, the demand is variable, and in automotive propulsion particularly, the speed and load must vary.

For reciprocating spark-ignition engines, it is usual to **throttle** or restrict the air supply to avoid over-lean mixtures at lower loads and to use maximum possible air only when the maximum load is required. The major areas of operation are summarized in **Table 3.1**, which will be explained further in the following paragraphs.

Table 3.1

Range of Operation	Governing Factor	Fuel-to-Air Ratio
Idling	Dilution of mixture by products of combustion	Rich
Normal power	Economy	Slightly lean
Maximum power	Full utilization of air	Rich

During a period of **idling** or no power demand, the air supply is throttled. The clearance volume gases, essentially inert, make up a large fraction of the charge at the end of the suction period. In addition, because the pressure in the cylinder is extremely low during suction, exhaust gases are drawn back into the cylinder during the valve-overlap period when both inlet and exhaust valves are open. The result is that a chemically correct mixture of air and fuel would be so diluted by inert gases that combustion would be erratic or impossible. A rich mixture (more fuel than the chemically correct amount for the oxygen available) must be supplied. An idling or stand-by situation is uneconomical in any case, but the necessity for a rich mixture makes it even more wasteful.

In the **normal power** range, which might be very roughly from 25 to 75 % of maximum possible indicated mean effective pressure, the prime consideration is usually for maximum fuel economy. Because a mixture of fuel and air is never completely homogeneous, and because temperature gradients within a cylinder

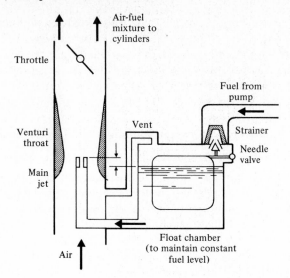

FIG. 3.19 An elementary carburetor.

interfere with uniform combustion, the chemically correct (stoichiometric) mixture of fuel and air will not burn completely and some fuel will be wasted. For this reason an excess of air, 5 or 10 % above theoretically correct, is supplied in order to avoid incomplete combustion.

The chemically correct air-to-fuel ratio for most gasolines is between 0.066 and 0.068 pounds of fuel per pound of air. The specific weight of fuel vapor is greater than that of air, and in addition, a high fraction of the fuel might still be liquid during induction of the charge. Thus the volume of air that can be supplied is the factor limiting the **maximum power** output for a given engine. To assure utilization of all oxygen that can be supplied, a rich fuel-to-air ratio is required.

No extremely simple device can meet the varying requirements of the usual spark-ignition engine. The modern automotive engine carburetor combines many refinements, limited examples of which will be described briefly here. All diagrams are highly schematic for the purpose of showing principles only. While most carburetors are actually of the down-draft type in which the air moves from top to bottom, it will be simpler to illustrate the carburetor components with an up-draft arrangement.

Figure 3.19 shows an elementary carburetor with a throttle, air venturi, fuel float chamber, and a single fuel jet. The principle of operation is familiar from fluid mechanics: a flow of air through the venturi will result in a reduction in static pressure at the throat to a value below the pressure just ahead of the venturi; the pressure difference will drive the liquid fuel from the float chamber, in which a constant level is maintained, the rate of fuel flow depending on both the pressure difference and the jet-orifice size. At a given engine speed, the suction above the venturi and thus the air flow will depend on the throttle opening.

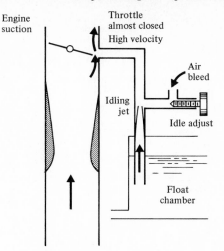

FIG. 3.20 An idling jet.

At the very low rates of air flow required for idling, when the throttle is nearly closed, the venturi pressure drop is not great enough to initiate any fuel flow. For this reason, a separate **idling jet** must be added to the basic carburetor. An example is shown in **Fig. 3.20**. When the throttle is practically closed, the full manifold suction operates on the outlet to this jet. In addition, the very high velocity past the throttle plate increases the suction locally. Fuel can therefore be lifted the additional height to the discharge point, but only at very low rates of air flow. As the throttle is opened, the main jet gradually takes over while the idle jet becomes ineffective.

The mass rate of air flow is established by the mean throat velocity and the air density, while the mass rate of flow of the constant-density fuel depends on the square of the air velocity and the air density. This means that as the throttle is opened and the suction at the venturi is increased, the mass rate of fuel flow increases more rapidly than the mass rate of air flow, and the single simple jet tends toward richness at greater flow demands. To provide a more nearly constant fuel-to-air ratio (about 0.064) in the normal power range, some type of **compensating device** is added to the basic carburetor. One example of this is the unrestricted air bleed through a compensating jet as shown in **Fig. 3.21**. The fuel supply to the compensating well is restricted by an orifice so that the level of fuel (h) in the well decreases as the air and fuel flow are increased. The compensating jet thus tends toward leanness as the main jet tends toward richness, the sum of the two remaining constant. At even higher rates of air flow, when the compensating well has been emptied, air is bled through the compensating jet to continue the leanness effect and, incidentally, to assist in fuel atomization.

As the maximum power range of operation is approached, some device must allow richer mixtures (about 0.080 lb fuel/lb air) to be supplied despite the compen-

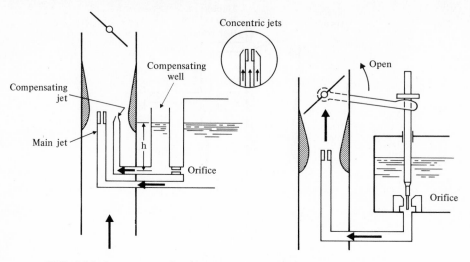

FIG. 3.21 A compensating jet.   FIG. 3.22 A meter rod economizer.

sating leanness. Such a device is the **meter rod economizer** shown in **Fig. 3.22**. The name economizer might be misleading; it stems from the fact that such a device provides a rich uneconomical mixture at high load demand without interfering with economical operation in the normal power range. This particular example of an economizer simply provides a large orifice opening to the main jet as the throttle is opened beyond a certain point. The rod may be tapered or stepped. Other examples provide for the opening of auxiliary jets through some linkage to the throttle movement, or through a spring action when manifold vacuum is lost as the throttle is opened.

When it is desired to accelerate the engine rapidly, a simple carburetor will not provide the rich mixture that is necessary. Rapid opening of the throttle will be immediately followed by an increased air flow, but the inertia of the liquid fuel will cause at least a momentarily lean mixture just when richness is desired for power. The delay in fuel delivery is aggravated by manifold wetting. One method of overcoming this lag in fuel flow is the **accelerating pump** shown in **Fig. 3.23**. Instead of the mechanical linkage shown, some carburetors have a pump plunger held up by manifold vacuum. Whenever that vacuum is reduced by the rapid opening of the throttle, a spring forces the plunger down in a pumping action identical to that of the pump illustrated. The plunger is raised again against the spring force when the throttle is partly closed and negative gauge pressure is reestablished.

During a **cold starting** period, at low cranking speeds and before the engine has warmed up, much richer than usual mixtures must be supplied simply because a large fraction of the fuel will remain liquid even in the cylinder, and only the vapor fraction can provide a combustible mixture with the air. The most common means of obtaining this rich mixture is by the use of a **choke**, which is a butterfly-

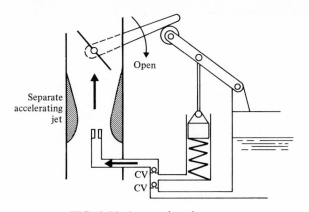

FIG. 3.23 An accelerating pump.

type valve placed ahead of the venturi. This restriction of air flow means that high negative pressure will exist at the main jet to allow the vent pressure from ahead of the choke to force a large rate of fuel flow. The choke must be opened as the engine warms up to prevent continued running with excess fuel, one result of which, aside from poor economy, is the possible dilution of lubrication oil with gasoline. An alternative to the choke in starting is the provision of auxiliary fuel jets that are opened manually or automatically only as required.

There are many other refinements that can be and are added to some carburetors. Aircraft engine carburetors, for instance, commonly employ movable or variable venturi sections and float chamber-pressure control to compensate for variations in air density with altitude. These and other refinements will not be illustrated here.

An alternative to the conventional carburetor is inlet-port or cylinder **fuel injection** [A.4]. Injection of fuel directly into the cylinder during suction and compression requires an intermittent action, which is a disadvantage, even though the timing need not be as precise as for a compression-ignition engine. Alternatively, a steady stream of fuel at moderate pressure directed toward the inlet valve port of each individual cylinder can provide, without precise timing, good fuel distribution among cylinders and rapid metering response to load demand. Furthermore, it is easily adaptable to reciprocating aircraft engines in which an even fuel distribution is difficult and variation in angle of flight requires additional complications in a float chamber. The fuel supply can still be controlled in injection carburetion, as in conventional carburetion, by the air flow through a venturi restriction in the air stream. The pressure drop controls the main fuel valve opening by means of a diaphragm. Idling and starting may still be assisted by auxiliary jets at the throttle valve, or by additional controls to the fuel supply valve. While the conventional carburetor is highly developed and refined, it is anticipated that fuel-injection carburetion will be affected by numerous innovations in the future.

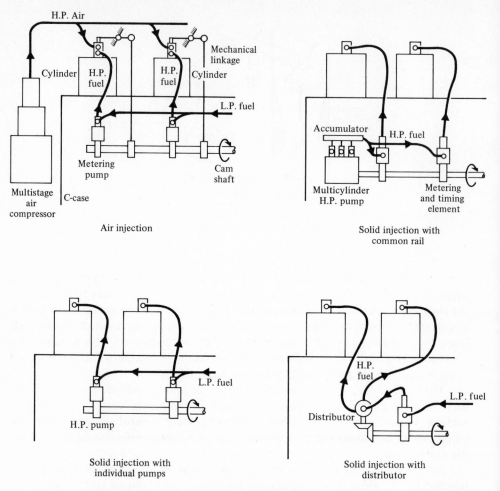

FIG. 3.24 Some fuel injection systems for diesel engines.

Diagrams of **Fig. 3.24** illustrate four possible means of fuel distribution and injection for multicylinder **compression-ignition** engines. The **air-injection** system which uses compressed air to assist in the atomization of the liquid fuel as they are injected together into the cylinder is no longer a common method. The losses and complexities of a high-pressure air compressor and the need for a separate timing mechanism are avoided with other systems.

The injection of liquid fuel directly into the cylinder or precombustion chamber without primary air atomization is known as **solid injection**. The earliest successful solid injection system was the **common rail-** or **accumulator-**type in which the high-pressure pump is separate from the metering and timing mechanisms for the individual cylinders. The **individual-pump** and the **distributor-pump** systems are more

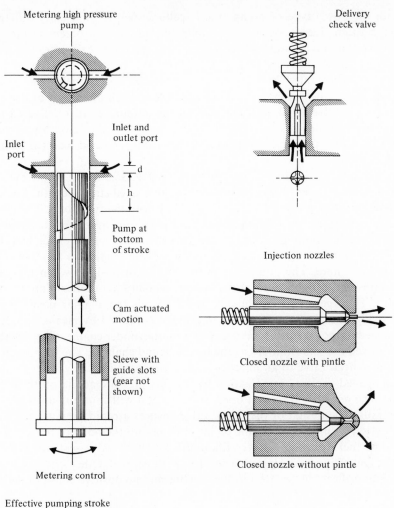

Metering high pressure
pump

Delivery
check valve

Inlet
port

Inlet and
outlet port

d

h

Pump at
bottom
of stroke

Cam actuated
motion

Sleeve with
guide slots
(gear not
shown)

Injection nozzles

Closed nozzle with pintle

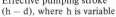
Metering control

Closed nozzle without pintle

Effective pumping stroke
$(h - d)$, where h is variable

FIG. 3.25 A Bosch-type fuel pump and delivery system.

common in modern diesel engines. In both of these, the pump that produces the high pressure also meters and times the fuel injection. The individual pump system may have the **unit injector** design in which the pump for each cylinder is mounted directly above the injection nozzle with no long lead lines between. An advantage of the distributor system for multicylinder engines is that only one metering device is necessary for all the cylinders.

No attempt will be made here to describe a large number of the ingenious fuel pumping and metering devices developed by various manufacturers. Only two will be illustrated. Several injection systems operate on some variation of the principle shown for the Bosch type of pump; the Excello system is sufficiently different to hint at the alternatives possible. Some appreciation of the problem of direct fuel injection for compression-ignition engines is realized if one calculates the amount of fuel to be delivered per cylinder per cycle to, say, a small four-cylinder, four-stroke diesel engine, rated 20 horsepower at 2000 rpm, with a fuel consumption of 0.5 lb/bhp-hr. Further, it must be noted that the smooth operation of the engine will be very sensitive to slight differences in the amount of fuel delivered to the various cylinders, and, finally, that the engine must also idle smoothly at no load with less than one-quarter the fuel consumption for the rated load (to overcome friction).

The **Bosch-type** injection pump illustrated in **Fig. 3.25** is one of the individual pump types: its plunger is cam operated through a fixed stroke length. The pump starts to deliver fuel when the plunger top closes both the fuel ports. The pressure drops and delivery ceases when the right-hand port is uncovered by the helical surface. The effective pumping stroke is (h-d). This effective stroke is varied by rotation of the plunger within the pump body so that h is made smaller for less fuel delivery and longer for more. The plunger is in the shutdown position when the vertical slot is opposite the right-hand port, and no pressure above the plunger can be developed. The sleeve with guide slots that controls the angular position of the plunger (while the plunger moves up and down) is rotated by means of a gear or rack attached to the "throttle" linkage. For a multicylinder engine, all guide sleeves must be moved the same amount as the load demand changes. The fuel is delivered through the check valve to an injection nozzle mounted in the cylinder head. The nozzle is opened against spring action by the fuel pressure acting on the exposed shoulder. The nozzle will remain open and injection will continue until the pump pressure falls. The design of the check valve ensures a rapid pressure fall to give clean injection with no dribble; the collar moving into the body of the valve withdraws some fuel from the delivery line until the check valve seats and traps the remaining fuel. Closed-type **injection nozzles** are illustrated. Other systems use open nozzles with orifices always open to the fuel line and no positive cutoff of delivery.

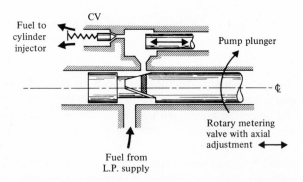

FIG. 3.26 An Excello-type distributor fuel pump.

In the **Excello** distributor-type system shown in **Fig. 3.26,** the metering element is separate from the pump plunger. The plunger operates through a fixed stroke, but is only effective when the port at the bottom of the pump chamber is closed by the shaded portion of the wedge-shaped segment of the rotating metering shaft. That shaft, though rotating continuously, can be moved to the left or right either manually or by a governor to allow longer or shorter effective pumping strokes. For a multicylinder engine, one pump plunger operates for each cylinder, and all plungers are located symmetrically around the shaft centerline so that the same metering element controls fuel delivery to all cylinders. The plungers are operated by a swash-plate drive.

3.2.8 *Stirling Cycle Reciprocating Engine*
In Section 3.4 of Chapter 1, it was noted that an engine operating on the so-called Stirling cycle could theoretically achieve the highest possible thermal efficiency between two particular source and sink temperatures. No discussion was introduced

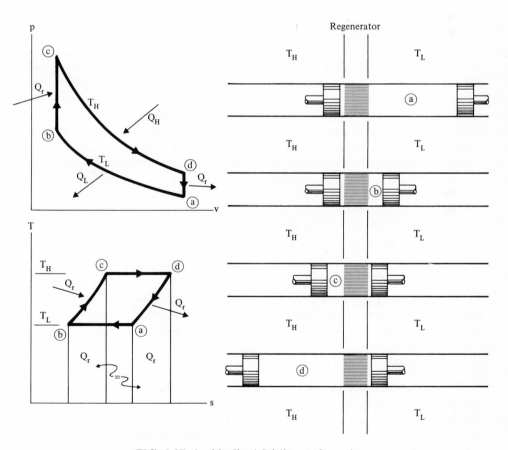

FIG. 3.27 An idealized Stirling cycle engine.

of exactly how such a cycle could be effected in practice except that internal heat transfer by means of a regenerator would be necessary. The "hot air" external-combustion engine, patented by Stirling in 1816, was an attempt to achieve the ideal cycle of two isothermal and two constant volume lines. It was, however, a cumbersome machine with many losses, and though it competed successfully with the steam engine in certain applications, it was eventually made obsolete by the internal-combustion engines. Recently, interest in such an engine has revived [A.5, 10, 11, 21, 22, 23]. The possibility of a quiet engine that can use any kind of fuel, or any other source of heat that can be transferred to the working fluid, is appealing for many applications today. Improved knowledge of heat transfer and the reduction of mechanical and fluid friction losses along with the possibility of using a working fluid other than air promise to make an approach to the Stirling cycle practical in several fields. The reversed Stirling cycle, like the reversed Carnot and Brayton cycles, can be used for refrigeration. (See Chapters 4 and 5.)

The **ideal cycle** and an imaginary means of achieving it with pistons having discontinuous motion are shown in **Fig. 3.27**. In this schematic representation of the working cycle, there are two spaces: in one of these, the working fluid can be compressed isothermally while heat is transferred from it to a sink at temperature T_L; in the other, the fluid can be expanded isothermally while heat is transferred to it at T_H. The two spaces are connected by a regenerator of negligible passageway volume, but with sufficient heat capacity so that it can become alternately a heat source and a heat sink.

The four different piston position diagrams identified as *a*, *b*, *c*, and *d* correspond to the four points similarly marked on the *p-v* and *T-s* diagrams. Between *a* and *b*, the fluid is compressed isothermally at T_L to its minimum volume. From *b* to *c*, the fluid is passed from right to left through the regenerator from which it receives sufficient heat to bring it to temperature T_H. Between *c* and *d*, the fluid is expanded isothermally at T_H. To return the cycle to its starting point, the fluid at its maximum volume is passed through the regenerator to which it transfers sufficient heat to lower the fluid temperature to T_L.

The imaginary cycle just described depends on several idealizations:

a) Perfect frictionless reversible heat transfer to an infinite sink at T_L and from an infinite source at T_H.

b) One-hundred percent effectiveness of the regenerator, wherein the fluid receives exactly the heat at minimum volume that it lost at maximum volume.

c) Discontinuous motion of the two pistons so that one remains stationary while the other moves, except for the operations *bc* and *da* during which the pistons move together to maintain constant volume.

Discontinuous motion would be possible but awkward. Several mechanisms can be devised without it to pass the working fluid back and forth through a regenerator and approach the Stirling cycle. The most promising of these involves the use of two pistons, a power piston and a displacer piston, in one cylinder. A **practical version** of this developed by the Philips Research Laboratories, the Netherlands,

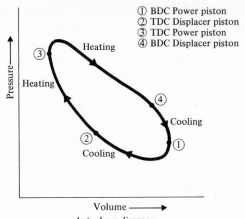

① BDC Power piston
② TDC Displacer piston
③ TDC Power piston
④ BDC Displacer piston

Actual p-v diagram

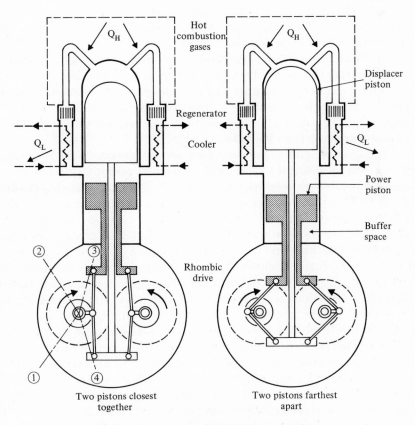

FIG. 3.28 A practical Stirling cycle engine.

with the later cooperation of the General Motors Research Laboratories, is shown schematically in **Fig. 3.28**. Each piston operates as a slider-crank mechanism with an offset centerline. The two motions are linked together through the "rhombic" drive mechanism shown. (A rhombus is a four-sided figure with straight sides, all of equal length.) An alternative mechanism uses two cranks nearly 90° apart on the same shaft to produce a similar motion.

Inspection of Fig. 3.28 will show that the motion of the displacer piston does not change the total volume of the working fluid at all. Its only purpose is to move the working fluid to the hotter side or the cooler side alternately. It is subjected to the same pressures above and below except for pressure differences from friction losses in the fluid being displaced. The total volume of the fluid then is affected only by the position of the lower piston, and the net work is the cyclic integral of $(p\, dV)$ for that piston. The trapped gas in the buffer space does no net work, but its presence minimizes leakage past the lower piston.

The two crank positions shown in Fig. 3.28 are the positions for minimum and maximum distances between the lower face of the displacer piston and the upper face of the power piston. These are not the positions for minimum and maximum total volume of working fluid. Maximum volume occurs at crank position ①, the outer dead center for the power piston, and minimum volume occurs at position ③, the inner dead-center position.

Net heat transfer to the working fluid occurs from crank position ② to crank position ④; during this part of the cycle the displacer piston is moving downward, forcing a larger fraction of the working fluid into the heater section. From crank position ④ back to crank position ②, fluid is being displaced out of the heater section, through the regenerator, and into the cooler section so that the net direction of heat transfer is from the fluid.

The resulting pressure-volume diagram for the working fluid is greatly distorted from the ideal Stirling cycle. Still, maximum operating efficiencies exceed those for comparable spark-ignition and compression-ignition internal-combustion engines [A.5]. The cycle work per pound of working fluid is represented by the area of the pressure-volume diagram if the base is specific volume. The work per cubic inch of power piston displacement is increased in practice by pressurizing the whole system so that mean pressures are 1000 psi or over. By this means, the obviously greater bulk of the engine compared to a conventional internal-combustion engine is offset to some extent; the external-combustion space and the heat exchangers still represent a disadvantage where space and weight are critical. Power regulation can be accomplished by adding working fluid to or withdrawing it from the total volume. This would make it possible, along with heat-source control, to maintain a constant top temperature despite load changes. If this complication is not practical for some applications, then part load operation becomes much less efficient.

Even at the high charge pressures (not to be confused with high indicated mean effective pressures), sealing is not as difficult as might be expected, because the displacer piston has almost equal pressure above and below, and the power piston operates entirely within the colder region. The crankcase must be closed, and can be hermetically sealed along with a generator. The working fluid for a land-operated engine can be hydrogen, which gives the highest thermal efficiency because of its low friction loss (atomic weight 1.008) and good heat transfer. However, it diffuses in time through metal walls, and make-up gas must be added continuously. The next best fluid is helium (atomic weight 4.003), which is considered suitable for space flight applications. The value of γ is not significant, because compression and expansion are isothermal rather than adiabatic.

The heat source for a more or less conventional application might be the combustion of almost any fuel, the hot products transferring heat to the tubes of the heater section. For most efficient burning and heat transfer, air would probably be supplied by a blower. The heat removal from the cooler section could be accomplished by means of cooling water circulated through a heat exchanger for heat transfer to the surroundings. Because less heat is lost to the exhaust gases and more to the coolant than in a conventional spark-ignition engine, the "radiator" would have to be larger. For applications in space travel, or where fuel is uneconomical, solar heating is possible. With this energy source, a thermal storage fluid is circulated between the engine's heater section and the focal area of the energy collector. Heat rejection requires that another fluid be circulated to a radiator for rejection to space.

3.3 ROTARY COMPRESSORS AND ENGINES

Rotary machines are similar to reciprocating machines in one important respect: both are **positive displacement** machines. This means that in one shaft revolution or one cycle, a solid member moves through a measurable volume so that an exact volume of gas is displaced or an exact change of volume of gas is allowed. The displacement is easily visualized in a simple reciprocating piston device. For some designs of rotary machines, it may be more difficult to visualize the displacement volume. Nevertheless, the same concept and definition of volumetric efficiency,

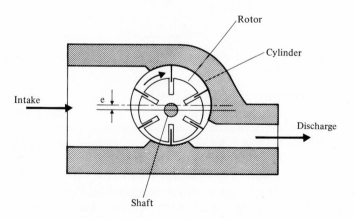

FIG. 3.29 A rotating vane type of rotary compressor.

Eq. (3.3a), can be applied to all positive displacement machines. Although many different designs are practical, only a few typical arrangements will be described. Most compressor designs are used for vacuum pumps as well.

A schematic section through a **rotating vane** type of rotary **compressor** is shown in **Fig. 3.29**. The rotor shaft is eccentric from the cylinder bore by the distance e. Vanes, which are free to move in slots in the rotor, are kept in contact with the cylinder wall by centrifugal force, sometimes augmented by springs or oil pressure. The inlet and discharge ports, which are sometimes a series of slots, do not cover the full rotor and vane widths, so that the outer edge of the vane is always in contact with a surface. It can be seen that, as the rotor turns, a definite volume of gas is trapped between two vanes and forced into a smaller volume before it is discharged.

The principal difficulty in this type of machine and in many other rotary machines is the prevention of leakage back from the high-pressure to the low-pressure side. This is particularly difficult at either end of the rotor where close tolerances are difficult to maintain, especially since temperatures vary. Because unbalanced forces are not so great, the speed of shaft rotation is usually higher than that allowable for reciprocating machines of the same displacement. Therefore rotary machines can usually handle higher volume rates of flow for the same physical size.

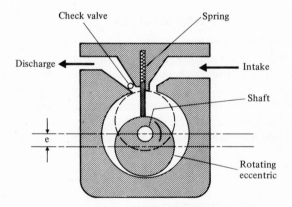

FIG. 3.30 A rotating eccentric type of rotary compressor.

The **rotating eccentric** type of rotary compressor, shown schematically in **Fig. 3.30**, is similar in many respects to the previously described design. The cylinder bore, however, is concentric with the shaft. When the eccentric is in the upper (broken line) position, it traps a definite volume of inlet gas as it passes the intake port. That gas is then forced into a progressively smaller volume on its way to the check-valve at the outlet port. The vane which can slide vertically in a slot in the casing provides a seal near the top. In order to prevent leakage at the outer edge of the eccentric as well as at the ends, there must be close fits between the rotating and stationary parts.

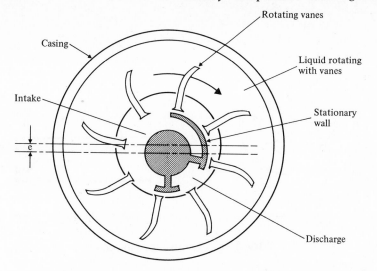

FIG. 3.31 A liquid piston type of rotary compressor.

Another rotary compressor that depends on eccentricity for its action is the **liquid piston** type shown in **Fig. 3.31**. The casing is eccentric from the center of shaft rotation. The diagram does not show that the rotating vanes are all connected to a shaft and rotate together. A liquid fills most of the casing and is forced to rotate with the vanes. Near the bottom of the diagram, the spaces between vanes are practically full of liquid. Because of casing eccentricity, the liquid recedes from the vane roots as the top of the diagram is approached, and is then forced back toward the center of vane rotation as the bottom of the diagram is again approached. In this way, the liquid acts as a piston, first allowing gas in between the vanes and then expelling the same gas. The intake and outlet passages are axial and are separated by the stationary wall shown. This design avoids solid-to-solid contact in rotor rotation.

A compressor commonly used in two-stroke internal-combustion engine air supply is the **Roots** type. It is usually referred to as a **blower** because it does not operate well at high-pressure ratios. It is used in a three-lobe configuration as well as in the two-lobe design shown in **Fig. 3.32**. The intake gas is trapped between the casing and the extremities of two lobes as the rotors turn. It is then discharged to the outlet side. The act of compression actually takes place by the backflow of some of the higher pressure gas into the space so that the rotors work against full discharge pressure instead of gradually building up pressure as in polytropic compression. The additional work of compression is noted on the pressure-volume diagram. The inevitable pulsation of delivery pressure is lessened by the three-lobe design, and sometimes further, by a helical shape in the axial direction. The two rotors do not actually contact each other, but are kept in correct relative position by gears at the end of each shaft. The rotor lobes are cut in the shape of cycloidal gear teeth

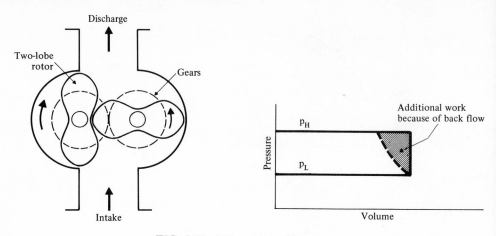

FIG. 3.32 A Roots type blower.

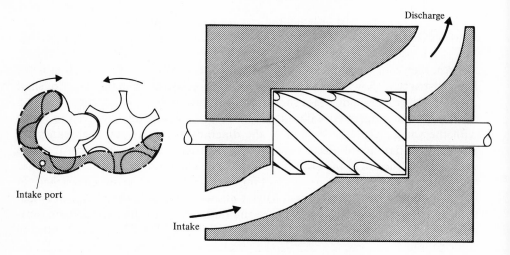

FIG. 3.33 A screw type (Lysholm) compressor.

to prevent interference while maintaining a practically constant clearance between the two.

A more gradual compression action is achieved in the **screw** type (sometimes called Lysholm) compressor show in **Fig. 3.33.** The shape of the intake port (shaded) is shown in the end view. As the rotors turn, they trap a definite volume of intake air as the lobes pass the port opening. When the two rotors mesh, the air is gradually compressed and delivered toward the opposite end. As in the Roots-type blower, the rotors do not actually contact each other, but are timed by separate gears (not shown). The screw type machine has been suggested for a reversible compressor-expander by Whitehouse, et al, *Power Engineering*, Jan., 1968.

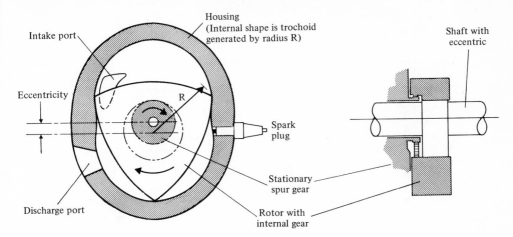

FIG. 3.34 A rotating combustion (Wankel) engine.

The one rotary-type internal-combustion engine considered important at this time is the **rotating combustion** (Wankel) design illustrated in **Fig. 3.34**. Again there is eccentricity, but this time it is between the center of shaft rotation and the center of a three-lobed rotor. An eccentric on the engine shaft is the driven arm of an epicyclic (planetary) gear train for which the other two members are a fixed central gear on the housing and an internal gear with which it mates on the rotor. The three-cornered rotor turns at one-third crankshaft speed, its endpoints at radius R generating the trochoid shape of the housing. Three cycle processes proceed simultaneously in the three separate chambers. For the position shown in Fig. 3.34, intake and compression of charge will take place in the uppermost chamber, combustion and expansion in the lower right chamber, and exhaust from the lower left. The design produces a very compact power plant capable of reasonably high shaft speeds. Several "cylinders" can be located axially on the same shaft to produce relatively smooth power output. The same problems of mechanical rubbing and end sealing are present as in most of the rotary compressors described.

3.4 CENTRIFUGAL AND AXIAL-FLOW COMPRESSORS AND TURBINES

Design problems and analysis of centrifugal and axial-flow machines, both compressors and turbines, are primarily problems in fluid mechanics. However, their overall performance as part of power plant cycles must be evaluated in thermodynamic terms. For these machines which are examples of steady-flow, steady-state systems in which the heat transfer per unit mass of working fluid is negligible, the energy terms are related by means of the first law of thermodynamics as given in the development of Eq. (1.34d):

$$h_1 + \frac{\bar{V}_1^2}{2g_c} = h_2 + \frac{\bar{V}_2^2}{2g_c} + w. \tag{3.13a}$$

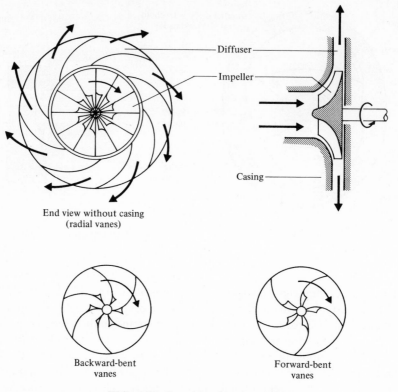

End view without casing
(radial vanes)

Backward-bent
vanes

Forward-bent
vanes

FIG. 3.35 Centrifugal compressors.

In terms of stagnation properties,

$$w = h_{o1} - h_{o2}. \tag{3.13b}$$

For a turbine, the work would be positive, and for a compressor, it would be negative. The same equations can be applied to subelements of these compressors and
turbines. Some fluid mechanics aspects will be examined very briefly.

3.4.1 Centrifugal Flow Compressors

A schematic diagram of a centrifugal compressor is given in **Fig. 3.35**. The impeller
rotates with the shaft while the diffuser guide vanes are fixed to the casing. The
compressed gas must be discharged through a scroll casing (not shown) which
surrounds the diffuser. For multistage compression, the discharge from the diffuser
of the first stage is ducted toward the center of rotation again to enter the impeller
of the next stage, which is mounted on the same shaft. By this means, quite high
pressure ratios can be achieved in centrifugal machines, which are usually compact
and well balanced. The whole work of compression is performed in the impeller.
While there is a velocity decrease and therefore a static pressure rise in the sta-

tionary diffuser, there can be no change in stagnation enthalpy there. At the entrance to the impeller, the vanes are curved forward to provide, as nearly as possible, a tangential relative velocity between the incoming gas and the rotating surfaces. The contour of the vanes beyond that may be radial, backward-bent, or forward-bent as shown in Fig. 3.35.

The velocities of the fluid leaving the impeller are shown in **Fig. 3.36**. The velocity of the fluid relative to the impeller ($\bar{W}$) will not be exactly parallel to the vane, but will lag somewhat on the average. When the circumferential velocity of the impeller ($\bar{V}_i$) is added to the velocity of the fluid relative to that circumference, the absolute velocity ($\bar{V}$) is obtained. (Here, absolute velocity will be that relative to "stationary" components of the compressor such as the casing. Of course, the whole compressor may have some velocity relative to the ground as in an aircraft power plant.) The absolute velocity can be divided into two mutually perpendicular components, the tangential velocity ($\bar{V}_t$) and the radial velocity ($\bar{V}_r$).

A useful parameter in the study of centrifugal compressors is the **slip factor** (SF) defined as

$$\text{SF} = \bar{V}_t / \bar{V}_i.$$

The example of a forward-bent vane shown in Fig. 3.36 gives a slip factor greater than 1.0. For a radial vane, it would be slightly less than 1.0, and for a backward-bent vane it would be considerably less than 1.0.

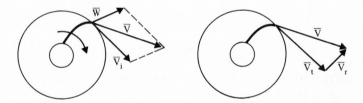

FIG. 3.36 Velocity components of fluid at impeller tip.

The **torque** input required at the shaft to drive the compressor is given by the rate of change of angular momentum imparted to the fluid. For a small element of fluid of mass "m," the torque would be

$$d\,\text{Torque} = \frac{d(rm\bar{V}_t)}{g_c dt},$$

where r is the radius. The total mass rate of flow is

$$\dot{m} = \int \frac{dm}{dt}.$$

From radius r_1 to r_2, the total torque would be

$$\text{Torque} = \frac{\dot{m}}{g_c}\,(r_2\bar{V}_{t2} - r_1\bar{V}_{t1}). \tag{3.14}$$

The usual units are

$$m \text{ lb}_m/\text{sec}, \quad r \text{ ft}, \quad V \text{ ft/sec}, \quad g_c \frac{\text{lb}_m}{\text{lb}_f} \cdot \frac{\text{ft}}{\text{sec}^2}$$

so that torque would be $\text{lb}_f - \text{ft}$. The required **power** input is simply

$$P = T\omega \text{ ft lb}_f/\text{sec}, \tag{3.15a}$$

where ω is the speed of angular rotation, radians per second. For the solid impeller,

$$\omega = \frac{\bar{V}_{i1}}{r_1} = \frac{\bar{V}_{i2}}{r_2}.$$

Thus

$$P = \frac{\dot{m}}{g_c}(\bar{V}_{i2}\bar{V}_{t2} - \bar{V}_{i1}\bar{V}_{t1}) \text{ ft lb}_f/\text{sec}. \tag{3.15b}$$

The **work** input per unit mass rate of flow is

$$-w_C = \frac{1}{g_c}(\bar{V}_{i2}\bar{V}_{t2} - \bar{V}_{i1}\bar{V}_{t1}) \text{ ft lb}_f/\text{lb}_m. \tag{3.16a}$$

Now, at the inlet to a centrifugal compressor where the fluid flow can usually be considered axial in direction, the last term in the bracket is zero. At the outlet, the slip factor can be used to relate the circumferential velocity of the impeller to the tangential component of the fluid velocity:

$$SF = \frac{\bar{V}_t}{\bar{V}_i}$$

$$-w_C = \frac{1}{g_c}(SF \ \bar{V}_i^2) \text{ ft lb}_f/\text{lb}_m. \tag{3.16b}$$

The specific work can be expressed as well in Btu/lb_m through the energy conversion factor (J). From Eq. (3.13b),

$$-w_C = h_{o3} - h_{o1} = \frac{SF \ \bar{V}_i^2}{Jg_c} \text{ Btu/lb}_m. \tag{3.17}$$

Here sections ① and ③ can be the inlet to the impeller and the outlet from the diffuser since, as already stated, there is no work or change of stagnation enthalpy beyond the impeller. In a compressor, the temperature changes are relatively small and the specific heat of a gas can usually be taken as constant. For a gas compressor,

$$h_{o3} - h_{o1} = c_p(T_{o3} - T_{o1})$$

and

$$\frac{T_{o3}}{T_{o1}} = \frac{SF \ \bar{V}_i^2}{c_p Jg_c T_{o1}} + 1. \tag{3.18}$$

For **reversible adiabatic** compression,

$$P_{o3}/P_{o1} = (T'_{o3}/T_{o1})^{\gamma/(\gamma-1)}.$$

Thus the work for an ideal compressor (reversible and adiabatic) is given by

$$-w'_C = c_p(T'_{o3} - T_{o1}) = c_p T_{o1}(P_{o3}/P_{o1})^{(\gamma-1)/\gamma} - 1. \qquad (3.19)$$

This is essentially the same as Eq. (3.9a) which was for a reciprocating, ideal, reversible adiabatic compressor with suction and delivery work included. The conversion factor (J) does not appear here because the units of Eq. (3.9a) would be ft lb_f/lb_m, while they would be Btu/lb_m for Eq. (3.19). Stagnation properties are used for the centrifugal compressor, because kinetic energy change cannot, in general, be ignored as it was for the reciprocating compressor.

The definition of isentropic efficiency of compression was given by Eq. (1.37a):

$$\eta_{Cs} = \frac{w'_C}{w_C} = \frac{h'_{o3} - h_{o1}}{h_{o3} - h_{o1}}. \qquad (3.20)$$

When this is combined with Eq. (3.18), the ideal temperature ratio can be expressed in terms of the actual work as

$$\frac{T'_{o3}}{T_{o1}} = \frac{\eta_{Cs} \, SF \, \bar{V}_i^2}{c_p J g_c \, T_{o1}} + 1 \qquad (3.21a)$$

and

$$\frac{P_{o3}}{P_{o1}} = \left[\frac{n_{Cs} \, SF \, \bar{V}_i^2}{c_p J g_c \, T_{o1}} + 1 \right]^{\gamma/(\gamma-1)}. \qquad (3.21b)$$

An ideal impeller with radial blades would have a compression efficiency and a slip factor both equal to 1.0.

3.4.2 Axial-Flow Compressors

The pressure rise in an axial-flow compressor is developed in quite a different way from that in a centrifugal-flow compressor. Its pressure rise results from the "lifting" action of an airfoil section and is therefore much smaller per stage than the pressure rise possible in a centrifugal machine. **Figure 3.37** shows three stages of an axial-flow compressor, each stage consisting of a set of stationary guide vanes followed by a rotor disc with blades at the outer edge. In each set of rotating blades, the velocity direction of the moving fluid is changed to create a pressure rise. In each set of stationary guide vanes, the fluid is redirected to have a relative velocity nearly tangential to the moving blades it will enter in the next rotor.

A highly simplified, two-dimensional, fluid flow description, similar to that given in Reference [B.4], will be presented to show how pressure rise is accomplished and the effect of lift and drag coefficients on stage efficiency. **Figure 3.38** shows the fluid flow between two adjacent rotor blades spaced a distance s apart. The height of each blade is $(r_b - r_a)$ and is called Δr. The direction of relative velocity between fluid and rotor is changed from $\bar{W}_1$ to $\bar{W}_2$. On the same figure, the forces on a blade are shown first for frictionless flow and then for flow with friction. Frictionless flow cannot create a drag force so that the resultant force (R) is a lift force (L), by definition perpendicular to the mean relative velocity direction

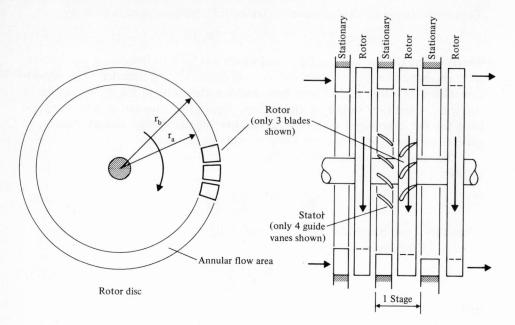

FIG. 3.37 An axial-flow compressor (schematic).

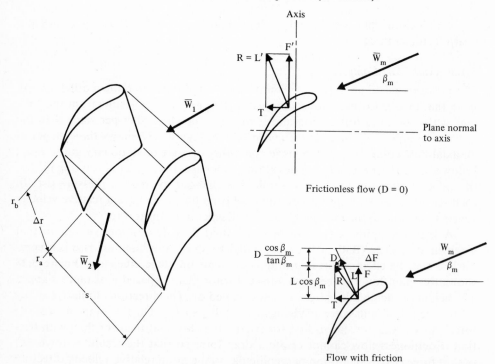

FIG. 3.38 Flow between two rotor blades, and effect of friction on pressure rise.

($\bar{W}_m$). The component of this force (T) in a tangential direction resists the torque input driving the compressor shaft. The component (F) in the axial direction results from the pressure rise from the inlet to the outlet. For each blade, this acts on an area ($s\Delta r$) normal to the axis.

When fluid friction is considered, the drag force (D) is added to the lift force, so that the resultant is no longer perpendicular to $\bar{W}_m$. For the same input torque, i.e., the same tangential force (T), the pressure force (F) is reduced, as shown, by

$$\Delta F = \frac{D}{\sin \beta_m}. \tag{3.22a}$$

The ideal or frictionless pressure rise is

$$(p_2 - p_1)' = \frac{(L + D \cot \beta_m) \cos \beta_m}{s\Delta r}. \tag{3.22b}$$

The actual pressure rise with friction is

$$(p_2 - p_1) = \frac{F' - \Delta F}{s\Delta r}. \tag{3.22c}$$

If the stage efficiency (η_{Cst}) is defined as

$$\eta_{Cst} = \frac{(p_2 - p_1)}{(p_2 - p_1)'}, \tag{3.23a}$$

then

$$\eta_{Cst} = 1 - \frac{D}{(L + D \cot \beta_m) \cos \beta_m \sin \beta_m}. \tag{3.23b}$$

Now by definition of the coefficient of lift (C_L) and the coefficient of drag (C_D),

$$L = C_L \tfrac{1}{2} \rho \bar{W}_m^2 A; \qquad D = C_D \tfrac{1}{2} \rho \bar{W}_m^2 A$$

and

$$\frac{L}{D} = \frac{C_L}{C_D},$$

where A is any useful area, say the projected area normal to $\bar{W}_m$. Then the efficiency expression can be simplified:

$$\eta_{Cst} = 1 - \frac{2}{\sin 2 \beta_m [(C_L/C_D) + \cot \beta_m]}. \tag{3.24a}$$

If β_m is nearly 45° as is likely, $\sin 2\beta_m$ is nearly 1.0. Then

$$\eta_{Cst} \cong 1 - \frac{2}{(C_L/C_D) + 1}. \tag{3.24b}$$

For good design, C_L should be much larger than C_D, and

$$\eta_{Cst} \cong 1 - 2\frac{C_D}{C_L}. \tag{3.24c}$$

The above development is admittedly quite a crude approximation, but still gives an insight into the means of producing a pressure rise and the effects of lift and drag coefficients as they are normally defined for an airfoil section. The actual

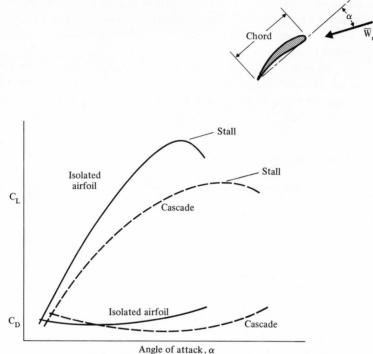

FIG. 3.39 The effect of angle of attack on coefficients of lift and drag for a particular airfoil shape either isolated or in a cascade.

values of C_L and C_D are affected greatly by two factors: the **angle of attack** of the fluid and whether the airfoil is **isolated** or in a **cascade**. A cascade is a group of airfoils parallel to each other and close enough so that the flow around each is affected by the presence of the adjacent airfoils. The closeness is indicated by the **solidity ratio,** which is the ratio of the chord length to the spacing (s), as defined on the diagrams. For normal values of solidity ratio in an axial-flow compressor, these effects are shown in **Fig. 3.39**. Although the ratio C_L/C_D is less for a cascade, the flow is better controlled, being essentially a flow between boundaries rather than around an immersed body. The point of **stall** where separation occurs is not so critical, and variations in the angle of attack do not affect performance as severely. While the physical angle of the blade relative to the shaft centerline does not change at a particular point on a particular blade, the angle of attack is affected greatly by the speed of rotation and the rate of fluid flow.

The **work** of ideal adiabatic compression for an axial-flow machine is expressed exactly the same way that it was for a centrifugal compressor in Eq. (3.19). That expression was based on Eq. (3.13b) with no consideration of the manner of producing the pressure rise. Similarly, the **isentropic efficiency** of the compressor can be expressed as in Eq. (3.20), which was based on the definition of isentropic compression efficiency given in Eq. (1.37a).

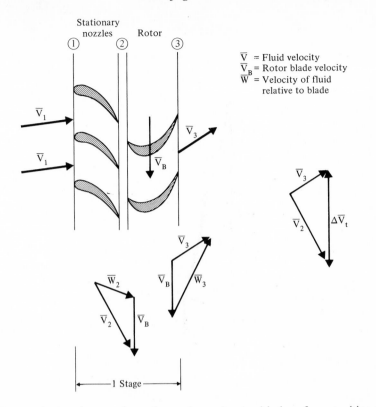

FIG. 3.40 Velocity changes through nozzles and rotor blades of one turbine stage.

3.4.3 Gas Turbines

Like air compressors, gas turbines may be axial-flow, radial-flow, or mixed-flow types. Although much of the following discussion will apply to any one of these types, the axial-flow configuration will be used for illustration: it is the most widely applied. One stage of an axial-flow turbine, much like a compressor stage, is a combination of a row of stationary blades or nozzles and a row of rotating blades at the periphery of a rotor disc. Because the gas is flowing from a high-pressure area to a low-pressure area in a turbine, separation of fluid flow from blade surfaces is not so serious a problem as in a compressor where the gas is forced to flow against a pressure difference. The direction of flow can be turned through a greater angle, and therefore more work per unit mass of gas can be obtained in one stage. Furthermore, gas temperature and therefore acoustic velocity will be higher so that higher relative velocities can be tolerated for the same limiting Mach number. Thus a single gas turbine stage can produce more power than several compressor stages can absorb, and greater pressure drop per stage is possible.

Three stationary blades, producing in effect two nozzles, and two rotor blades are shown for one stage of a turbine in **Fig. 3.40**. Absolute and relative gas velocities are shown in the vector triangles at the nozzle outlet ② and at the rotor outlet ③.

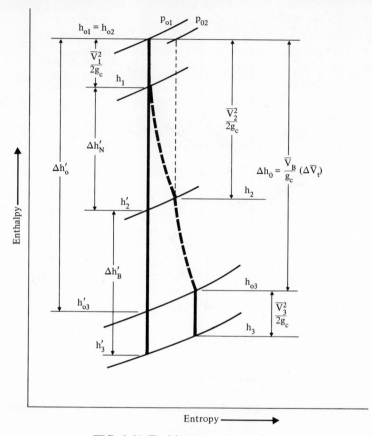

FIG. 3.41 Turbine stage, general case.

Despite the fact that the gas velocity changes in both direction and magnitude in the nozzles, no useful work is done because all surfaces are stationary. The kinetic energy is increased at the expense of static enthalpy, but the stagnation enthalpy must remain constant. All useful work, eventually available at the shaft, is done by the gas as it goes through the moving blades.

From a consideration of change of angular momentum, the work of a compressor was expressed by Eq. (3.16a). When this is applied to the rotor of an axial-flow turbine for which the impeller tangential velocity is constant at some mean radius, the turbine work can be written

$$w_T = \frac{\bar{V}_B}{g_c}(\bar{V}_{t2} - \bar{V}_{t3}), \tag{3.25}$$

where $\bar{V}_{t2}$ and $\bar{V}_{t3}$ are the tangential components of gas velocity at sections ② and ③ respectively, the difference being shown as $\Delta\bar{V}_t$ in Fig. 3.40.

In the discussion of component efficiencies for a gas-turbine power plant, Section 3.10 of Chapter 1, the expansion process through the whole turbine was shown on enthalpy-entropy coordinates in Fig. 1.52. The expansion through one stage can now be shown, **Fig. 3.41**, with nozzle and rotor blade effects in sequence. This is identified as the general case, since no particular relationship is noted between Δh_B and Δh_N; that will be specified presently.

It has already been noted that there can be no change in stagnation enthalpy through the stationary nozzles, so that

$$h_{o1} = h_{o2}.$$

There is, however, an increase in velocity because of a decrease in pressure, so that the ideal change in static enthalpy through the nozzles is

$$\Delta h'_N = h_1 - h'_2.$$

If there is a further drop in pressure through the rotor blades from p_2 to p_3, the ideal change in static enthalpy from ② and ③ is

$$\Delta h'_B = h'_2 - h'_3.$$

The overall ideal drop in stagnation enthalpy is the maximum possible work of the turbine stage per unit mass of fluid:

$$w'_{Tst} = h_{o1} - h'_{o3} = \Delta h'_o. \tag{3.26a}$$

Unless $\bar{V}_3$ is equal to $\bar{V}_1$, this is not the sum $\Delta h'_N$ plus $\Delta h'_B$.

The endpoints of actual gas expansion through the nozzle and then through the rotor blades are joined by broken lines on the diagram. The actual work per unit mass of fluid for one stage is

$$w_{Tst} = h_{o1} - h_{o3}. \tag{3.26b}$$

The stage efficiency, defined in the same way as turbine efficiency, is

$$\eta_{Tst} = \frac{h_{o1} - h_{o3}}{h_{o1} - h'_{o3}} = \frac{\Delta h_o}{\Delta h'_o}. \tag{3.26c}$$

There are various types of turbine stage designs. For a purely **impulse** turbine stage, there is no change, ideally, in the magnitude of the velocity vector entering and leaving the rotor blades. The whole tangential force is created by a change in direction of that velocity. For this situation, $\bar{V}_3$ would equal $\bar{V}_2$ and p_3 would equal p_2 so that $\Delta h'_B$ would be zero.

For a **reaction** stage, there is a change in magnitude as well as direction from $\bar{V}_2$ to $\bar{V}_3$. Thus there is a change in static enthalpy through the rotor. To describe a type of turbine stage, the term **percent reaction** is used, and is defined as

$$\% \text{ Reaction} = \frac{\Delta h'_B}{\Delta h'_B + \Delta h'_N}. \tag{3.27}$$

Obviously then, a purely impulse stage is a zero-percent reaction stage. In one multistage gas turbine, the type of stage is not necessarily the same throughout. In fact, even in one stage, the type may change radially along the blade, being say 50 % reaction at the tip and more nearly impulse at the root.

PROBLEMS

1. A reciprocating engine has a 3-in. bore and a 4-in. stroke and runs at 800 cycles/min. It produces an indicator diagram $\frac{1}{2}$ sq in. in area and $2\frac{1}{2}$ in. long, with an indicator having a spring scale of 400 psi/in. What is the indicated horsepower per cylinder?

2. A four-stroke compression-ignition engine, for which fuel is injected during the compression stroke, takes in ambient air at 60 F and 14.7 psia with a volumetric efficiency of 80% at 1900 rpm. Its fuel-to-air ratio is 0.058 with a fuel of heating value 19,000 Btu/lb. The six cylinders have a 4-in. bore and a 4.5 in. stroke.

   a) For an assumed overall thermal efficiency of 28%, estimate the horsepower output and the brake specific fuel consumption.

   b) What is the brake mean effective pressure?

3. An air compressor compresses 1500 cfm of free air (60 F and 15 psia) and delivers it at 105 psia. Its isentropic compression efficiency is 82%.

   a) Calculate the required horsepower using Eq. (3.9a), where γ is assumed to be constant at 1.40.

   b) Check your answer using the Air Tables.

4. A two-stage air compressor operates between 15 psia and 270 psia with intercooling at the optimum intermediate pressure. Given that intercooling brings the air back to the ambient temperature of 80 F, and both stages have isentropic compression efficiencies of 100%, what is the overall isothermal compression efficiency?

5. Since intercooling has obvious advantages in gas compressors, why do we seldom find more than two or three stages of compression in practice?

6. A reciprocating compressor has a clearance volume that is 12% of the displacement volume. By what percentage would the volumetric efficiency be changed ideally if the clearance volume were increased to 16% when the pressure ratio is 6 : 1 and the exponent (n) is 1.30?

7. A stationary, single-stage centrifugal compressor has an impeller diameter of 18 in., and runs at a speed of 15,000 rpm. The inlet air is at 60 F and 1 atm. If the slip factor is 0.95 when the pressure ratio is 4 : 1, what is the isentropic compression efficiency?

8. In an axial-flow compressor, the axial component of velocity is constant at 400 fps, and the average air density is 0.075 lb/cu ft. In one row of moving blades, the relative velocity between air blades is changed from 30 degrees to 45 degrees with respect to a plane normal to the axis. What is the pressure rise across this row of blades?

9. Why is a high solidity ratio desirable in an axial-flow compressor although that decreases the coefficient of lift?

10. The inlet static temperature and pressure for a single-stage gas turbine are 1700 F and 100 psia. The outlet pressure is 22 psia. For this to be a 60% reaction turbine, what must the pressure be between the nozzles and the rotating blades? Assume thermodynamic properties as given in the Air Tables.

11. The stagnation pressure and stagnation temperature are 90 psia and 1800 R for gases with a velocity of 800 fps approaching a single-stage turbine. The turbine exhaust pressure is 18 psia. If half the static pressure drop takes place in the nozzles and half in the moving blades, what is the "percent reaction?" Assume thermodynamic properties as given in the Air Tables.

REFERENCES

A. *Articles and Separate Publications*

1. BARTHALON, M. E. and H. HORGEN, "French Experience with Free Piston Engines," *Mechanical Engineering*, May, 1957, pp. 428–31.
2. CAGI, "The Fundamentals of Compressed Air Power," *Compressed Air and Gas Inst.*, Cleveland, 1960, 16 pp.
3. DAVIS, C. W., E. M. BARBER, and E. MITCHELL, "Fuel Injection and Positive Ignition—a Basis for Improved Efficiency and Economy," *SAE Trans.*, 1961, pp. 120–131.
4. DOLZA, J., "A Discussion of the Basic Design and Operation of the General Motors Fuel Injection System," *G.M. Engineering Jl.*, July–Aug., 1957, pp. 2–7.
5. FLYNN, G., W. H. PERCIVAL, and F. E. HEFFNER, "Revival of 1816 Stirling Engine Cycle," *SAE Jl.*, Apr., 1960, pp. 42–51.
6. GAY, E. J. (Ed.), "Powerplants for Industrial and Commerical Vehicles—a Look at Tomorrow," *SAE Special Publication*, Apr., 1965, 48 pp. (Discussion Supplement, August, 1965, 19 pp.)
7. GLASSMAN, A. J., "Alkali-Metal Turbine Geometry as Affected by Blade-Speed Limitations," NASA TN D-2777, *NASA*, Washington, Apr., 1965, 16 pp.
8. HARVEY, R. J. and T. C. ROBINSON, "Steam Puts Gleam on Portable Powerplants for Remote Area Operation," *SAE Jl.*, Dec., 1964, p. 68.
9. JOHNSON, I. A. and R. O. BULLOCK, Editors, "Aerodynamic Design of Axial-Flow Compressors," NASA SP-36, *NASA*, Washington, 1965, 508 pp.
10. KIRKLEY, D. W., "Determination of Optimum Configuration for Stirling Engine," *Jl. Mech. Eng. Science*, Vol. 4, No. 3, Sept., 1962, pp. 204–212.
11. KOHLER, J. W. L. and C. O. JONKERS, "Fundamentals of the Gas Refrigerating Machine," Vol. 16, No. 3, 1954, p. 69, and "Construction of the Gas Refrigerating Machine," Vol. 16, No. 4, p. 105, *Philips Tech. Rev.*, 1954.
12. LUNDQUIST, W. G., "Dyna-Star Design Shrinks Piston-Engine Size," *SAE Jl.*, Mar., 1964, p. 66.
13. MEYER, W. E., "Small Engines on the March Abroad," *SAE Jl.*, Vol. 73, No. 6, pp. 32–34.
14. MEYER, W. E., "World Survey of Small Engine Developments," *SAE Paper*, 952A, Jan., 1965, 15 pp.
15. Miller, E. H. "Modern Nuclear Turbine Design," *General Electric*, Schenectady, Apr., 1964, 14 pp.
16. OLSEN, D. R., "Simulation of a Free Piston Engine with a Digital Computer," *Trans SAE*, Vol. 66, 1958, p. 668.
17. STARKMAN, E. S., "What Engine Will Power the Car of Tomorrow?" *SAE Jl.*, Vol. 72, No. 4, 1964, pp. 34–39.
18. STRONG, R. E., "200 MW Steam Injection Peaking Plant," *Gas Turbine*, Sept.–Oct., 1966, pp. 40–41.
19. TABOR, H. Z. and L. BRONICKI, "Tiny Turbine + Right Working Fluid = High Efficiency," *SAE Jl.*, Vol. 73, No. 6, pp. 52–55.
20. VINCENT, E. T., "Charts Help Map Performance of Turbocharged CI Engines," *SAE Jl.*, Dec., 1963, pp. 66–69.

21. WALKER, G., "Operating Cycle of Stirling Engine with Particular Reference to Function of Regenerator," *Jl. Mech. Eng. Science*, Vol. 3, No. 5, Dec., 1961, pp. 394–408.

22. WALKER, G., "An Optimization of Principal Design Parameters of Stirling Cycle Machines," *Jl. Mech. Eng. Science*, Vol. 4, No. 3, Sept., 1962, pp. 226–240.

23. WELSH, H. W. and D. S. MONSON, "Adapting Stirling Engine to 'One-Year-In-Space' Operation," *SAE Jl.*, Dec., 1962, pp. 44–51.

24. WITHROW, L. and G. M. RASSWEILER, "Slow Motion Shows Knocking and Non-Knocking Explosions," *SAE Trans.*, Vol. 39, 1936, p. 297.

B. *Books*

1. ARMSTRONG, L. V. and J. B. HARTMAN, *The Diesel Engine*. New York: Macmillan, 1959.

2. CHURCH, E. F., *Steam Turbines* (third edition). New York: McGraw-Hill, 1950.

3. DRIGGS, I. H. and O. E. LANCASTER, *Gas Turbines for Aircraft*. New York: Ronald, 1955.

4. DURHAM, F. P., *Aircraft Jet Power Plants*. Englewood Cliffs: Prentice-Hall, 1951.

5. DUSINBERRE, G. M. and J. C. LESTER, *Gas Turbine Power* (second edition). Scranton: International Textbook, 1958.

6. FRAAS, A. P., *Combustion Engines*. New York: McGraw-Hill, 1948.

7. HELDT, P. M., *High-Speed Combustion Engines*. Philadelphia: Chilton Co., 1951.

8. JUDGE, A. W., *Small Gas Turbines—and Free Piston Engines*. London: Chapman and Hall, 1960.

9. KING, R. O. and Associates, *The Combustion of Fuel Vapor and Gases*, Vol. 1, Defense Research Board of Canada, 1947–1957.

10. LEE, J. F., *Theory and Design of Steam and Gas Turbines*. New York: McGraw-Hill, 1954.

11. LICHTY, L. C., *Combustion Engine Processes*. New York: McGraw-Hill, 1967.

12. OBERT, E. F., *Internal Combustion Engines—Analysis and Practice*, Scranton: International Textbook, 1950.

13. TAYLOR, C. F., *The Internal Combustion Engine in Theory and Practice*, Vol. 1. Cambridge, Mass.: M. I. T. Press, 1960.

14. TAYLOR, C. F., and E. S. TAYLOR, *The Internal-Combustion Engine* (second edition). Scranton: International Textbook Co., 1961.

15. VINCENT, E. T., *The Theory and Design of Gas Turbines and Jet Engines*. New York: McGraw-Hill, 1950.

16. ZUCROW, M. J., *Aircraft and Missile Propulsion*, Vols. I and II. New York: John Wiley, 1958.

CHAPTER 4

REFRIGERATION AND HEAT PUMP CYCLES

4.1 INTRODUCTION

In engineering, the term **refrigeration** must imply more than cooling or keeping something cool. We are not concerned here with the effect that can be achieved simply by the convective or radiant heat transfer to ambient conditions. What we will investigate are various means of transferring heat from some substance or space at one temperature and disposing of that same energy elsewhere at a higher temperature. The process may be continuous or intermittent.

The so-called **Clausius** statement of the **second law of thermodynamics** is: "It is impossible to construct a device that operates in a cycle and produces no effect other than the transfer of heat from a cooler to a hotter body." The implication of this statement is that a system that does produce the transfer of heat from a cooler **source** to a hotter **sink** requires the input of some additional work or energy. Thus this chapter will be concerned with various systems in which the input of work or energy will accomplish such a refrigeration effect.

But the purpose of the system may not be primarily to cool or maintain the low temperature of the source of heat. Instead, the purpose may be to deliver heat energy to some body or sink in order to raise its temperature or maintain its temperature at a usefully high level. Such a system is called a **heat pump**. Thermodynamically, there is no difference between a refrigerator and a heat pump. The difference is entirely in the point of view, that is, in the assignment of a purpose to the process. In many applications, the same machine may at one time operate as a refrigerator and at another time as a heat pump. In some situations, both the cooling effect at one temperature level and the heating effect at another may be desired, and the system then operates simultaneously as a refrigeration machine and as a heat pump.

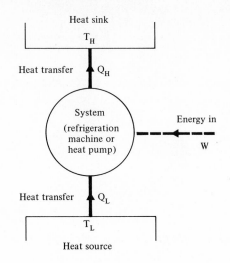

FIG. 4.1 A thermodynamic system acting as a refrigeration machine or heat pump.

Figure 4.1 shows the thermodynamic system just described. The absolute temperature of the source is T_L degrees, and the heat transferred from the source is the refrigeration effect, Q_L. The energy rejection is to the sink at temperature T_H degrees, and this amount of heat delivery is Q_H. Both effects are accomplished by the input of work, W. For continuous operation, application of the **first law of thermodynamics** to the system shows that, over a particular period of time,

$$Q_H = Q_L + W. \tag{4.1a}$$

Where a working fluid is being considered in the system, q and w are conventionally used to represent energy per unit mass of the fluid. It is often convenient to consider energy transfer per unit of time. For this, a dot over a symbol will represent a time rate. By definition, the rate of work is power. Thus the above equality can be written

$$\dot{Q}_H = \dot{Q}_L + \dot{W} = \dot{Q}_L + P. \tag{4.1b}$$

4.1.1 *Units*

The common unit of refrigeration is the **ton**. This was originally intended to be the rate of heat transfer necessary to freeze one ton of water at 32 F in 24 hours. In America, the short ton (2000 lb mass) is used. If the latent heat or enthalpy of fusion of liquid water to ice is taken as 144 Btu/lb (actual value 143.35 Btu/lb), then the exact definition is obtained: one ton is equivalent to the removal of 288,000 Btu/day, which is more conveniently applied as **12,000 Btu/hr** or **200 Btu/min**.

It should be noted that the **British ton** of refrigeration is about 19 % larger than the American ton since it is defined as equal to a cooling rate of 1 Kilogram-calorie/sec (about 238 Btu/min). This was taken because it is not greatly different from the required rate of heat removal to freeze one long ton (2240 lb) or one metric ton (10^6 grams $= 2205$ lb) of ice in a day. On the European continent, the unit of refrigeration is commonly the **frigorie**, which is equivalent to a heat removal rate of about 50 Btu/min.

The term ton as a unit of refrigeration is a reminder of one of the early commercial applications of refrigeration, the manufacture of ice. Before that industry was founded, the prime source of cooling for food storage and shipping was natural ice. Just as James Watt coined the term horsepower because his engines were to compete economically with animal power, so the refrigeration industry needed a unit that would convey a meaningful quantity to the potential market of the nineteenth century. Certainly today, ratings in Btu per minute or per hour are more meaningful to the engineer.

4.1.2 Coefficient of Performance, Efficiency, and the Carnot Cycle

The term **thermal efficiency** is not easily applicable to a refrigeration machine or a heat pump. For a power cycle, the thermal efficiency is the power output divided by the energy input; the ratio must always be less than unity. For the system of Fig. 4.1, we have seen that the heat delivered to the sink must be greater than the work energy supplied in accordance with the first law of thermodynamics. In place of thermal efficiency, the term **coefficient of performance** (COP) is used. This might be defined as "what you want to achieve, divided by what you have to pay for to achieve it." The personal pronoun is used on purpose since, for a given situation, the numerical value of the COP will depend on the subjective judgement of what was desired.

If the system of Fig. 4.1 is to operate as a refrigeration machine, the desired effect is the removal of heat energy at the rate $\dot{Q}_L$ from the source. As a heat pump, the desired effect is the delivery of heat energy at the rate $\dot{Q}_H$ to the sink. In either situation, the power supplied ($P = \dot{W}$) must be "paid for." Thus, from the point of view of refrigeration, the coefficient of performance would be

$$\text{COP}_R = \frac{\dot{Q}_L}{\dot{W}} = \frac{Q_L}{W}. \tag{4.2a}$$

Considered as a heat pump, the same system would have the following coefficient of performance:

$$\text{COP}_{HP} = \frac{\dot{Q}_H}{\dot{W}} = \frac{Q_H}{W}. \tag{4.2b}$$

Because the sum ($Q_L + W$) is equal to Q_H, these two values of COP are easily related:

$$\text{COP}_{HP} = \frac{Q_L + W}{W} = \text{COP}_R + 1.0. \tag{4.2c}$$

This is true for both ideal and actual systems so long as the systems are adiabatic, except for the heat transfers Q_L and Q_H.

The term **horsepower per ton** for a refrigeration machine is analogous to the term "brake specific fuel consumption" (pounds of fuel per brake horsepower-hour), which is applied to fuel consuming prime movers. Both terms represent attempts to relate the required input to the desired output or effect. Horsepower per ton is an inverse indication of the COP, just as "bsfc" is an inverse indication of the brake thermal efficiency. It is easily shown for a refrigerator that

$$HP/ton = \frac{4.72}{COP_R}.$$

When the term "efficiency" is applied to a refrigeration system, the only significant meanings are **cycle efficiency** (η_{cy}) and **relative efficiency** (η_r), sometimes called **reduced efficiency**, introduced in Section 1.1.2 for power cycles. The cycle efficiency for a real refrigeration system is the ratio of the actual COP value to the value of COP for a particular ideal cycle to which an approach is attempted. The relative or reduced efficiency, on the other hand, is the ratio of the COP value for some ideal cycle to the COP value for a reversible cycle working between the same principal temperature levels.

As noted in Section 1.1.1, it can be shown as a corollary of the second law of thermodynamics that any two reversible power cycles that receive heat energy only at one temperature level and reject heat only at another lower temperature level must have the same thermal efficiency. It can similarly be shown that any two reversible refrigeration cycles receiving and rejecting heat energy at two particular temperature levels must have identical values of coefficient of performance. Thus

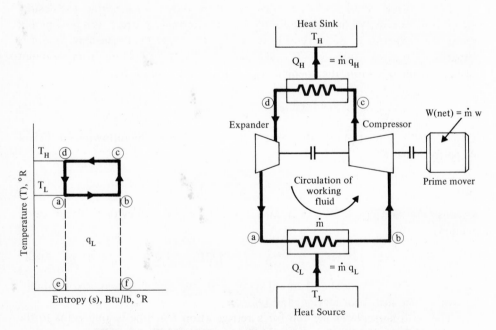

FIG. 4.2 The ideal Carnot refrigeration or heat-pump cycle.

relative efficiency can be referred to any reversible cycle so long as heat transfer for the reference cycle takes place at only two temperature values. Meaningful cycle temperatures must be selected for the comparison.

The simplest cycle to study for this purpose is the **Carnot cycle. Figure 4.2** shows the Carnot cycle on temperature-entropy coordinates where, because the two isentropic and two isothermal processes in this ideal cycle are reversible, enclosed areas represent energy quantities. Also shown is a schematic diagram of the same cycle. Note that the net work input must be the difference between the work of compression from **b** to **c** and the work of expansion from **d** to **a**. Heat is assumed to be transferred isothermally to the hypothetical working fluid from **a** to **b** (with a resultant increase in the fluid's entropy); the heat transferred at T_L degrees Rankine is represented by the area **abfea**, which is q_L Btu per pound of working fluid. As the fluid is compressed isentropically from **b** to **c**, the temperature is increased to T_H. Heat is now transferred from the working fluid isothermally at T_H until a

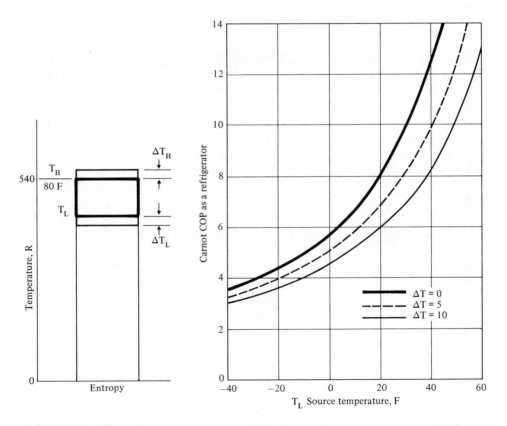

FIG. 4.3 The effect of source temperature and heat exchanger temperature difference on the COP of the ideal Carnot cycle as a refrigerator.

state point **d** is reached from which isentropic expansion will return the fluid to **a**. The heat energy rejected at T_H is represented by the area **cdefc**, which is q_H Btu per pound, and the work done *on* the working fluid is represented by the area **abcda** which is w Btu per pound.

Inspection of the T-s diagram for the Carnot cycle shows that the coefficient of performance can be expressed as the ratio of absolute temperatures. For the Carnot cycle operating as a refrigerator:

$$\text{COP}_R = \frac{q_L}{w} = \frac{T_L}{T_H - T_L}. \tag{4.3a}$$

As a heat pump, the Carnot cycle COP is:

$$\text{COP}_{HP} = \frac{q_H}{w} = \frac{T_H}{T_H - T_L} \tag{4.3b}$$

$$= \frac{T_L}{T_H - T_L} + 1.0, \tag{4.3c}$$

which is the COP as a refrigerator plus unity. Compare Eq. (4.2c).

In all real heat exchangers, some temperature difference is required. For source and sink temperatures T_L and T_H, the refrigerating fluid must actually work in a cycle between temperatures $(T_L - \Delta T_L)$ and $(T_H + \Delta T_H)$. This enlarged temperature span then reduces the COP of the cycle. **Figure 4.3** shows a Carnot cycle operating with a variable source temperature T_L and a sink temperature fixed by ambient air or cooling water temperature at 80 F. The curves of Fig. 4.3 show the effect of source temperature on the ideal cycle COP_R with no temperature differences in the heat exchanger, and they show the effect of 5-degree and 10-degree differences added to the same cycle. The plotted values of COP are those used when we view the system as a refrigerator. For the system as a heat pump, all values of COP would be increased by unity.

4.1.3 Means of Refrigeration

There are many ways in which a refrigeration effect can be achieved. Vapor cycles utilizing either mechanical compression or chemical absorption are the most common. Those two plus steam-jet and gas-cycle systems will be dealt with in this chapter. Cryogenics, the name given to the study of extremely low temperatures, usually deals with systems that are different in many respects from the refrigeration devices described here. Thus cryogenics is discussed in Chapter 5. From the point of view of commercial application, a relatively new, but potentially very important means of refrigeration, is the thermoelectric device. While the overall thermodynamic evaluation of that type of device allows a close parallel to the cycles of this chapter, a discussion of the basic principles of thermoelectricity can be grouped more logically with a discussion of other direct energy-conversion devices. For this reason, the thermoelectric heat pump is dealt with in Chapter 7.

4.2 MECHANICAL VAPOR COMPRESSION

The Carnot cycle, which has been discussed, cannot be achieved by a real working fluid. It is approached best in a cycle that utilizes constant-pressure vaporization and constant-pressure condensation of the fluid. In these two processes, most of the heat transfer, at the lower and at the higher temperatures respectively, is accomplished isothermally. The relatively high latent heats of vaporization and condensa-

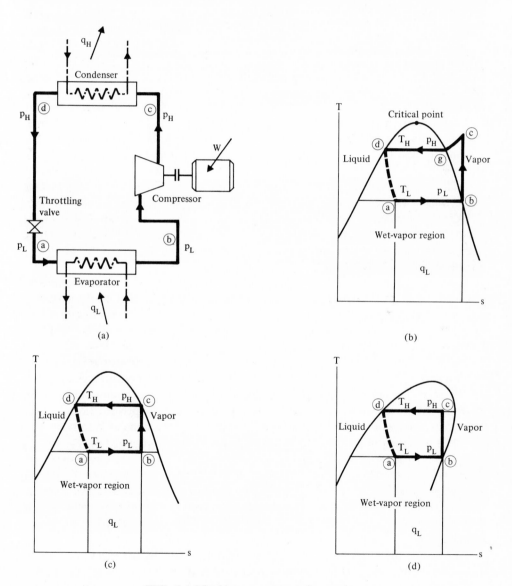

FIG. 4.4 Ideal vapor-compression cycles.

tion for most fluids allow large refrigeration effects for modest rates of circulation. Isentropic compression is approached in modern compressors. On the other hand, appreciable work cannot usually be recovered by the expansion of a very wet vapor through a machine, and so the expansion is allowed to take place in a throttling device such as a valve or a fixed restriction in the flow passage.

Figure 4.4(a) shows the simplest possible mechanical vapor-compression system for refrigeration or heat pumping. There are only two pressure levels in this system (friction loss being ignored): vapor at **b** is **compressed** from p_L psia to p_H at **c**; condensed liquid at **d** is **throttled** with no work output from p_H to p_L at **a**.

The existence of the two pressure levels allows evaporation of the **refrigerant** (practical working fluid) at a relatively low temperature, and condensation of the same fluid at a higher temperature. Heat must be transferred to the fluid to produce the evaporation, and it is in this low-pressure heat exchanger that the refrigeration effect is achieved. Heat is transferred from the **secondary fluid** which passes through the **evaporator**. Another fluid, the **cooling medium** passing through the **condenser**, carries away the sum of the heat energy absorbed by the working fluid in the evaporator and the work of compression supplied in the mechanically driven compressor.

The temperature-entropy diagram of Fig. 4.4(b) shows a **dry-compression** cycle, where the isentropic compression starts on the saturated vapor line at point **b**. At the end of compression, the working fluid is superheated vapor at **c**. In the condenser, the first heat removal is equivalent to the superheat from **c** to **g**. However, for most refrigerants, the greatest part of the heat rejected is in the condensation process from **g** to **d**.

A closer approach to the shape of the Carnot cycle is seen in the **wet-compression** cycle of Fig. 4.4(c), where the isentropic compression terminates on the saturated vapor line at **c**. The disadvantage of this cycle is that the fluid compressed is in the wet vapor region. A mixture of liquid and vapor would cause mechanical problems for most compressors, and the deviation from isentropic compression may be larger than for the dry-compression cycle. Some fluids exhibit a reversal from negative to positive slope for the saturated vapor line on T-s coordinates as shown in Fig. 4.4(d). Isentropic compression from a dry, saturated point could then be entirely within the wet vapor region. The absence of a "superheat horn," while being an apparent thermodynamic advantage, is a mechanical disadvantage.

The **throttling process** is irreversible and cannot be represented on T-s coordinates. It is usually considered to be adiabatic when accomplished in a short distance as in a valve or orifice. Also, no external work is done. Therefore the state at the endpoint is known: application of the first law of thermodynamics to a steady-flow process with no work done and no heat transfer shows that the enthalpy of the fluid after throttling (at point **a**) is the same as it was before throttling (at point **d**).

Consider the diagram of **Fig. 4.5**(a). Enthalpy can be taken as zero at the arbitrary datum point **H**. Enthalpies at any other point will be given by the areas under constant pressure lines from **H** to the point. Calculations will show that the liquid saturation line is very nearly the same as a constant pressure line on this diagram; the enthalpy at point ② of saturated liquid at p_L is very nearly the enthalpy of liquid at T_L and p_H. Thus the area under the line *H-2-d* (which includes the shaded triangle) is equal to the area under the line *H-2-a* (which includes the shaded rectangle). The unshaded area under the two lines is common to both, and therefore the triangular area is equal to the rectangular area. This shows clearly that the horizontal displacement of point ***a*** depends on the slope of the line *2-d*. The area *a345a* has been called the **throttling loss**, but this is not a desirable term

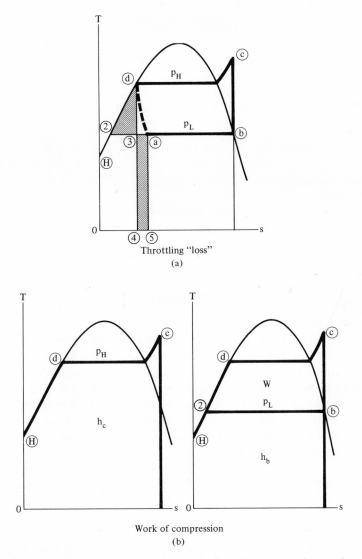

FIG. 4.5 Enthalpy and work in an ideal vapor-compression cycle.

since there is no loss of energy in this "constant enthalpy" process. There is, however, a reduction in the potential refrigeration effect by the amount of the shaded rectangle as compared to the heat removal possible with a Carnot cycle which would have isentropic expansion from *d* to ③ and evaporation from point ③.

The **net work** required in a mechanical vapor-compression cycle is not affected by the left side of the *T-s* diagram, because no work is recoverable from the expansion process as, ideally, it would be in the Carnot cycle. When the steady-flow energy equation is applied to the adiabatic compression process, it is seen that all the work done on the fluid must result in a change of enthalpy of the fluid:

$$w = h_c - h_b \text{ Btu/lb.} \tag{4.4}$$

This is illustrated in the two *T-s* diagrams of Fig. 4.5(b). The enthalpy at point *b* is the area under the constant pressure line *H-2-b*. Similarly, the area under the constant pressure line *H-d-c* is the enthalpy at point *c*. The net work input for the cycle, then, is the area *bcd2b*, regardless of the exact point from which throttling takes place. Compare this discussion with that of the Rankine power cycle in Section 1.2.3.

On the other hand, the point from which throttling takes place does govern the amount of refrigeration effect possible per pound of working fluid circulated. The refrigeration effect is the difference in enthalpy of the end-state points in the constant pressure evaporation process *a–b*. Thus

$$q_L = h_b - h_a. \tag{4.5a}$$

But

$$h_a = h_d,$$

so

$$q_L = h_b - h_d \text{ Btu/lb.} \tag{4.5b}$$

4.2.1 Pressure-Enthalpy Diagram

A thermodynamic property chart that is little used outside of the field of vapor-compression refrigeration is the **pressure-enthalpy diagram** for which a typical shape is shown in **Fig. 4.6**. Its main advantages are that heat quantities are dealt with as linear distances and that pressure lines are straight. The vapor-compression cycle operates between two pressures; when the relatively small pressure drops resulting from fluid friction are neglected, the principal points in the cycle are easily located on one or the other of the two pressure lines. Since enthalpy is the same at the beginning and the end of the throttling process, the line *d–a* is readily drawn on this diagram, and *a* is immediately located from *d*. In the idealized dry-compression cycle shown in Fig. 4.6, the quantities q_L, w, and q_H are easily determined, and the fact that q_H is the sum of the other two energy quantities is more readily apparent than on the *T–s* diagram.

The properties of vapors commonly used for refrigeration are conveniently presented on *p–h* coordinates with constant specific volume, constant temperature, and constant entropy lines shown in the vapor region, and constant dryness lines

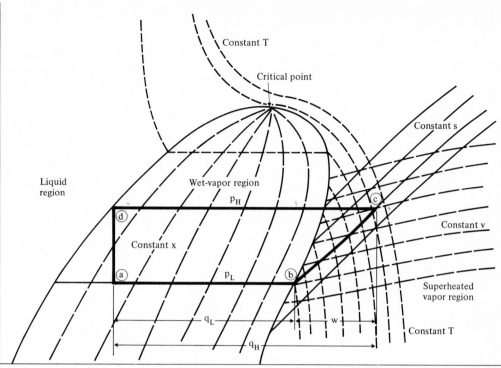

FIG. 4.6 A typical *p-h* diagram for a refrigerant with the ideal vapor-compression cycle shown.

shown in the wet-vapor region. A logarithmic scale is used for the pressure ordinate to compress the chart but retain accuracy at the lower values. Also useful is the fact that, because of the logarithmic scale, equal pressure ratios are represented by equal vertical distances. The enthalpy abscissa, while usually linear, frequently has a line of discontinuity in the wet-vapor region; the scale is often changed to provide greater accuracy where it is needed on the right-hand side.

4.2.2 Subcooling and Superheating

The idealized cycle discussed, where throttling and compression start at the liquid and vapor saturation lines respectively, has disadvantages in practice. In an actual system, if the fluid leaving the condenser were brought only to the liquid saturation line, any friction loss in the piping would result in throttling into the wet-vapor region. The resulting vapor formation and increase in specific volume would reduce the capacity of the piping or, what is more important, of the throttling device.

Subcooling of the liquid at p_H, before the throttling valve, is shown on the left side of both the *T-s* diagram and the *p-h* diagram of **Fig. 4.7**. The diagrams

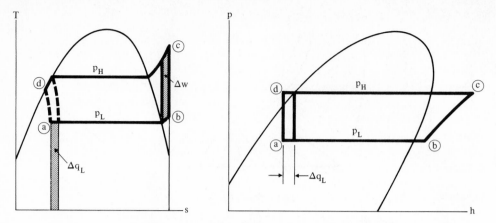

FIG. 4.7 Subcooling and superheating in an ideal vapor-compression cycle.

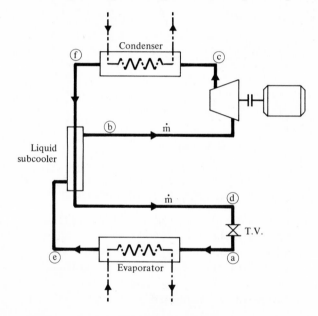

FIG. 4.8 Liquid subcooling with suction vapor.

show the possible increase in refrigeration effect resulting from the subcooling of the high-pressure liquid. Whether this whole difference becomes available as a net increase in q_L will depend on how the subcooling is achieved. As will be seen, the increase on the left side of the diagram may be obtained by an equal reduction on the right side.

On the other side of the diagrams, point *b*, from which compression starts, is shown in the superheated vapor region. It is usually important to avoid the admission of liquid phase fluid to the compressor, and several degrees of superheat will ensure against local entrainment of liquid. The principal disadvantage of **superheating** the compressor suction vapor is that the increased specific volume will reduce the capacity of the compressor in terms of mass rate of circulation. Also, the top temperature of the cycle at *c* will be raised. The *T-s* diagram shows more conclusively than does the *p-h* diagram that the work of compression from *b* to *c* must be increased by the superheating of the suction vapor.

It is clear that, in a simple cycle such as that shown in Fig. 4.4(a), the temperature at *d* cannot be below the condenser coolant temperature, and the temperature at *b* cannot be above the source temperature. To provide for both **subcooling** and **superheating**, a heat exchanger is often added to the system as shown schematically in **Fig.4.8**. When subcooling is achieved in this manner, the heat transfer into the working fluid between *e* and *b* (which is equal to the heat transfer out from *f* to *d*) is not part of the refrigeration effect.

4.2.3 Volumetric Efficiency, Capacity, and Power

The cooling capacity of a refrigeration system depends on the rate of circulation of the refrigerant which, for a positive-displacement type of compressor, depends on the speed of rotation, the displacement, and the **volumetric efficiency** (η_v). The effects of clearance volume, fluid friction losses, and cylinder temperature on compressor volumetric efficiency were discussed in Chapter 3. It is interesting to note that, particularly in some small units, the capacity of the system at any given speed can be altered by a change in the clearance volume in compressors which are otherwise identical. The change can be made simply by boring out the cylinder head or, more conveniently, the piston crown.

By definition,

$$\eta_v = \frac{\dot{m}(v_b)}{(\text{Piston displacement}) \times (\text{Cycles per minute})}. \tag{4.6a}$$

Thus

$$\dot{m} = \frac{\eta_v(\text{PD})N}{v_b} \text{ lb/min}, \tag{4.6b}$$

where v_b is the specific volume (cu ft/lb) at suction conditions, $\dot{m}$ is the rate of circulation of the working fluid (lb/min), PD is the piston displacement (cu ft/cycle), and N is the number of cycles per minute.

Both the system capacity and the power required are directly related to the rate of circulation. For a single compressor system and with the notation of Fig. 4.8,

$$\text{Capacity} = \frac{\dot{m}(h_e - h_a) \text{ tons}}{200}, \tag{4.7a}$$

$$\text{Power} = \frac{\dot{m} \times w}{42.4\,\eta_m}, \tag{4.7b}$$

and

$$\text{hp/ton} = \frac{200}{42.4} \times \frac{w}{(h_e - h_a)\eta_m}, \tag{4.7c}$$

where w is work in Btu per pound, and η_m is the **mechanical efficiency.**

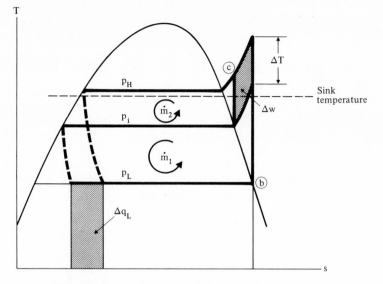

FIG. 4.9 Staging of an ideal dry-compression cycle.

4.2.4 Multistage Vapor Compression

For all fluids, the saturated liquid line has a positive slope on *T-s* coordinates. Most fluids that are used as refrigerants exhibit a negative slope for the saturated vapor line, although some, such as Refrigerant 113, exhibit a reversal to a positive slope below some point. (See also Section 1.2.7.) Thus, for all fluids, there is an increasing throttling loss per pound as evaporation pressure is lowered while the throttling point stays fixed. Also, for most fluids, the vapor temperature at the end of compression to a given condensation pressure is increased more and more as the evaporation temperature and pressure are lowered. **Figure 4.9** shows how both these deviations from the Carnot cycle can be reduced by the use of two-stage compression with intercooling. It is also evident that the required rate of circulation of refrigerant through the evaporator is reduced for a given refrigeration rate or tonnage.

In Chapter 3, it was shown that the optimum interstage pressure for ideal gas compression, as given by Eq. (3.11b), is

$$p_i = \sqrt{p_H p_L}.$$

For vapor compression, the optimum value is likely to be slightly different, depending on the fluid used and on the overall pressure ratio.

For a refrigeration or heat-pump cycle, the condensation temperature is governed by the available sink temperature. Intercooling by an external coolant therefore cannot in general be carried to a low enough temperature. **Figure 4.10** shows a feasible system for staging with a **closed heat exchanger** between the high-pressure and low-pressure cycles. Such an arrangement is called **cascading.** The

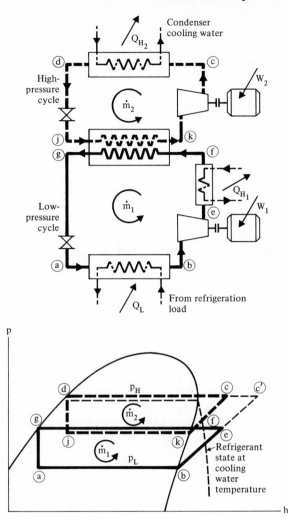

FIG. 4.10 Staging of a vapor-compression cycle with a closed heat exchanger.

refrigeration effect of the upper stage is the heat transfer from the lower stage, except for a small amount of superheat removal that may be possible to an external coolant. The common heat exchanger is sometimes referred to as a **cascade con-denser**. The *p-h* diagram in this particular example implies that both cycles utilize the same type of refrigerant. Overlapping of the cycles is necessary to provide a temperature difference in the cascade condenser, where the two fluids are separated by a solid barrier.

As long as the same refrigerant is used in both the high-pressure and the low-pressure cycles, there is really no need to separate the two in a closed-type heat exchanger. Heat transfer can be improved, and the necessity for a temperature

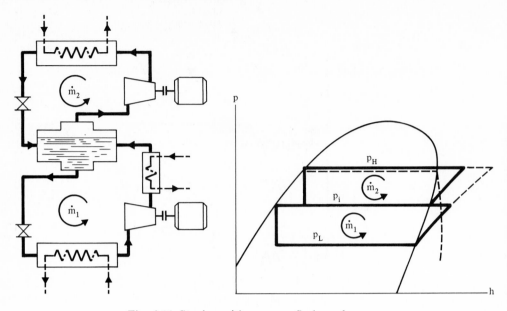

Fig. 4.11 Staging with an open flash cooler.

difference can be eliminated by the use of an **open-type heat exchanger** or **flash cooler,** in which the fluids entering at f and at j mix intimately. This revision is shown in **Fig. 4.11.** The use of open heat exchangers is actually more common.

In the systems of either Fig. 4.10 or Fig. 4.11, $\dot{m}_1$ is established by the required refrigeration rate, and $\dot{m}_2$ is determined by an energy balance on the common heat exchanger. Consider an energy balance on the flash cooler:

$$\text{In} \qquad\qquad \text{Out}$$
$$(\dot{m}_1 h_f + \dot{m}_2 h_j) = (\dot{m}_1 h_g + \dot{m}_2 h_k).$$

Thus

$$\frac{\dot{m}_1}{\dot{m}_2} = \frac{h_k - h_j}{h_f - h_g}. \qquad (4.8a)$$

The required capacity of the system determines the mass rates of flow:

$$\dot{m}_1 = \frac{\text{Required tons} \times 200}{h_b - h_a} \text{ lb/min}. \qquad (4.8b)$$

With intimate mixing in the open flash cooler (where energy absorption is accomplished by evaporation), the fluid of the upper cycle is indistinguishable from that of the lower cycle and both are in equilibrium at p_i. This allows the ratio $\dot{m}_1/\dot{m}_2$ and the pressure p_i to vary slightly as required by the conditions of operation.

4.2.5 Multiple-Fluid Cascading

A major advantage of a cascade system with a closed-type heat exchanger is that the types of fluid in the two stages need not be the same. When wide temperature ranges must be spanned, extremes in pressure can be avoided if fluids are chosen

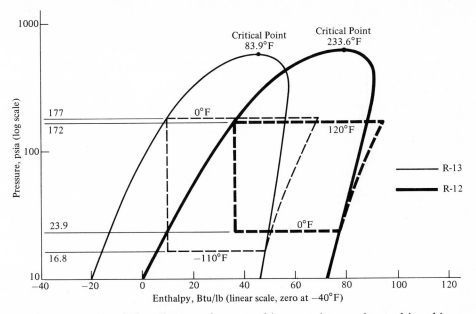

FIG. 4.12 Pressure-enthalpy diagrams for two refrigerants that can be used in a binary cascade system.

for reasonable evaporation and condensation pressures in the two or more temperature ranges. The equipment arrangement for a two-stage **binary** system can be the same as was shown for a one-fluid system in Fig. 4.10. The cooling water heat exchanger in the lower cycle would probably not be used. **Figure 4.12** shows the *p-h* diagrams for Refrigerants 12 and 13, superimposed on each other. (The designation of refrigerants by number will be discussed in Section 4.3). It is seen that a cascade system with these two refrigerants can span the temperature range from −110 F to +120 F, with an intermediate temperature of zero F; the lower pressures would both be above atmospheric pressure, while the higher pressures would be well below 200 psia. No overlap in temperatures is shown, but that would be necessary for a temperature difference in the common heat exchanger; pressures would be only slightly altered. This binary-fluid cascade system is analogous in many ways to the binary-fluid power cycle discussed in Chapter 1. An example of three-fluid cascading that has been used for the production of extremely low temperatures and gas liquefaction will be given in Section 5.3.2.

4.2.6 Controls for Vapor Compression Systems

Although the schematic presentations and descriptions of vapor compression systems so far have indicated that the throttling or adiabatic pressure-drop operation takes place in some sort of a valve, a simple manually set valve would not in general be practical. A fixed restriction or some automatically adjusting device are the alternatives. Various types of throttling devices are illustrated in **Fig. 4.13**.

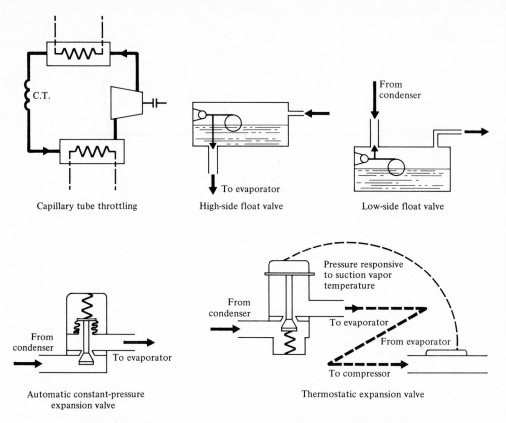

FIG. 4.13 Throttling devices for vapor-compression systems.

A pressure drop or progressive throttling must accompany the flow of any real fluid in a pipe. Where a fixed restriction is suitable or even desirable, as in hermetically sealed household refrigerators, a **capillary tube**, which is a very inexpensive component, can take the place of the throttling valve. In a fine bore tube, the pressure drop can be predicted quite accurately if the fluid viscosity is known. Thus the bore and length can be set at the time of manufacture. Subsequent system control then cannot be accomplished through the use of a throttling device, but must be achieved by on-off motor control. Since the tube remains open when the motor stops, the pressures can equalize throughout, and the motor will restart against a minimum load. Its greatest advantages are simplicity and economy. The tube must be protected from blockage by any foreign substance in the system. Proper selection of length and bore must be made with care, sometimes by trial and error.

A simple control that can respond to the rate of vaporization in the evaporator is the **"low-side" float valve**. This requires operation with a flooded evaporator which ensures fully wetted surfaces for good heat transfer. On the other hand, there is still no control of the pressure and thus temperature of evaporation. A "**high-side**" **float valve**, which maintains a constant liquid level in the condenser, responds to the rate of condensation, but its action is only indirectly affected by evaporator conditions.

The pressure of evaporation can be controlled by an **automatic constant-pressure expansion valve** actuated by the evaporator pressure or simply by the pressure immediately after throttling. That pressure, working on a bellows or diaphragm, against an adjustable or a fixed spring force, will act to shut off the valve as the evaporator pressure goes up, and it will allow the valve to open as the pressure goes down. This is reasonably satisfactory for practically constant-load operation, but for appreciable changes in load, it is actually operating in the wrong direction. A decrease in cooling load will give a decreased rate of vaporization and a drop in evaporator pressure; the constant-pressure expansion valve, in trying to raise the pressure, will allow an increased mass rate of refrigerant flow for the decreased load. Conversely, an increased load and evaporator pressure will produce a decreased mass rate of flow.

Variable load operation is handled best by a **thermostatic expansion valve**, which attempts to maintain a constant number of degrees of superheat in the suction vapor. A balance is required between an adjustable spring force plus valve discharge pressure, operating on one side of a diaphragm or bellows, and capillary tube pressure from a feeler bulb, located on the compressor suction line, operating on the other side. The liquid and vapor in the feeler bulb, which may or may not be the same fluid as the refrigerant, will exert a saturation pressure corresponding to the suction temperature. For a given evaporator pressure, any number of degrees of superheat in the discharge line can be achieved by an adjustment of the spring force. The exact number of degrees of superheat may not be achievable under very different running conditions, because the pressure-temperature relationship for the feeler-bulb fluid will not be linear. However, the fact that some superheat will be required at all times is sufficient in practice. This type of control ensures that the internal surfaces are wetted through most of the evaporation coil with only a small fraction of the coil being used to provide superheat. Where appreciable pressure drop occurs in large coils, the evaporator pressure may be taken by a separate equalizing line either from the middle of the coils or from the discharge line near the feeler-bulb location.

Where it is necessary to maintain better control over the number of degrees of superheat, particularly at low temperatures where the pressure-temperature response of most refrigerants is small, a **differential temperature** expansion valve can be used. This device (not illustrated) is not actuated directly by evaporation pressure. It utilizes two feeler bulbs: one senses evaporation temperature and the other senses suction temperature. Both contain the same fluid, which is chosen for good response in the temperature range anticipated. Their pressure forces are opposed and a spring force on one side only establishes the temperature differential, which is the number of degrees of superheat.

Control of **condenser cooling water** is desirable for two reasons: (1) to maintain a constant condenser pressure and thus condensation temperature, and (2) to adjust the cooling water consumption to the minimum amount required as the load

(a) Condenser cooling water control

(b) Manual shut-off valve

(c) Dehydrator

(d) Strainer

(e) Electric solenoid valve

(f) Thermostatic expansion valve

(g) Manual shut-off valve (optional)

(h) Cold chamber electric thermostat

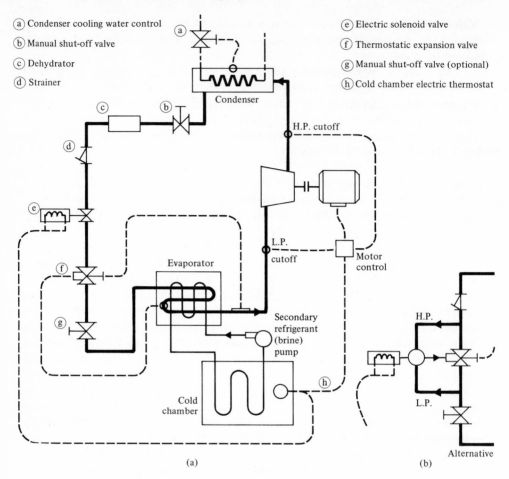

FIG. 4.14 A schematic diagram of controls for a vapor-compression system.

changes. Usually, a **modulating valve** is operated by condenser pressure acting to open it, and a spring force acting to close it. If precise control is not required, an **on-off solenoid valve**, electrically operated in conjunction with the compressor motor, may suffice.

The **electric motor** of a refrigeration system may have several controls. The cold-chamber temperature may actuate an on-off switch through a temperature sensitive element such as a bimetallic coil. To protect the system and the motor, cut-out switches may be actuated by motor overload, by low suction pressure, and by high condenser pressure. Where the characteristics of the refrigerant allow, the low pressure cut-out should operate at something above atmospheric pressure so that, in case of a leak in the system, no air will be drawn in and the leak can be located before all the refrigerant is lost. However, this is not important for small sealed units.

Figure 4.14 (a) shows a schematic layout of a vapor compression refrigeration system with major components and typical controls, including a thermostatic expansion valve. The alternative arrangement shown in Fig. 4.14 (b) avoids the necessity of having the whole refrigerant flow pass through the solenoid valve for a large system. The same three-way solenoid valve could control several thermostatic expansion valves serving parallel evaporators. For part-load operation, solenoid valves can be used to cut out individual coils in a bank of coils progressively as the load decreases. Important accessories, not shown in Fig. 4.14, include a safety valve or blow-out disc on the high-pressure side, a charging valve on the compressor-suction side, and an oil separator and return line often required after compressor discharge.

When the air is cooled by **direct expansion coils,** that is, with the air passing directly over the evaporator coils, usually provision must be made for **defrosting.** This can be avoided in comfort air conditioning if some control prevents the evaporator temperature from going below 32 F. When the outside air is the heat source for a heat pump, frosting of the coils must be expected at times with a vapor-compression cycle.

Defrosting can be accomplished in several ways. One common and effective means is by **reversing the cycle**, using special valves, so that the evaporator becomes the condenser and vice versa. Usually, the interruption of normal operation can be tolerated for sufficient time to clear the coils of all ice formation. Also, **electric heating** can be used with the metal coils themselves acting as conductors. The coils can be **sprayed** at intervals with either water or one of several liquid absorbents with a great affinity for water; these are usually solutions of lithium chloride, calcium chloride, or lithium bromide in water. An alternative to defrosting, but one that is often not practical, is to **dehumidify** all air before it passes over the evaporator coils. (Dehumidification with solid adsorbents and liquid absorbents is discussed under Air Conditioning in Section 6.4.) The beginning and end of the defrosting cycle can be timed automatically, past experience establishing the period by necessity or convenience, or the cycle can be controlled by the temperature differential between the air and the refrigerant, since that difference will increase with the addition of an ice-heat barrier.

4.3 REFRIGERANTS FOR VAPOR-COMPRESSION SYSTEMS

Sixty-nine different substances are listed in **Table 4.1** as refrigerants. These are the substances that have been assigned standard designations or refrigerant numbers by the American Society of Heating, Refrigerating and Air-Conditioning Engineers. (The designations were originally assigned by the American Society of Refrigerating Engineers, which combined with the American Society of Heating and Air-Conditioning Engineers to form the ASHRAE in 1959). While all these have been used as refrigerants or can be advantageously employed under certain conditions, only a dozen or so are very important commercially at the present time.

Some explanation of the system of numerical designation is required, particularly for the **halocarbons,** or **halogenated hydrocarbons**. These have been syn-

Table 4.1

ASHRAE STANDARD DESIGNATION OF REFRIGERANTS*

Refrigerant number	Chemical name	Chemical formula
Halocarbon compounds		
10	Carbontetrachloride	CCl_4
11	Trichloromonofluoromethane	CCl_3F
12	Dichlorodifluoromethane	CCl_2F_2
13	Monochlorotrifluoromethane	$CClF_3$
13Bl	Monobromotrifluoromethane	$CBrF_3$
14	Carbontetrafluoride	CF_4
20	Chloroform	$CHCl_3$
21	Dichloromonofluoromethane	$CHCl_2F$
22	Monochlorodifluoromethane	$CHClF_2$
23	Trifluoromethane	CHF_3
30	Methylene chloride	CH_2Cl_2
31	Monochloromonofluoromethane	CH_2ClF
32	Methylene fluoride	CH_2F_2
40	Methyl chloride	CH_3Cl
41	Methyl fluoride	CH_3F
50*	Methane	CH_4
110	Hexachloroethane	CCl_3CCl_3
111	Pentachloromonofluoroethane	CCl_3CCl_2F
112	Tetrachlorodifluoroethane	CCl_2FCCl_2F
112a	Tetrachlorodifluoroethane	CCl_3CClF_2
113	Trichlorotrifluoroethane	CCl_2FCClF_2
113a	Trichlorotrifluoroethane	CCl_3CF_3
114	Dichlorotetrafluoroethane	$CClF_2CClF_2$
114a	Dichlorotetrafluoroethane	CCl_2FCF_3
114B2	Dibromotetrafluoroethane	$CBrF_2CBrF_2$
115	Monochloropentafluoroethane	$CClF_2CF_3$
116	Hexafluoroethane	CF_3CF_3
120	Pentachloroethane	$CHCl_2CCl_3$
123	Dichlorotrifluoroethane	$CHCl_2CF_3$
124	Monochlorotetrafluoroethane	$CHClFCF_3$
124a	Monochlorotetrafluoroethane	CHF_2CClF_2
125	Pentafluoroethane	CHF_2CF_3
133a	Monochlorotrifluoroethane	CH_2ClCF_3
140a	Trichloroethane	CH_3CCl_3
142b	Monochlorodifluoroethane	CH_3CClF_2
143a	Trifluoroethane	CH_3CF_3
150a	Dichloroethane	CH_3CHCl_2
152a	Difluoroethane	CH_3CHF_2
160	Ethyl chloride	CH_3CH_2Cl
170*	Ethane	CH_3CH_3

* Reprinted by permission from ASHRAE Guide and Data Book, 1965–66 and ASHRAE Standard 34–37, ASA B79.1–1960.

Table 4.1 (*continued*)

ASHRAE STANDARD DESIGNATION OF REFRIGERANTS

Refrigerant number	Chemical name	Chemical formula
218	Octafluoropropane	$CF_3CF_2CF_3$
290*	Propane	$CH_3CH_2CH_3$
Cyclic organic compounds		
C316	Dichlorohexafluorocyclobutane	$C_4Cl_2F_6$
C317	Monochloroheptafluorocyclobutane	C_4ClF_7
C318	Octafluorocyclobutane	C_4F_8
Azeotropes		
500	Refrigerants 12/152a 73.8/26.2wt %‡	CCl_2F_2/CH_3CHF_2
501	Refrigerants 22/12 75/25wt %	$CHClF_2/CCl_2F_2$
502	Refrigerants 22/115 48.8/51.2wt %	$CHClF_2/CClF_2CF_3$
Miscellaneous organic compounds		
Hydrocarbons		
50	Methane	CH_4
170	Ethane	CH_3CH_3
290	Propane	$CH_3CH_2CH_3$
600	Butane	$CH_3CH_2CH_2CH_3$
601	Isobutane	$CH(CH_3)_3$
1150†	Ethylene	$CH_2{=}CH_2$
1270†	Propylene	$CH_3CH{=}CH_2$
Oxygen compounds		
610	Ethyl ether	$C_2H_5OC_2H_5$
611	Methyl formate	$HCOOCH_3$
Nitrogen compounds		
630	Methyl amine	CH_3NH_2
631	Ethyl amine	$C_2H_5NH_2$
Inorganic compounds (Cryogenic)		
702	Hydrogen (normal and *para*)	H_2
704	Helium	He
720	Neon	Ne
728	Nitrogen	N
729	Air	$0.21O_2, 0.78N_2, 0.01A$
732	Oxygen	O_2
740	Argon	A

* Methane, ethane, and propane appear in the halocarbon section in their proper numerical order, but these compounds are not halocarbons.

† Ethylene and propylene appear in the hydrocarbon section to indicate that these compounds are hydrocarbons, but are properly identified in the section unsaturated organic compounds.

‡ Carrier Corporation Document 2-D-127, p. 1.

(*continued*)

Table 4.1 (*continued*)

ASHRAE STANDARD DESIGNATION OF REFRIGERANTS

Refrigerant number	Chemical name	Chemical formula
Inorganic compounds (noncryogenic)		
717	Ammonia	NH_3
718	Water	H_2O
744	Carbon dioxide	CO_2
744A	Nitrous oxide	N_2O
764	Sulflur dioxide	SO_2
Unsaturated organic compounds		
1112a	Dichlorodifluoroethylene	$CCl_2{=}CF_2$
1113	Monochlorotrifluoroethylene	$CClF{=}CF_2$
1114	Tetrafluoroethylene	$CF_2{=}CF_2$
1120	Trichloroethylene	$CHCl{=}CCl_2$
1130	Dichloroethylene	$CHCl{=}CHCl$
1132a	Vinylidene fluoride	$CH_2{=}CF_2$
1140	Vinyl chloride	$CH_2{=}CHCl$
1141	Vinyl fluoride	$CH_2{=}CHF$
1150	Ethylene	$CH_2{=}CH_2$
1270	Propylene	$CH_3CH{=}CH_2$

thetically produced, particularly for their refrigerating properties, by the substitution of a halogen for one or more of the hydrogen atoms in methane, ethane, or propane (CH_4, C_2H_6, C_3H_8). The halogens are fluorine, chlorine, bromine, iodine, and astatine, but only the first three are used. The last digit in the designation (the one on the right) indicates the number of fluorine atoms in the molecule of the compound formed. The second digit from the right is one more than the number of hydrogen atoms remaining. The third digit from the right is one less than the number of carbon atoms in the molecule, so that a blank indicates a basic methane structure, a 1 indicates an ethane structure, and a 2 indicates a propane structure.

Table 4.2 gives certain properties of selected refrigerants arranged in order of their increasing boiling-point temperatures at atmospheric pressure. The boiling points give some indication of the evaporation temperatures at which the various refrigerants are useful, and under certain conditions determine the type of compressor that might be used. Atmospheric-pressure boiling temperature does not by itself indicate practical temperature ranges for a cycle. **Figure 4.15** (a) shows four refrigerants that could be selected to cover four parts of the temperature range from −150 F to +140 F, if each refrigerant in its own range is to have an evaporation pressure greater than atmospheric and a condensation pressure less than 100 psig. Figure 4.15 (b) shows that two refrigerants could cover the temperature range from −110 F to +120 F, with compressor suction pressures above atmospheric

Table 4.2
PHYSICAL PROPERTIES OF SELECTED REFRIGERANTS ARRANGED IN ORDER OF INCREASING BOILING POINT [§]

Refrigerant number	Chemical formula	Molecular weight	B. P. at 1 Atm., F	Critical Temp., F	Critical pressure, psia	Latent heat at B. P., Btu/Mol	Current use status	Safety group*
729	(Air)	28.97	−317.8	−221†	547†	—	Gas cycle	1
1150	C_2H_4	28.03	−155.0	48.8	731.8	5,791	Limited	3
13	$CClF_3$	104.47	−114.6	83.9	561	6,670	Yes	1
744	CO_2	44.01	−109.3‡	87.8	1071.1	10,837‡	No	1
13B1	$CBrF_3$	148.9	−72.0	152.6	575	7,606	Yes	1
290	C_3H_8	44.06	−44.2	202	661.5	8,006	Limited	3
22	$CHClF_2$	86.48	−41.36	204.8	721.9	8,787	Yes	1
717	NH_3	17.03	−28.0	271.4	1657	10,036	Yes	2
500	(Azeotrope)	99.29	−28.0	221.1	631	8,786	Yes	1
12	CCl_2F_2	120.93	−21.62	233.6	596.9	8,591	Yes	1
764	SO_2	64.06	14.0	314.8	1141.5	10,697	No	2
114	$CClF_2CClF_2$	170.93	38.39	294.3	474	10,085	Yes	1
21	$CHCl_2F$	102.93	48.06	353.3	750	10,720	No	1
11	CCl_3F	137.38	74.48	388.4	635	10,765	Yes	1
113	$CCl_2FC ClF_2$	187.39	117.63	417.4	495	11,828	Yes	1

* Safety Group Designation (ASA Standard B9.1)
1. Negligible toxicity, nonflammable
2. Toxic or flammable or both
3. Highly flammable and explosive

† See also Table 5.1
‡ Sublimation from solid to vapor (see Section 4.4).
§ Reprinted by permission from ASHRAE Guide and Data Book, 1965–66.

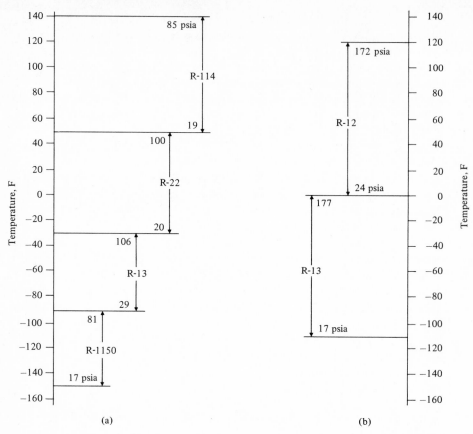

FIG. 4.15 Saturated temperature and pressure ranges for selected refrigerants.

and delivery pressures no higher than 165 psig. Suction pressures can be below atmospheric to reduce evaporation temperature, but leakage into the system would then become a possibility. The moderate delivery pressures given are well below practical limits if heavier equipment can be used. In general, however, the deviation from an ideal Carnot cycle is greater, i.e., the cycle efficiency is lower, as the pressure range for the compressor is increased.

Centrifugal compressors operate most satisfactorily when the suction volumes are large. This would occur either for large specific volumes of the suction vapor or for large tonnage per compressor. Because of speed limitation, reciprocating compressors cannot handle very large volume rates; multiple cylinders and multiple compressors are used, but this becomes uneconomical beyond a point. For the halocarbons listed in Table 4.2, the specific volumes of saturated vapor at 1 atm pressure vary from 2.2 to 3.5 cu ft/lb; their latent heats per mole (which is essentially per unit of volume) all have the same order of magnitude. Thus if these various refrigerants are used at refrigerating temperatures close to their atmospheric boiling point, the specific volumes do not dictate the type of compressor to be used. Only when temperature ranges are more or less fixed, as they would be, say, for air-conditioning applications, can it be said that the boiling-point temperature dictates the type of compressor; when that is so, the refrigerants with the higher atmospheric boiling temperatures, such as R.113, R.11, and R.114, are likely to be used with centrifugal compressors, and R.22, R.500, and R.12 are likely to be used with reciprocating compressors for small tonnage applications. Any of the halocarbon refrigerants might require a centrifugal compressor for large capacity operation. Rotary compressors, though essentially constant displacement machines, do not have the same speed limitations as do reciprocating compressors. They are used in some small systems because of their simplicity, and in other situations to span the gap between situations that demand reciprocating machines and those that demand centrifugal machines.

The composition of the molecule determines the atmospheric boiling point. For the basic methane structure (for instance, R.11, R.12, R.13, R.21, and R.22) it is seen that the boiling point decreases as the number of fluorine atoms increases, and it increases as chlorine atoms are added. The same is seen to be true for the basic ethanes (only R.113 and R.114 are listed), but these temperatures are higher since the heavier ethane has a higher boiling point than methane.

Propane, despite its flammability, is used as a refrigerant, but usually in applications such as petroleum refining, where similarly hazardous fluids are being handled. Because of its low boiling point, its high specific volume (6.7 cu ft/lb for saturated vapor at one atmosphere), and the danger of positive compression, it is usually compressed in a centrifugal machine.

Carbon dioxide, once a popular refrigerant because of its safety, has largely been replaced by other fluids except in self-refrigeration for the production of solid carbon dioxide or "dry ice." (See Section 4.4). Its high ratio of critical pressure to critical temperature indicates the necessity for reciprocating compressors. High-pressure ratios are also required for ammonia, which is still being used in some systems despite its toxicity. Also sulphur dioxide, popular in domestic refrigerators some years ago, has little in its favor now. These three compounds and air are included in Table 4.2, largely for the sake of comparison.

The halogenated hydrocarbons are, with certain exceptions, safe refrigerants; they are nontoxic and nonflammable. Although their latent heats per pound are low compared to ammonia, their latent heats per mole are comparable, so that their volume rates of circulation are not excessive. The superheat horn on the *T-s* diagram, which indicates deviation from the Carnot cycle, is usually small in the proper temperature range; in fact, for R.113 and R.114, an isentropic compression line, starting with dry saturated vapor, moves into the wet vapor region and the superheat horn fails to exist.

Certain desirable properties can be achieved by the **azeotropes**, which are homogeneous mixtures of two halocarbons. An example is R.500, which is a mixture of R.12 and R.152a in proportions 73.8 % to 26.2 %. It has the same atmospheric boiling point as ammonia, and is just sufficiently different from R.12 to produce the same capacity in a direct driven compressor with a 50 cps motor that would be achieved by R.12 with a 60 cps motor. Conversely, substitution of R.500 for R.12 in a constant-displacement compressor would increase the capacity and power by about 20 % (R.500 is produced under a patent of the Carrier Corporation).

Many refrigerants designated for use in vapor-compression cycles can also be used as **secondary refrigerants** to carry the cooling effect without changing phase. Methylene chloride (dry) and R.11, for instance, are useful at low temperatures when conventional **brines** such as glycol or salt solutions become too viscous.

Many nonthermodynamic properties besides flammability and toxicity are important in the selection of refrigerants. Discussion of such things as affinity for water and oil, viscosity, conductivity, and corrosiveness will be found in the ASHRAE Guide and Data Books, and the literature of the various refrigerant manufacturers should be consulted.

4.4 THE PRODUCTION OF SOLID CARBON DIOXIDE

The manufacture of solid carbon dioxide, or "dry ice," provides an excellent example of multistage compression, flash intercooling (already discussed), and the phenomenon of **sublimation**, which is evaporation occurring directly from the solid to the vapor state. While much less important commercially than it was prior to the extremely wide-spread use of small vapor-compression systems in refrigeration applications such as food transportation, "dry ice" still has important applications. It is useful as a safe expendable refrigerant. The dry gas sublimated from the solid provides very effective insulation, and is a preservative.

Solid carbon dioxide is usually produced in a system which uses the CO_2 itself as a refrigerant. The pressure-temperature relationship of this substance explains the necessity for staging; the location of the **triple point** at which solid, liquid, and vapor exist in equilibrium, accounts for the possibility of sublimation at, or somewhat above, normal atmospheric pressure. **Figure 4.16** (a) shows the division of a pressure-temperature diagram into its various phase regions; without scales, it is a typical shape for many substances. Figure 4.16 (b) is a *p-h* diagram for carbon dioxide, showing its significant values. It is apparent that if saturated liquid CO_2 is throttled from, say, 80 psia to atmospheric pressure, the resulting state is a mixture of solid and vapor, where over 50 % of the mixture is solid. Saturated liquid at 80 psia is well below zero F (about -68 F). At a reasonable sink temperature (say cooling water over 65 F), the saturation pressure of CO_2 is over 800 psia. Several stages of compression are required to span this large pressure range. Although an external refrigeration system with a different refrigerant certainly could be used to condense carbon dioxide at about -68 F, the temperature range is great and staging might still be required; furthermore, very effective heat transfer is accomplished in carbon dioxide flash coolers.

A schematic diagram of a practical system is shown in **Fig. 4.17**, and typical pressures are noted. Water-cooled heat exchangers might be used immediately after the first and second compressors to bring the vapor near cooling water

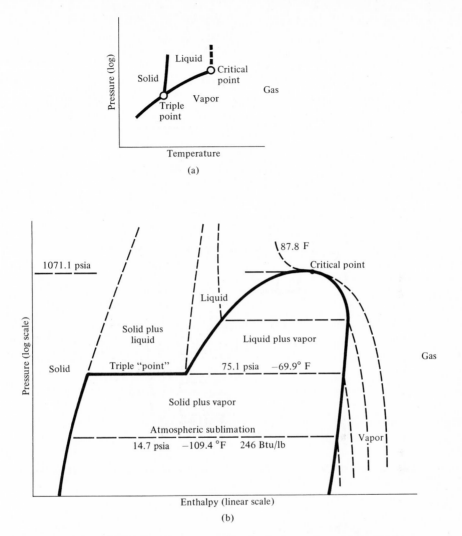

FIG. 4.16 Solid, liquid, vapor, and mixed phases for carbon dioxide.

temperature before the flash coolers, and thus reduce circulation rates in the higher stages.

The solid CO_2 is removed from the snow chamber and compressed into blocks for handling. The make-up gas to replace the solid produced is manufactured in various ways, such as in the partial combustion of charcoal in a limited atmosphere, in a fermentation process, or as a by-product of some process such as the hydrogenation of petroleum. The snow chamber may operate above atmospheric pressure to reduce the pressure range of the cycle, in which case the make-up gas would be supplied by external compressors.

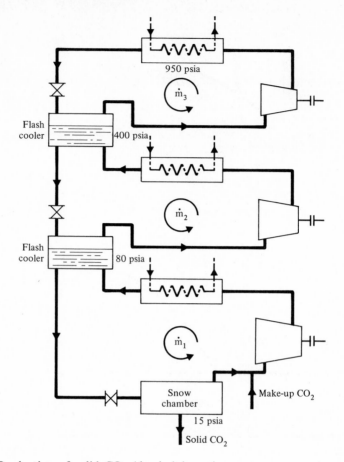

FIG. 4.17 Production of solid CO_2 (dry ice) in a three-stage compression system with flash coolers.

4.5 STEAM-JET REFRIGERATION

Water vapor can be used as a refrigerating fluid, but reference to the steam tables will show that to achieve the modest refrigeration temperature of 40 F, an evaporation pressure of $\frac{1}{4}$ in. Hg abs is required, and the specific volume of saturated vapor at that pressure is over 2400 cu ft/lb. Water, therefore, is not commonly used in ordinary mechanical vapor-compression systems, but it can be used when a high vacuum can be produced by other means. One method of producing a high vacuum is chemical absorption, which will be described in Section 4.6. Another method is steam-jet ejection, which is described here. Both jet-ejection and absorption systems have the advantage of using mostly "low-grade" energy and only relatively small amounts of shaft work. Still, water as a refrigerant is limited to such applications as air conditioning, where temperatures as high as 40 F are quite satisfactory.

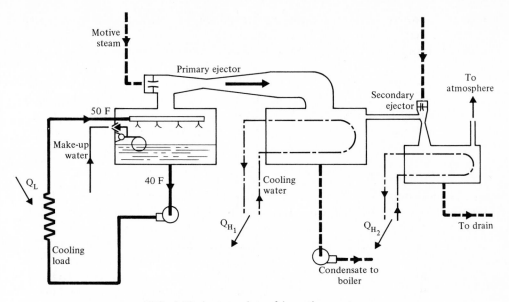

FIG. 4.18 A steam jet refrigeration system.

Figure 4.18 presents a schematic arrangement for water chilling, where the evaporation of a relatively small amount of water in a **flash cooler** reduces the temperature of the main body of water. The chilled liquid is then pumped as the refrigeration carrier to the cooling-load heat exchanger. The required high vacuum is produced by a **steam-jet ejector**; the steam is supplied to the ejector nozzle at any pressure greater than 2 atm abs. The resulting high velocity, as the steam is expanded through the nozzle, must be accompanied by a low static pressure. The evaporated water from the flash cooler is entrained with the motive steam, and both are discharged to a condenser.

As in the condenser of a steam power plant, the vacuum can be maintained only if gases, primarily air, originally dissolved in the steam and make-up water and then released at the very low pressure, are continuously removed. That is accomplished by a **secondary ejector** which removes both the gases and some water vapor to a secondary condenser. Although Fig. 4.18 shows only the two stages, three ejection stages may be required before the gases can be released to the atmosphere because of the large pressure ratio involved.

Such a system does not have a high COP, and the equipment is large and must be tightly sealed. Its main advantage is that it can utilize the energy in the steam directly without the necessity of expensive "high-grade" energy (shaft power) except for the circulation and condensate pumps. Consequently, it is also relatively free from vibration. An additional inherent advantage is that no separate secondary fluid need be used to distribute the refrigeration effect, because the remaining liquid water chilled directly by evaporation can be used. Steam-jet refrigeration might be economical where excess steam capacity is available, as it would be for the summer months in many steam plants. Its advantages and economy must be compared with those of absorption refrigeration to be discussed in the next section.

There is no thermodynamic reason why fluids other than water could not be used for jet-ejection evaporation, so that different temperatures could be achieved and extreme vacuums could be avoided. However, unless the jet fluid is the same as the refrigerating fluid, separation would be necessary for recovery of the refrigerant, and it is not usual that surplus high-pressure fluid other than steam is available.

4.6 ABSORPTION REFRIGERATION

The greatest disadvantage of a vapor-compression refrigeration system is that it requires the expenditure of expensive, "high-grade" energy in the form of shaft work to drive the compressor; a concommitant disadvantage is the vibration of the rotating machine, whether it is the reciprocating, rotary, or centrifugal type. The elimination of the necessity for shaft work (except perhaps for liquid pumps) has been the prime reason for the economic success of the various **absorption** and **adsorption** vapor refrigeration systems. These systems function on the principle that many substances can attract and hold large quantities of the vapors of other substances at a relatively low temperature; the vapors are driven off when energy is added and the temperature is raised. When the vapor is retained on the surface or in the capillary pores of a solid, the process is called **adsorption**; and the heat that is transferred to keep the process isothermal is the latent heat of condensation of the vapor that is adsorbed, plus the relatively small heat of wetting. When there is a chemical reaction between the two substances, or when the vapor goes into solution, the process is called **absorption**, and, in addition to the latent heat of the vapor, the heat of chemical reaction or heat of solution (either of which may be positive or negative) must be removed. Practical combinations of refrigerant and absorbent require that large quantities of one be easily absorbed into the other. This implies an **attraction** between the molecules of the two substances, and thus a **positive heat release** (exothermic reaction) accompanying the formation of the solution. Consequently, in actual absorption refrigeration systems, the latent heats of vaporization from solution are considerably higher than those for vaporization alone at the same temperatures

To illustrate the feasibility and principle of operation of an absorption refrigeration system, a schematic diagram of a simple system is shown in **Fig. 4.19**. This could be an **aqua/ammonia** system in which the refrigerating fluid (R) is ammonia and the absorbent or carrier (C) is water; or it might be a **lithium bromide/water** or **lithium chloride/water** system, where water is the refrigerating fluid (as it was seen to be in steam-jet refrigeration). In the latter case, the liquid carrier (C) is a strong solution of lithium bromide or lithium chloride in water, since lithium salt alone is a solid at the temperatures used. (These **hygroscopic brines** are used to dehumidify air as described in Chapter 6, as well as to absorb water in a refrigeration system).

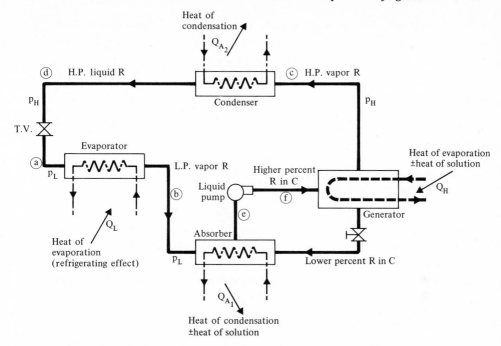

FIG. 4.19 A simple absorption refrigeration system.

In the operation of the absorption system, the condenser, throttling valve, and evaporator perform the same functions as in a vapor-compression system: high-pressure vapor at c is condensed to liquid at d from which condition it is throttled to wet vapor at a; the evaporation of the refrigerant at pressure p_L and temperature T_L from a to b produces the refrigeration effect. At this point, the two types of refrigeration systems lose their similarity. The vapor-compression machine is removed; instead, the cool vapor (R) at low pressure is absorbed into the carrier (C). In the **absorber**, the latent heat of condensation plus or minus the heat of solution must be removed by cooling water or some other medium. The strong solution of R in C is removed by the **liquid pump**. (In an aqua/ammonia system, the stronger solution is less dense and will rise to the top.) It is in the liquid pump that the pressure is raised. Because the change of specific volume for the liquid is so much smaller than it would be for a vapor over the same pressure range, the pump work is practically negligible compared to the work of vapor compression.

In the **generator**, the refrigeration vapor is separated from the carrier by the addition of energy which must be the heat of vaporization from liquid at f to vapor at c, again plus or minus the heat of solution. From c, the cycle is repeated. For the process to be continuous, the weak solution left in the generator by the removal of vapor must be returned to the absorber through some control that will maintain the pressure difference between the "high side" and the "low side" of the system.

4.6.1 Theoretical COP of an Absorption System

It is simple to derive an expression for the maximum possible coefficient of performance of an absorption system. **Figure 4.20** shows the energy flows to and from a simple absorption system such as that illustrated in Fig. 4.19. It is assumed that the entire refrigeration effect (Q_L) is accomplished by the isothermal transfer of heat at temperature T_L, that the energy supply in the generator (Q_H) is received isothermally at T_H, and that all energy rejection in both the absorber and the condenser is accomplished as a result of heat transfer at ambient temperature T_A. The work of the pump will be assumed to be negligible.

By the first law of thermodynamics, the net energy input to the system must be zero, so that

$$Q_L + Q_H = Q_{A1} + Q_{A2} = Q_A. \qquad (4.9a)$$

The second law dictates that the net entropy change in the system is either equal to zero in the ideal case, or greater than zero if any irreversibilities are present. The **system** must be taken to include the sources and sinks of the heat flows as well as the apparatus of the refrigeration cycle; the latter, operating on a continuous cycle, repeatedly returns to its own starting point and must have a zero entropy change. Consequently,

$$\Delta S_L + \Delta S_H + \Delta S_A \geq 0.$$

It has been assumed that each of the heat transfers is isothermal. Thus

$$-\frac{Q_L}{T_L} - \frac{Q_H}{T_H} + \frac{Q_A}{T_A} \geq 0, \qquad (4.9b)$$

$$\frac{Q_L}{T_L} + \frac{Q_H}{T_H} \leq \frac{Q_L + Q_H}{T_A}, \qquad (4.9c)$$

$$Q_L\left(\frac{T_A - T_L}{T_L}\right) \leq Q_H\left(\frac{T_H - T_A}{T_H}\right). \qquad (4.9d)$$

Therefore the maximum possible coefficient of performance of an absorption refrigeration process, which is the ratio of the refrigeration effect to the heat supplied, is given by

$$\frac{Q_L}{Q_H} \leq \left(\frac{T_L}{T_A - T_L}\right)\left(\frac{T_H - T_A}{T_H}\right). \qquad (4.9e)$$

This result might have been anticipated. It is seen that the maximum possible COP is the product of the ideal COP of a refrigerator working between T_L and T_A and the ideal thermal efficiency of an engine working between T_A and T_H. The absorption refrigeration system includes not only the evaporator and condenser, which provide the refrigeration or heat pumping effect, but also the absorber and generator, which make the operation possible. The expressions for the coefficient of performance of a vapor compression system used in Sections 4.1 and 4.2 were developed without consideration of the efficiency of the prime mover.

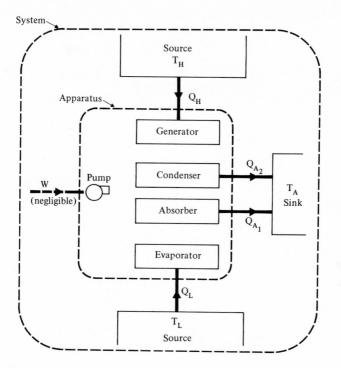

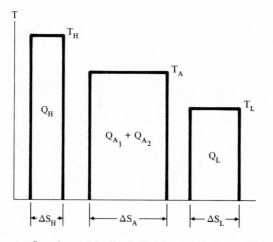

FIG. 4.20 Energy flow in an idealized simple absorption refrigeration system.

4.6.2 Lithium Bromide/Water System

Two types of absorption-refrigeration plants have been used extensively: the ammonia/water and the lithium bromide/water systems. The latter is simpler because the refrigerating fluid can be separated relatively easily in pure form from the carrier fluid, and it will be described first. Commercially, it is the more important of the two at the present time.

The simple absorption system of Fig. 4.19 was made up of four separate components in addition to the liquid pump. **Figure 4.21** shows a schematic diagram of a **practical LiBr/water system** in which the evaporator and absorber, both at the lower pressure, are combined in one shell, and the condenser and generator, at the higher pressure, are combined in a second shell. A liquid-liquid heat exchanger is added to improve the COP. Also shown is a pressure-concentration diagram to help explain the operating cycle. A detailed chart of the pressure-concentration diagram can be found in the appendix.

From the bottom of the lower vessel, a relatively weak solution (high percentage of water) is pumped by pump P_1 through the heat exchanger to enter the condenser-generator at condition ②. Heat is transferred to the fluid from steam or hot-water coils in the **generator** portion. The solution is heated to condition ③ at which temperature and pressure water vapor will be separated from it at the container pressure of say 2.8 in. Hg abs. The concentration of LiBr in the remaining solution is thus increased from condition ③ to condition ④, and the stronger solution (less water) is returned to the lower vessel through the heat exchanger. A three-way valve between lines 1–2 and 4–5 can be used to control the actual rate of liquid circulation between the upper and lower chambers in response to chilled water temperature variation. The water vapor that was driven from the solution in the upper shell is **condensed** at shell pressure when it comes in contact with the cooling water coils above. It is prevented from returning to solution by the condensate pan, which directs the liquid through a separate line to the upper part of the lower shell. A U-tube seal or restriction in this line can maintain the pressure difference between the two shells.

The refrigeration effect is achieved when the returning water flashes to wet vapor at the lower pressure and proceeds to take up energy equivalent to its heat of vaporization. This it receives in the **evaporator** portion of the shell by heat transfer from the **chilled water** line that circulates to the actual refrigeration load. The purpose of pump P_2 is simply to keep the liquid water spraying over the chilled water coils to improve the heat transfer. The low pressure of about $\frac{1}{4}$ in. Hg abs in the bottom shell is possible because, at the lower temperature, more water is rapidly taken into the solution of LiBr and water. The concentration of that solution is changed from condition ⑤ to condition ①. Heats of condensation and solution are removed by the circulation of cooling water through the part of this shell that acts as the **absorber.** The spray in the absorber section brings the solution into intimate contact with water vapor to aid absorption, and facilitates heat transfer with the cooling water coils.

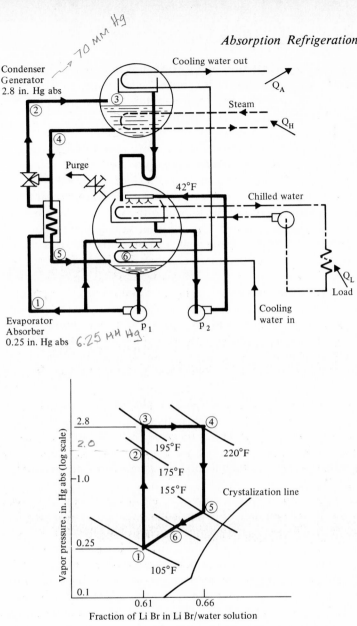

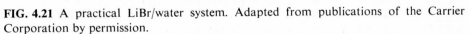

FIG. 4.21 A practical LiBr/water system. Adapted from publications of the Carrier Corporation by permission.

The extremely low pressure in the **evaporator-absorber** could not be maintained indefinitely without the operation of a purge system to remove noncondensible gases. This can be accomplished by jet ejectors, as in the steam-jet refrigeration system already described, by a more complicated bubble column, which will not be described, or by a vacuum pump operated occasionally.

The most serious limitation on this system is imposed by the fact that water is the refrigerant. The temperatures are above 32 F (actually above 40 F ,with a minimum pressure of $\frac{1}{4}$ in. Hg abs) which makes this unsuitable for many refrigeration applications. However, it is very satisfactory for air conditioning and for process water cooling. Water, rather than glycol or brine, can be circulated as the secondary fluid when all the temperatures are above 32 F, and frosting of heat-transfer surfaces is not possible. Some systems actually circulate the primary water from pump P_2 to the cooling load, but the use of a secondary fluid is usually preferable to reduce contamination and leakage problems. The existence of two temperature levels makes it practical to put the same cooling water through both the absorber and the condenser in series, with a consequent saving in total cooling water flow.

Lithium bromide/water systems are manufactured in sizes from 50 to 1000 tons. Where low pressure steam is the energy source, steam rates somewhat less than 20 lb/hr/ton are achievable. The resulting COP is considerably less than unity, but cheap energy can make such a figure economically competitive with much higher COP values for vapor-compression machines. In the absorption system, the pump horsepower per ton is very small. Minimum vibration may, in some applications, be an even greater consideration than economy.

4.6.3 Temperature-Concentration and Enthalpy-Concentration Diagrams for Ammonia and Water

The factor that makes an ammonia/water absorption system more complicated than a lithium bromide/water system is that both ammonia and water will exert vapor pressures at the temperatures involved. Being in solution, each will have a vapor pressure less than it would have alone at the same temperature. Still, as energy is added at some particular total pressure, both ammonia and water will be vaporized.

In the discussion to follow, ammonia and water are used as an example of a **homogeneous binary mixture**. This particular example is used because it is found in an important class of refrigeration systems. Also, it is easier to comprehend the phenomena involved when they are applied to particular fluids whose properties are somewhat familiar. The principles apply to any similar solution.

Separately, ammonia and water have different boiling temperatures for any given pressure. A solution of the two has a boiling temperature somewhere between the two individual values. (This is not true for solutions exhibiting an **azeotropic** point, for which the boiling temperature at a given pressure may be either above or below the higher and the lower of the two values of the pure substances. An azeotropic mixture vaporizes as though it were a pure substance, with no change in concentration value for either the liquid or the vapor during the vaporization process. See Bosnjakovic [B.3], p. 108.) As would be expected, the **concentration** of the solution determines the boiling temperature at a particular total pressure. The vapor that is driven from the liquid solution at saturation conditions when energy is added does not, initially, have the same proportions of the two components as the original liquid solution had. In the case of an ammonia and

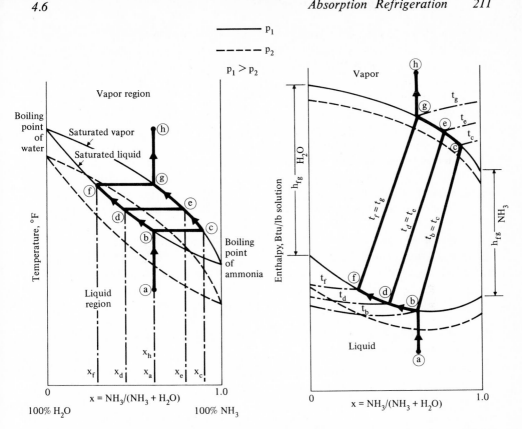

Fig. 4.22 Vaporization of a solution of ammonia and water at pressure p_1.

water solution, ammonia, having the lower boiling point for the existing pressure, will vaporize preferentially. The resulting vapor will have a higher proportion of ammonia than did the original liquid solution. The result of this will be to reduce the concentration of ammonia in water for the remaining liquid. From this, two things follow: first, the boiling temperature of the remaining liquid will rise (tending to approach the boiling temperature of water alone at the prevailing total pressure); second, subsequent vapor (being produced from a weaker solution of ammonia in water) will have a lower ammonia concentration than that of the first vapor to be driven off. When an amount of energy equivalent to the full enthalpy of vaporization has been added, the proportions of the two substances in the vapor must be exactly the same as the original proportions in the liquid, if neither substance has actually been removed during the evaporation process.

The way in which a binary mixture is evaporated by the addition of energy at constant total pressure, as described above, is illustrated in steps in the two diagrams of **Fig. 4.22**. The change from subcooled liquid at *a* to superheated vapor at *h* can be followed on both the temperature-concentration (*t-x*) and the enthalpy-concentration (h-x) diagrams. On both the t-x and the h-x diagrams, the locations of the saturated liquid and the saturated vapor lines depend on the total vapor pressure. The process is described for one pressure (p_1). Saturation lines for a different pressure (p_2) are shown also. (For a discussion of the extension of the temperature-concentration diagram of a two-component solution up to and beyond the critical point, see Fig. 5.2 and the accompanying explanation of the "contact point" and the "plait point" for nitrogen and oxygen.)

As energy is added to a liquid solution of ammonia and water at concentration x_a, the temperature will rise until the saturated liquid temperature of that concentration is reached at point **b**. Further heat transfer to the solution will drive off some vapor, and the first vapor to form at t_b will have a composition x_c. For more vapor to be formed, the temperature of the remaining liquid solution must be raised.

Points **d** and **e** represent the conditions of liquid and vapor, respectively, in equilibrium at some higher temperature and the same total pressure before evaporation of the liquid has been completed. Note that liquid at concentration x_d is in

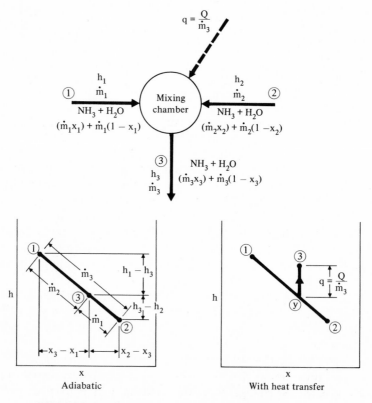

FIG. 4.23 Mixing of two streams of aqua/ammonia.

equilibrium with vapor at concentration x_e. As energy is being added, the concentrations of both the liquid solution and the vapor mixture are being reduced. When an amount of energy equivalent to the whole enthalpy of evaporation for the given solution has been added, the final concentration of the saturated vapors is x_g, which is the same as x_a. The last bit of liquid to vaporize would have had a concentration x_f. As more energy is added to the vapors, the temperature will rise, but the concentration value will remain fixed.

If heat were to be transferred from superheated vapor at h, the process described would be reversed. Condensation would begin at g with the liquid formed initially having a concentration value x_f. Finally, liquid at b would be subcooled at constant concentration to a. For the heating or cooling and change of phase processes, the t-x diagram gives no indication of the amount of energy to be added or removed. On the h-x diagram, the vertical distances do represent energy in or out at constant pressure. The distance from b on the saturated liquid line to g on the saturated vapor line represents the "latent heat" or enthalpy of vaporization per pound of solution.

At the beginning of this subsection, it was noted that an investigation of the properties of ammonia and water solutions will serve as an example for other homogeneous binary mixtures. The uses of the enthalpy–concentration chart, illustrated in what follows, apply as well to other mixtures besides ammonia and water. Consider the mixing of two streams of different concentration of ammonia and water as shown in **Fig. 4.23**. A mass and energy balance will give the following equations:

$$\begin{array}{ccc} \textbf{In} & & \textbf{Out} \\ \dot{m}_1 + \dot{m}_2 & = & \dot{m}_3 \quad \text{lb/min,} \\ (\text{NH}_3)_1 + (\text{NH}_3)_2 & = & (\text{NH}_3)_3 \quad \text{lb/min,} \\ \dot{m}_1 x_1 + \dot{m}_2 x_2 & = & \dot{m}_3 x_3 \quad \text{lb/min,} \\ \dot{m}_1 h_1 + \dot{m}_2 h_2 + Q & = & \dot{m}_3 h_3 \quad \text{Btu/min.} \end{array}$$

For adiabatic mixing, the heat-transfer term (Q) drops out:

$$\dot{m}_1 h_1 + \dot{m}_2 h_2 = \dot{m}_3 h_3 = (\dot{m}_1 + \dot{m}_2) h_3 \quad \text{Btu/min.}$$

Thus

$$\frac{\dot{m}_1}{\dot{m}_2} = \frac{h_3 - h_2}{h_1 - h_3}. \tag{4.10a}$$

Similarly,

$$\frac{\dot{m}_1}{\dot{m}_2} = \frac{x_3 - x_2}{x_1 - x_3} = \frac{x_2 - x_3}{x_3 - x_1}. \tag{4.10b}$$

These simple equations indicate two facts: first, point ③ on the h-x diagram is on a straight line between points ① and ② because the differences in x-values are proportional to the differences in h values; second, the distance of point ③ along the line from one of the points compared to the distance from the other point is

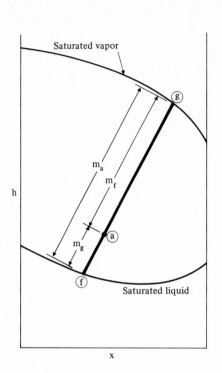

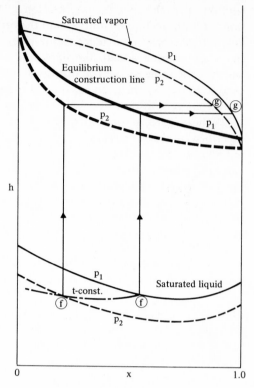

FIG. 4.24 A binary fluid mixture of saturated liquid and vapor in equilibrium.

FIG. 4.25 The use of an *h-x* chart for ammonia and water with equilibrium construction lines.

inversely proportional to the ratio of the respective mass rates of flow. The latter point is demonstrated on the first *h-x* diagram of Fig. 4.23; if a scale is selected so that the distance $\overline{1\text{-}2}$ represents the mass rate of flow $\dot{m}_3$, then $\dot{m}_1$ and $\dot{m}_2$ are represented by the distance $\overline{2\text{-}3}$ and $\overline{1\text{-}3}$ respectively.

For mixing with heat transfer, the *x*-value does not change from that for adiabatic mixing. The value of h_3 will be higher or lower than the adiabatic value as indicated by the vertical distance $\overline{y\text{-}3}$ on the second *h-x* diagram.

A binary mixture of **saturated liquid and saturated vapor in equilibrium** could be considered an example of adiabatic mixing. For equilibrium, the liquid and vapor would be at the same temperature. In **Fig. 4.24**, points *f* and *g* represent the conditions of the saturated liquid and the saturated vapor, respectively. Point *a* gives the condition of the mixture. The previous considerations of adiabatic mixing show that *a* is in a straight line between *f* and *g*. Thus isothermals in the wet vapor region must be straight lines. Also, as shown on the diagram, the dryness fraction m_g/m_a would be given by the ratio of the distance $\overline{f\text{-}a}$ to $\overline{f\text{-}g}$. It should be easy to remember which segment of the line represents the amount of vapor, because the smaller the amount of vapor, the closer the point **a** must be to point **f**.

In Fig. 4.22, it was shown that an *h-x* diagram for a homogeneous binary mixture such as ammonia and water would have a separate pair of saturated vapor and saturated liquid lines for each equilibrium pressure to be plotted. The specific heat of liquid is not much affected by pressure, so that isothermals in the liquid region apply to all pressures in the ranges usually dealt with. A set of isothermals in the wet vapor and the superheated vapor regions would, however, apply to only one pressure. On a practical diagram, therefore, they are omitted, and to connect saturated liquid and saturated vapor points that are in equilibrium with each other, **equilibrium construction lines** are plotted for each pressure. A constant *x*-line from the saturated liquid point will intersect a constant *h*- line from the equilibrium saturated vapor point at the construction line. **Figure 4.25** illustrates the use of this type of chart for two different pressures and one particular temperature. A chart with scales is given in the appendix.

4.6.4 Separation of a Binary Mixture by Heat Transfer at Constant Pressure

It was stated that the principal difficulty in an ammonia/water absorption-refrigeration system not found in a lithium bromide/water system is the need to separate the ammonia from the water, since the two will be driven off together from the liquid in the generator. The fact that this separation can be achieved by heat exchange can be explained with the aid of an *h-x* diagram for a simple system.

Figure 4.26 shows, schematically, a generator supplied with a relatively **strong aqua** (high percent of ammonia in water) at condition **a** from which it is desired to drive off ammonia vapor as pure as possible by heat transfer. The first *h-x* diagram of Fig. 4.26 shows that if Q_H Btu/min are added in the **generator** for a rate of liquid supply $\dot{m}_a$ lb/min, the vapor driven off will have a concentration x_d, and the remaining liquid will have a concentration x_c. (The points **c** and **d** on the *h-x* diagram must be found by trial and error with a straight edge through point **b**). For a refrigeration system, the purity of ammonia vapor at **d** is not great enough. If less energy were added to point **b′**, the purity might still not be high enough, and also very little vapor would be driven off per pound of liquid circulated; if more heat were added to point **b″**, the concentration of ammonia in the vapor would be even lower. The separation must be continued in a second step.

The vapor at condition **d** is brought in contact with cooling coils in a part of the system sometimes called a **dephlegmator**. When Q_A Btu/min are removed, the state point is lowered to **e**. Because water will condense preferentially, the concentration of the remaining vapor is increased to x_g while the condensate has a concentration x_f. (Again, points **g** and **f** must be found by trial and error on the chart). Variation in the energy quantities Q_H and Q_A can produce vapor of any desired concentration between x_a and very nearly 1.0. The limiting case in which $x_g = 1.0$ cannot quite be attained in such a simple system, since the isotherm *gef* would then be vertical and x_f would have to be 1.0 also. This is incompatible with an achievable value of x_d.

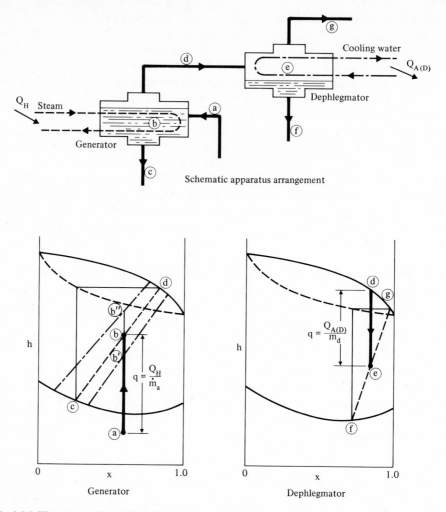

FIG. 4.26 The separation of a binary mixture by heat transfer in two continuous-flow processes.

4.6.5 *Ammonia/Water System*

A practical large ammonia/water system must be more complicated than the simple system of Fig. 4.19 for two reasons: the vapor supplied to the condenser must be dry (anhydrous) ammonia, and the economy must be improved by internal heat exchange. The simple separation system of Fig. 4.26 is made more effective and more efficient by the insertion of a purification or **rectifying column** between the generator and the dephlegmator. Also, the dephlegmator becomes a **reflux cooler** as its condensate is returned to the generator through the rectifying column. The use of single and double rectifying columns for the separation of oxygen and nitrogen from atmospheric air will be illustrated in Chapter 5, Section 4. For a

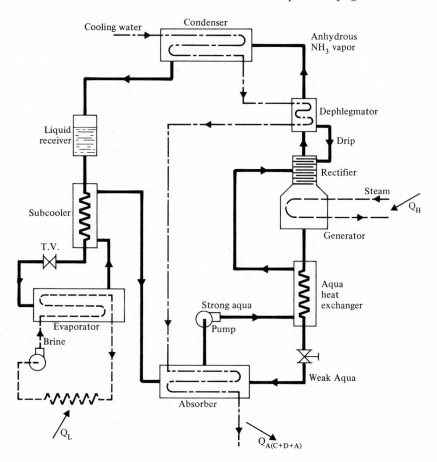

FIG. 4.27 A schematic industrial aqua/ammonia system.

detailed discussion of mass and energy balances and limiting conditions in reflux coolers and rectifying columns, the reader is referred to Bosnjakovic [B.3], Chapter 4.

Figure 4.27 is a schematic arrangement of a practical aqua/ammonia system. With dry ammonia vapor supplied to the condenser at a high pressure (say 160 psia), the condenser, subcooler-superheater, throttling valve, and evaporator operate in exactly the same manner as in an ammonia vapor-compression system. In the absorber, the cool vapor is taken into solution by weak aqua (low concentration of ammonia in water) returning from the generator, while enthalpy of condensation and enthalpy of solution are removed by heat transfer to the cooling water. The strong aqua is pumped from the absorber to the generator, passing through a liquid-liquid heat exchanger to be heated regeneratively by the warm, weak aqua moving from the generator to the absorber.

The strong aqua is actually introduced to the generator via the rectifying column. This liquid flow joins the condensate returning from the dephlegmator to come in intimate contact with the vapors rising from the generator. The heat transferred from low-pressure steam, hot water, or any other economical supply in the generator drives off vapor with too much water content. In the open counterflow process of the rectifying column, both mass and energy are transferred. The rising vapor becomes progressively drier, while the falling liquid carries additional condensed water back to the generator.

The cooling water required in the condenser, the dephlegmator, and the absorber need not be separate supplies. Because of the different temperature levels, the water is usually passed through all three in series.

The advantage of the ammonia/water system, which uses ammonia as the refrigerant, over any system such as lithium bromide/water, which uses water as the refrigerant, is the ability to refrigerate below 32 F. While this is not usually necessary for comfort air conditioning, the lower temperatures are commonly required in various industrial applications. This system in large ratings is adaptable to many process industries where relatively low-temperature energy supplies are available. At the present time, large ammonia/water systems, because of their complexity, do not compete economically with large lithium bromide/water systems where the latter are suitable; also, they often do not compete economically with large vapor-compression machines for lower temperatures now that centrifugal compressors have been improved. They compete with large centrifugal compressors only where shaft power is unusually expensive or low-grade energy unusually cheap.

One current application of the ammonia/water system that may grow in importance is the utilization of solar energy for the generator heat source of a refrigerator for food preservation and perhaps for comfort cooling. This is likely to be used in underdeveloped countries where it is often found that high solar intensities are coupled with a lack of electric power. An intermittently operated ammonia/water system can eliminate the liquid pump and can be matched to the solar cycle. Valves can be operated manually to allow low-pressure absorption during a cool night, and then vapor generation at high pressure for refrigeration during the day. For this application, other binary systems besides ammonia and water and lithium bromide and water have been successfully tried [A.5, A.6, and A.31].

4.6.6 Three-Fluid Absorption System

The two-fluid absorption system succeeded in replacing a vapor compressor requiring a large amount of shaft work with a liquid pump with practically negligible energy requirement compared to the refrigeration effect. By the addition of a third fluid, the pump has also been removed to eliminate completely all moving parts. The system is sometimes called the von Platen-Munters system after its Swedish inventors.

The three-fluid system is shown schematically in **Fig. 4.28**. The fluids most commonly used are ammonia (the refrigerant), water (the absorbent), and hydrogen (a neutral gas used to support a portion of the total pressure in part of the system).

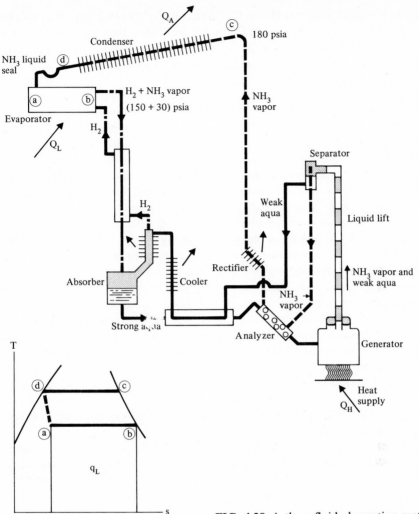

FIG. 4.28 A three-fluid absorption system.

This method of refrigeration is presently used only in domestic units where the coefficient of performance is less important than quiet trouble-free operation.

In the generator of Fig. 4.28, energy is added by heat transfer to a relatively strong aqua/ammonia solution. As vapor is accumulated at the top of the generator, a slug of liquid is forced a short distance up the narrow tube leading to the separator. Vapor follows this liquid into the tube, and, as vaporization continues, alternate slugs of liquid and vapor are forced to rise through the tube in a manner similar to that of a coffee "percolator."

While relatively weak aqua is allowed to drain by gravity from the bottom of the separator to the absorber, vapor with a high ammonia concentration is taken from the top. This vapor is forced to bubble through cool liquid in the **analyzer** and is further cooled, releasing water, in the rectifier. Dry ammonia vapor rises to the condenser. From the condenser, as from all finned elements shown, heat rejection is to the ambient air, which is usually circulated only by natural convection.

The unique feature of this system is encountered as the liquid ammonia passes through the liquid seal to the evaporator. Here the ammonia joins hydrogen, which is trapped on the "low" side, and following Dalton's law of partial pressures, is called upon to support only the difference in pressure between the total pressure in the whole system and that which hydrogen is able to support in that volume at the existing temperatures. The total pressure under operating conditions is usually around 180 psia, and that pressure is supported by the ammonia alone on the "high" side, including the condenser. Let it be assumed that the hydrogen is capable of supporting 150 psia. The ammonia then, as soon as it passes the liquid seal, is called upon to support (or is subjected to) only 30 psia. Mixing the ammonia with the hydrogen is equivalent to passing it through a throttling valve. At its low pressure of 30 psia, the ammonia, now in the wet vapor region, vaporizes completely to produce the refrigerating effect.

The cold ammonia vapor and hydrogen circulate by natural convection through a gas-gas heat exchanger to the absorber where the ammonia vapor comes in contact with the weak aqua from the separator. At the low temperature of the ammonia and hydrogen, absorption takes place so that hydrogen alone rises through the heat exchanger to the evaporator, while a strong aqua solution flows, by gravity, back to the generator.

The liquid-liquid heat exchanger preheats the strong aqua on its way to the generator and precools the weak aqua going to the absorber. The internal heat exchanges make the system more effective for its size and raise the coefficient of performance.

On the *T-s* diagram accompanying the schematic diagram, an attempt is made to follow the ammonia alone from its entrance into the condenser to its exit from the evaporator. The absorption and separation processes cannot be shown easily, and a gap must be left in the cycle between points *b* and *c*.

4.6.7 *Combination of Vapor-Compression and Absorption-Refrigeration Systems*

Many arrangements of different types of refrigeration equipment can be devised whenever more than one unit would be needed to carry the full refrigeration load. Also, several units are often preferred to one large unit so that those units not required can be shut down when the demand is less than full load.

A combination of one steam-turbine-driven, vapor-compression machine with two lithium bromide/water absorption machines has been suggested [A.7]. **Figure 4.29** shows such a combination with the exhaust from the turbine providing the generator heat for the two absorption systems. A turbine bypass is provided to allow operation of the absorption systems alone. Cooling water lines, absorber circulation, etc, are not shown. An overall steam rate of about 13 lb/hr/ton is claimed compared to 18 to 20 for good absorption systems.

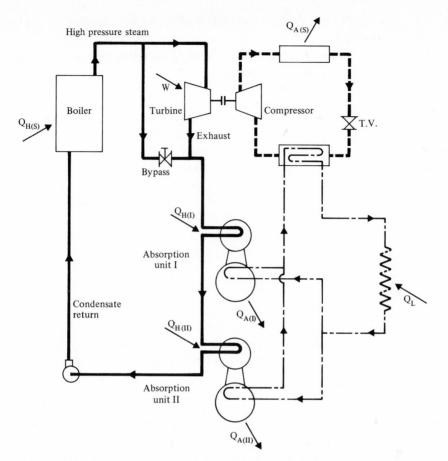

FIG. 4.29 A combination of a vapor-compression system with a steam turbine drive and two absorption refrigeration units.

4.7 GAS-CYCLE REFRIGERATION

Refrigeration can be accomplished by means of a gas cycle rather than a vapor cycle. Whereas the refrigeration effect per pound of fluid circulated in a vapor-compression cycle is equivalent to a large fraction of the enthalpy of vaporization, in a gas cycle it is only the product of the temperature rise of the gas in the low-side heat exchanger and the specific heat of the gas. Therefore a large cooling load requires a large mass rate of circulation. To keep the size of equipment down, the whole apparatus may be under pressure, which requires a closed cycle. The throttling valve used for the expansion process of a vapor-compression cycle is usually replaced by an expansion engine or turbine for a gas cycle, because useful work can thereby be recovered, and because work must be done by the gas for an appreciable temperature drop to be achieved.

4.7.1 The Simple Brayton Refrigeration Cycle (*Without Regenerative Heat Transfer*)

An idealized gas-refrigeration cycle, using air and operating between 100 and 300 psia is illustrated in **Fig. 4.30**. It is seen that with water cooling down to approximately 100 F, quite low temperatures are possible. (For reasonably accurate calculations, the air tables or the T-s diagram in the appendix can be used.) Note, however, that not all the refrigeration effect is achievable at the lowest temperature. A secondary fluid in the lower heat exchanger can be brought close to the minimum air temperature if a counter-flow arrangement is used.

The cycle shown is a **Brayton refrigeration cycle**. This cycle was presented as a power cycle in Chapter 1, and its efficiency was analyzed in Section 1.3.5. With the assumption of constant specific heat usually made for an ideal gas cycle, the Brayton cycle's coefficient of performance as a refrigerator is

$$\text{COP}_R = \frac{\text{Refrigeration effect}}{\text{Net work}}$$

$$= \frac{h_b - h_a}{(h_c - h_b) - (h_d - h_a)}$$

$$= \frac{T_b - T_a}{(T_c - T_b) - (T_d - T_a)}. \tag{4.11a}$$

This can be simplified because there is isentropic compression from **b** to **c** and isentropic expansion from **d** to **a**, and $p_a = p_b$ and $p_c = p_d$. Thus

$$\frac{T_b}{T_a} = \frac{T_c}{T_d},$$

and

$$\text{COP}_R = \frac{T_a(T_b/T_a - 1)}{T_d(T_c/T_d - 1) - T_a(T_b/T_a - 1)}$$

$$= \frac{T_a}{T_d - T_a}. \tag{4.11b}$$

To put this in terms of the pressure ratio from **b** to **c** or **d** to **a**,

$$\text{COP}_R = \left[\frac{T_d}{T_a} - 1\right]^{-1}$$

$$= \left[\left(\frac{p_H}{p_L}\right)^{(\gamma-1)/\gamma} - 1\right]^{-1}. \tag{4.11c}$$

The value of COP_R for the simple Brayton cycle as a refrigerator tends toward infinity as the difference $(T_d - T_a)$ approaches zero in the same way that the efficiency of the comparable power cycle would approach zero at these conditions. Equation (4.11c) does not, however, lead to the conclusion that the cycle COP_R can actually be increased indefinitely for refrigeration between two given temperatures simply by the choice of a small enough pressure ratio. For operation of the simple cycle of Fig. 4.30 between a source temperature T_b and a sink tempera-

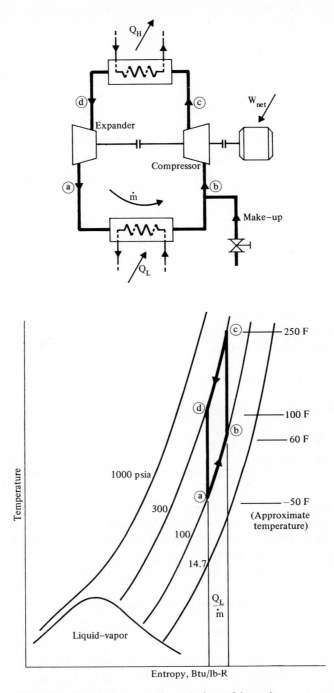

FIG. 4.30 An ideal closed air cycle for refrigeration.

ture T_d, the minimum pressure ratio for cooling is

$$\left(\frac{p_H}{p_L}\right)_{min} = \left(\frac{T_d}{T_b}\right)^{\gamma/(\gamma-1)}. \tag{4.12}$$

Obviously, however, the refrigeration per pound of fluid circulated would be zero at this pressure ratio. As the pressure ratio is increased, the required rate of circulation for a given tonnage is lowered. A compromise must be reached between the cycle efficiency and the rate of circulation. Losses and therefore deviation from the ideal COP will increase as the flow rate becomes greater. (Compare this with the discussion of the optimum compression ratio for a Brayton power cycle.)

It would be logical to compare the ideal Brayton cycle with the Carnot cycle operating between temperatures T_b and T_d. The **relative efficiency** or the ratio of the ideal Brayton COP to the ideal Carnot COP is

$$\eta_r = \left(\frac{T_a}{T_d - T_a}\right)\left(\frac{T_d - T_b}{T_b}\right)$$

$$= \frac{(T_d - T_b)}{(T_b/T_a)(T_d - T_a)}$$

$$= \frac{T_d - T_b}{T_c - T_b}. \tag{4.13}$$

For a given temperature range from *b* to *d*, the Brayton cycle COP approaches the Carnot cycle value as T_c approaches T_d (its minimum limit). That is to say again, the lower the pressure ratio, the higher the COP of the idealized cycle.

Actual cycle COP varies considerably from that of the ideal cycle, primarily because of inefficiencies in gas compression and expansion, and because of fluid friction and mechanical losses. For a large system, multistage compression might be worthwhile. A nonpressurized system is large and bulky, a pressurized system is heavy, and both are likely to create considerable vibration forces; but air-cycle refrigeration has the advantage of utilizing a completely nontoxic and a readily replaceable working fluid. For this reason it has been used widely in the past for such applications as shipboard cargo cooling. Make-up gas must be supplied under pressure for a pressurized system and must be dry if temperatures go below 32 F. Dry gas also helps to protect the equipment. Improved leakage control and the introduction of nontoxic synthetic refrigerants have decreased the importance of the air cycle in general-purpose refrigeration. Refrigeration applications in which the gas cycle has an inherent advantage are those in which the cool gas itself can be utilized. But even here, the advantages of direct cooling must be balanced against those of a separate vapor-compression refrigeration cycle. Compactness, serviceability, and other mechanical and economic considerations must be taken into account.

4.7.2 Aircraft Cabin Cooling

A very common application of gas-cycle refrigeration, where the working fluid is utilized and not usually recycled, is aircraft cabin cooling as illustrated schematically in **Fig. 4.31**. The very special circumstances of this application must be noted.

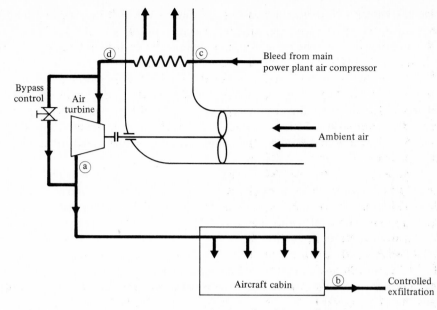

FIG. 4.31 Open cycle aircraft cabin cooling.

1. The compressed air is available and is a small percentage of the amount handled by the compressor of a turbojet or a supercharged aircraft engine. This in itself may not be a prerequisite as evidenced by the fact that the same system has been utilized for aircraft with other types of power plants; still, the refrigeration compressor shaft work is not a large percentage of the propulsive power being utilized.
2. Large amounts of cool ambient air are easily available for cooling the compressed air.
3. In addition to cooling, replacement of stale air is required in the cabin.
4. Pressurization of the cabin may be required.
These points are mentioned as examples of peripheral considerations that must often go into the selection of a type of refrigeration process.

More complex systems for aircraft cooling sometimes include evaporative cooling with an expendable refrigerant, or additional compressors and expansion turbines to augment the refrigeration effect [A.20]. In all aircraft applications, extremely high shaft speeds keep the size of the equipment remarkably small. The attendant noise would not, however, be acceptable in most other comfort cooling applications.

4.7.3 The Brayton Cycle with Regenerative Heat Transfer

It is possible to produce refrigeration with a Brayton gas cycle at very low temperatures, even with relatively small pressure ratios, when internal heat transfer by means of a **regenerative heat exchanger** is added to the cycle as shown in **Fig. 4.32**. The term regenerative heat exchanger, or simply regenerator, is used here as it was for power cycles in Chapter 1 to denote a heat-transfer device in which the working fluid at one point in the cycle exchanges energy by heat transfer with the same working fluid at another point in the cycle. Thermodynamically, it does not matter whether the hotter and colder fluids flow through continuously, as implied in Fig. 4.32, and are separated by some barrier, or whether the fluid at one time gives up heat energy to the walls or packing of the regenerator and at another time reverses its direction of flow to regain the same energy by heat transfer. In heat-transfer nomenclature, the counter-flow heat exchanger is often called a **recuperator**, and the reversed flow type is often called a **regenerator**. Either type may function as a regenerator in terms of power or refrigeration cycle analysis.

Because the cycle represented schematically in Fig. 4.32 can be used to achieve extremely low temperatures, it might be considered proper to reserve discussion of it until Chapter 5, in which we will deal with cryogenics. Nevertheless it is presented here for continuity with Sections 4.7.1 and 4.7.2. Reference to it will be made again in Chapter 5, Section 5.

To derive a simple expression for the coefficient of performance of the system of Fig. 4.32, we will again assume that the specific heat of the working fluid is constant. Further, we will assume that the **regenerator efficiency** is 100%, which means that T_d is equal to T_n. As a refrigerator, the coefficient of performance is

$$
\begin{aligned}
\text{COP}_R &= \frac{T_n - T_a}{(T_c - T_b) - (T_d - T_a)} \\
&= \frac{T_a(T_d/T_a - 1)}{T_b(T_c/T_b - 1) - T_a(T_d/T_a - 1)} \\
&= \frac{T_a}{T_b - T_a} \\
&= \left[\frac{T_b}{T_d} \left(\frac{p_H}{p_L} \right)^{(\gamma-1)/\gamma} - 1 \right]^{-1}.
\end{aligned}
\tag{4.14}
$$

If T_b is considered the ambient or sink temperature, and T_d (approximately equal to T_n) is the refrigeration temperature, then Eq. (4.14) differs from the COP of a Carnot cycle only by the inclusion of the pressure-ratio term. At a pressure ratio of unity, however, the refrigeration per pound is zero even for the ideal cycle, just as it would be for the minimum pressure ratio, Eq. (4.12), for the simple gas cycle of Fig. 4.30.

The relative efficiency of the regenerative cycle compared to a Carnot cycle between T_b and T_d is

$$
\eta_r = \frac{T_b - T_d}{T_b(p_H/p_L)^{(\gamma-1)/\gamma} - T_d}.
\tag{4.15}
$$

The fact that this goes to 1.0 as the pressure ratio goes to 1.0 simply corroborates the statement following Eq. (4.14).

Although the overall temperature range has been increased compared with the range for the system of Fig. 4.30, the refrigeration effect per pound of fluid circulated has been decreased. Since the work of expansion through a given pressure

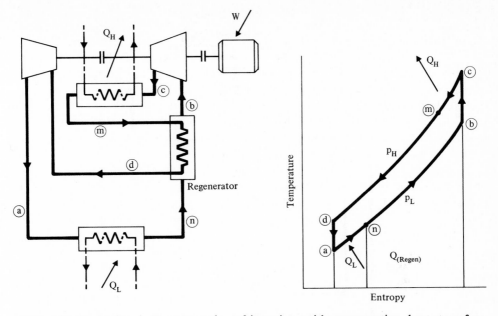

FIG. 4.32 Ideal closed Brayton-cycle refrigeration with regenerative heat transfer.

ratio is reduced as the temperature is lowered, the net work input for the cycle must be increased. The coefficient of performance, therefore, is not the same as for a simple cycle without regeneration.

In the discussion of simple gas-power cycles in Section 1.3, it was noted that gases with higher γ-values permit higher efficiencies (see Fig. 1.26). The same dependence on γ is noted for the coefficient of performance of a gas-refrigeration cycle. For a closed cycle, there is no need to use air ($\gamma = 1.40$) for the working fluid. Monatomic gases ($\gamma = 1.64$ to 1.67) would actually be superior. The most frequently used gas has been helium.

A **combined refrigeration and power cycle** using helium throughout has been suggested [A.15]. A schematic diagram of such a system along with a temperature-entropy diagram for the ideal cycles are shown in **Fig. 4.33**. The two turbines, one in the power cycle and one in the refrigeration cycle, operate at entirely different temperature levels and combine to drive the common compressor. The two rates of gas circulation ($\dot{m}_P$ and $\dot{m}_R$) can be so balanced that external shaft work is neither required nor available. The *T-s* diagram is idealized in that turbine and compressor inefficiencies are not shown; the pre-cooler and after-cooler are assumed to bring the respective gas streams to ambient temperature; and the two regenerators are shown to be 100 % efficient. With reasonable efficiencies assumed for all components, the optimum pressure ratio for the power cycle is found to be very close to that for the refrigeration cycle using helium; the former is approximately 2.0 and the latter approximately 1.8 [A.15].

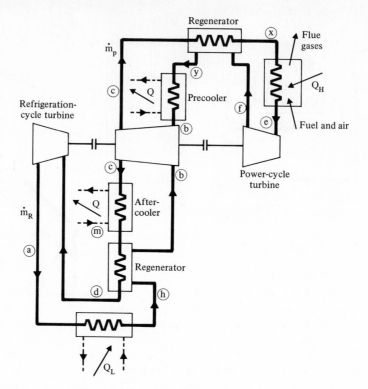

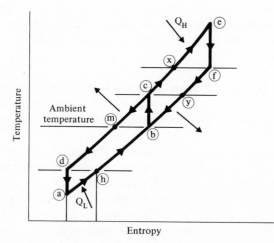

FIG. 4.33 Combined refrigeration and power Brayton gas cycles (working fluid:helium).

4.7.4 The Stirling Cycle

In Chapter 3, Section 2, the Stirling cycle was analyzed as a power cycle. It is composed ideally of two isothermal and two constant volume lines. The heat transfer at constant volume is internal, i.e., regenerative. The same cycle, reversed, can be used for refrigeration. However, machines attempting to operate on the Stirling cycle are not competitive with vapor systems for most applications at conventional refrigeration temperatures. On the other hand, such machines do achieve cryogenic temperatures in a single stage (as can the Brayton cycle with regenerative heat transfer). While it is likely that the Stirling cycle will find wider application in the future, its use at the present time is primarily in air liquefaction. Therefore a detailed discussion will not be undertaken here, but will be presented in Chapter 5, Section 5.2.

A similarly arbitrary distinction is made in the postponement of discussion of other primarily cryogenic gas cycles such as the Gifford-McMahon system and the Gifford pulse-tube device.

4.7.5 The Vortex Tube

Another device for producing a refrigeration effect with gas alone is the **vortex tube.** This was patented by Ranque [A.24] and systematically investigated by Hilsch [A.13]; either one or the other name is often used to identify the device. It is shown in **Fig. 4.34**.

Compressed gas is supplied at pressure p_a and temperature T_a in a tangential direction to produce a vortex within the tube. Part of the gas escapes through the larger exit tube at a rate controlled by the valve located 30 or more diameters downstream. The remainder, at a lower pressure, escapes through a concentric orifice into a tube leading in the opposite direction. It is found that the fraction of gas (x) escaping through the orifice is at a temperature T_b which is below T_a, while the fraction $(1 - x)$ is at a temperature T_c which is higher than T_a. The temperature differences produced depend on the pressure ratio p_a/p_b, on the fraction x, which is controlled by the valve, and on the type of gas used.

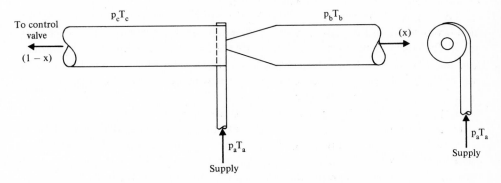

FIG. 4.34 A vortex tube (Hilsh or Ranque tube).

The temperature difference between the two escaping steams is attributable to two effects. First, there is a radial pressure gradient in any vortex because of the centrifugal force on the rotating fluid, so that the gas near the axis is at a reduced pressure while that at the outer wall is compressed. Second, the viscosity of the fluid prevents the establishment of a completely free vortex in which no energy is transferred from one annular layer to the next; the tendency of the viscosity to produce a solid or forced vortex transfers energy from the inner layers toward the outer layers as the latter retard the tangential velocity of the former. (A true "free vortex" would have infinite velocity and zero pressure at its axis.) Both the pressure differences and the viscous drag effect perform work on the fluid near the wall at the expense of energy in the fluid near the axis, so that gas escaping from the core is at a lower temperature than T_a while the remainder is at a higher temperature.

The expansion of the gas to p_b is neither adiabatic nor reversible, so that the temperature T_b is above the ideal value for isentropic expansion from p_a to p_b. Assuming that the cooler gas can produce useful refrigeration if it is heated again to T_a, the refrigeration effect per pound of gas supplied is

$$Q_L = (x)c_p (T_a - T_b). \tag{4.16}$$

This is less than that possible in the Brayton cycle between p_a and p_b, because only the fraction x is cooled, and because T_b is higher than the isentropic expansion temperature. In addition, there is no possibility of recovering the work of expansion as in the ideal Brayton cycle. The coefficient of performance of the vortex tube as a refrigerator is therefore comparatively low, and it is likely to be applied only where its simplicity is a distinct advantage. With pressure ratios as high as 10:1, temperature depressions $(T_a - T_b)$ as great as 60 C degrees (108 F degrees) have been reported by Hilsch for air.

4.8 HEAT PUMP APPLICATIONS

In the introduction to this chapter, it was emphasized that the refrigeration cycle can be used as a heat pump, i.e., with its prime purpose to deliver heat to a sink at high temperature rather than primarily to remove heat from a source at low temperature. The working cycles remain essentially the same: vapor-compression, absorption, and gas systems are used. In this section, only a few applications of heat pumps will be described. Innumerable arrangements can be devised in the application of basic refrigeration cycles to the purpose of heating. In all cases, the economics of the system must be tested against the possibility of using the prime power supply (electricity or fuel) or some alternative energy supply directly as heat.

Many heat sources are available in various locations and industries; large quantities of heat energy can be extracted from lakes and rivers, from wells, from the outside air, from waste liquids and gases, from suitably trapped solar energy, and even from freezing water. Wherever such heat sources are available, but the temperature of the source is too low for direct application, a heat pump may be feasible from the point of view of economics or convenience. Frequently in air conditioning and various processes in industry, heating and cooling are required simultaneously in different places. In such instances, a heat pump is often the most logical means of accomplishing both.

4.8.1 Heat Pumps for Air Conditioning

One common use of the heat pump is in heating buildings for comfort. Since the first cost of installing such equipment is high compared with conventional fuel heating systems, the heat pump becomes competitive particularly when the same building area at different times provides a cooling load of the same order of magnitude as the heating load. In such situations, the same apparatus can usually be used with only some modification in piping or ducting arrangements. Even if the cooling and heating loads are not quite equal, the peaks of either might be carried by some small standby system such as unit air conditioners and direct electric heating.

Heat pumps are classed according to the nature of their sources and sinks as air-to-air, liquid-to-liquid, air-to-liquid, liquid-to-air, or some combination of these. Only two examples are given here. **Figure 4.35** is a simplified schematic presentation of an air-to-air system, where the air-conditioning duct work is arranged so that the same refrigeration or heat pump system can be used for both summer cooling and winter heating. In the summer, all four air dampers are in the upper position, so that recirculated room air is passed over the evaporator coils for cooling; outside air, passing over the condenser, removes the energy picked up by the working fluid in the evaporator plus the work of the compressor. For operation of the system as a heat pump in the winter, all four dampers are moved to the lower

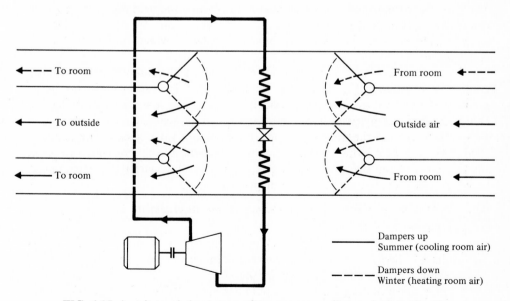

FIG. 4.35 An air-to-air heat pump for summer cooling and winter heating.

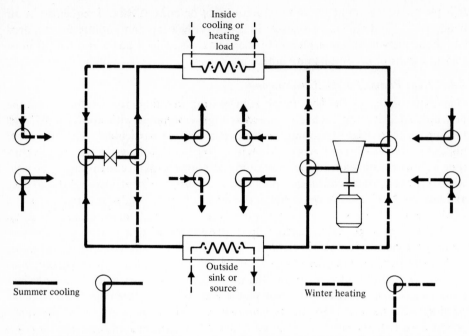

FIG. 4.36 A liquid-to-liquid heat pump for summer cooling and winter heating.

position; heat transfer from the outside air, even at relatively low temperatures, can then be accomplished at the evaporator coils while the heat rejection from the condenser coils is delivered to the recirculated room air. Controlled mixing of fresh outside air with recirculated room air can be provided for in the ducting arrangement.

A simplified schematic arrangement of a liquid-to-liquid system for summer cooling is presented in **Fig. 4.36**. Here, the flow paths of the source and sink fluids are not changed, but the flow of the working fluid is. For summer cooling, the discharge from the compressor goes to the external-fluid heat exchanger, which acts as a condenser, and the discharge from the throttling valve goes to the internal-fluid heat exchanger, which functions as the evaporator coil. For winter heating, the functions of the two heat exchangers are reversed. (Quarter-turn switch-over valves are shown schematically in the diagram.)

4.8.2 Utilization of Waste Heat of Prime Mover to Augment Low Temperature Source

For any prime mover, a large percentage of the energy supplied is rejected as "unavailable" energy; the heat of condensation in a steam turbine power plant is a good example. To recapture this energy alone through a heat pump that utilizes all the shaft power of the prime mover could not increase the available energy above the direct utilization of the initial source for heat transfer. But to add this recovered

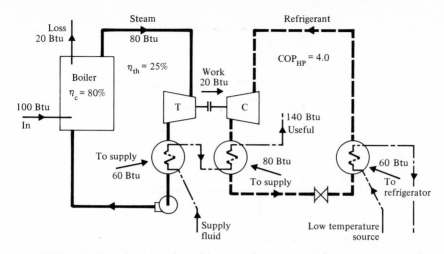

FIG. 4.37 Power cycle rejected heat used to augment heat pump source.

energy to the heat pumped from a separate low-temperature source can increase the net output to well above the original fuel energy input and justify the use of high-grade energy (shaft work) in a heating system.

An example of such a combination of power cycle and heat-pump cycle is shown in **Fig. 4.37**. With the efficiencies assumed as noted on the diagram, the delivered heating or process fluid absorbs more energy than could be obtained from even 100 % combustion efficiency in a steam heating plant or from the heat-pump cycle alone. More flexibility of operation might be obtained by using the turbine to drive an electric generator and an electric motor to drive the refrigerant compressor, although additional losses would be introduced. If the turbine were a larger one used to drive other machines besides the compressor, more energy, of course, would be available in the exhaust condenser.

4.8.3 *Water Distillation with a Heat Pump*

Often in an industrial situation, the fluid to be heated or cooled can be used as the working fluid in a vapor-compression cycle that achieves the desired effect. An example is the water distillation process shown in **Fig. 4.38**.

The supply water, presumably not pure enough for use, is preheated and then vaporized at atmospheric pressure. The saturated vapor is compressed to about two atmospheres, so that its condensation temperature is sufficiently high to vaporize water at 212 F with some temperature difference in the vaporization chamber coils. The final discharge at point f can be controlled by a valve to prevent the condensate from flashing to vapor until it is sufficiently subcooled in the heat exchanger. The *p-h* diagram shows the ideal cycle. The only energy input is $(h_d - h_c)$. The schematic diagram shown is highly simplified from such large-scale plants as the one described by Brennan [A.3].

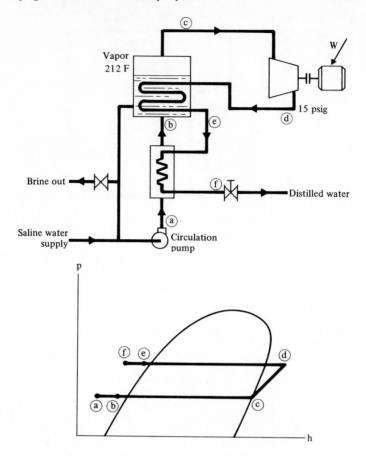

FIG. 4.38 Water distillation with a heat pump.

4.8.4 Recovery of Latent Heat of Fusion

For any liquid used as a source of heat, the mass rate of circulation could be reduced if the latent heat of fusion could be recovered. When water is cooled, only one Btu per pound is available for each degree F drop in temperature. The latent heat of fusion, though, is nearly 144 Btu/lb. In the heat-pump system shown schematically in **Fig. 4.39**, the source of heat energy is the liquid water below the ice of a northern lake. A commercial shaved-ice machine (which has a rotating drum inside of which a refrigerant can vaporize so that ice is formed on the outside) is used to extract the heat of fusion. (The ice is removed by scraper blades and taken by conveyors to a blower for disposal). The system is used to provide heated water. In addition to the heat transfer to the water supply in the condenser of the heat-pump cycle, heat energy is reclaimed from both the cooling water and the exhaust of the driving engine. Radiation and convective heat loss from the engine are

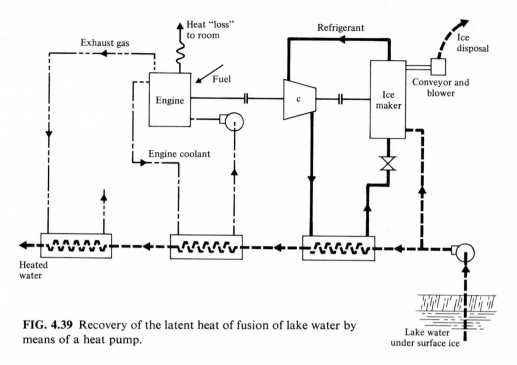

FIG. 4.39 Recovery of the latent heat of fusion of lake water by means of a heat pump.

useful in helping to heat the building. All components including pumps and blower are either belt or chain driven from the engine shaft. It is reported that energy equivalent to 170% of the higher heating value of the diesel fuel can be obtained as useful heat [A.1].

4.8.5 *Internal Heat Exchange in Mixture Separation Systems*

The separation of ammonia and water described for absorption refrigeration (Sections 4.6.4 and 4.6.5) is only an example of the general problem of binary-mixture separation. Bosnjakovic [B.3] points out that under certain conditions the energy rejected in the dephlegmator might be recoverable economically by means of a heat pump for use at the higher temperature in the generator. At the same time, the generator would act as the condenser for the vapor leaving the dephlegmator. Such a system is illustrated in **Fig. 4.40**. It must be remembered that the work (w) and heat transfer (q) noted on the h-x diagram are energy per unit mass of fluid leaving at g, not per unit mass initially vaporized in the generator. Whether or not the heat pump arrangement is warranted economically depends on the cost of shaft power for the compressor drive and on the temperature difference (and, therefore, the pressure difference) required for the particular separation process. Auxiliary generator heat supply may be required.

The Linde single-column and double-column systems for air separation, to be described in Chapter 5, Section 4, could be considered as further examples of

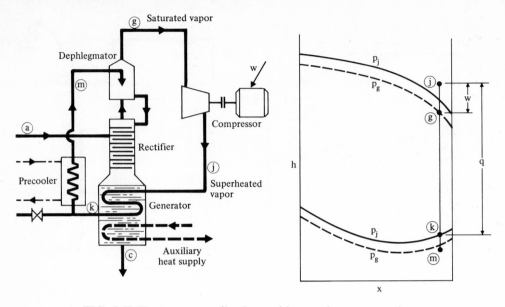

FIG. 4.40 Heat pump application to binary mixture separation.

the application of a heat pump to mixture separation. In these systems, the initial air supply is compressed and then cooled in a counter-flow heat exchanger before being supplied to the generator of the separation column. At the higher pressure, the supply air is condensed while the heat transfer from it vaporizes liquid oxygen and nitrogen at a lower pressure.

PROBLEMS

1. From the definition of entropy, use a temperature-entropy diagram to prove that the coefficient of performance of a Carnot cycle refrigerator is $T_L/(T_H - T_L)$.
2. From the point of view of desirable properties for a vapor-compression refrigerant, what is the significance of liquid-phase specific heat?
3. A vapor-compression refrigeration cycle using Refrigerant-12 operates between 100 psia and 10 psia. Saturated liquid is throttled, and the fluid evaporates to the dry saturated condition.
   a) How much refrigeration is effected per pound of fluid circulated, and at what temperature is refrigeration available?
   b) For the ideal cycle with isentropic compression, what is the coefficient of performance as a refrigerator?
4. In a vapor-compression refrigeration system, does the number of degrees of subcooling of the liquid in the condenser affect
   a) the work done per pound of fluid circulated?
   b) the work done per ton of refrigeration?

5. A Refrigerant-12 vapor-compression refrigeration system has an air-cooled condenser and a subcooling-superheating, liquid-vapor, counter-flow heat exchanger. Condensation, without subcooling, is at a saturation pressure corresponding to 10 F degrees above ambient air temperature of 90 F. Subcooling is to within 20 degrees of the evaporation temperature of 10 F. There is no superheating in the evaporator. Assume isentropic compression and ignore friction losses. Use chart values to calculate the coefficient of performance as a refrigerator. Sketch the thermodynamic cycle on both the *p-h* and *T-s* coordinates to identify significant state points.

6. For a Refrigerant-12 system, assume that the liquid before throttling and the suction vapor are both saturated. The reciprocating compressor clearance factor is 0.15. Assume that the exponent (*n*) for vapor compression is 1.3. By what percentage will the refrigeration capacity be changed if the condensation temperature is kept at 100 F, but the evaporation temperature is changed from 20 F to zero F?

7. A five-ton, two-stage, vapor-compression refrigeration system uses Refrigerant-12 with evaporation in the lower cycle at 5 psia, condensation in the upper cycle at 140 psia, and flash intercooling at 30 psia. Assume that there has been no subcooling of liquid before throttling and no superheating of vapor before compression.
a) What is the ideal horsepower requirement?
b) What is the ideal overall coefficient of performance?
c) Compare these values with the power and the coefficient of performance for ideal single-stage operation over the same range.

8. A vapor compression, cascaded, two-stage, multifluid refrigeration system cools a chamber to −80 F and heats condenser cooling water to +100 F. The two refrigerants are R-12 and R-13. Assume ideal "dry-compression" cycles with no subcooling of liquid and no superheating of suction vapor. Assume that a 10 degree minimum temperature difference exists in all heat exchangers, and that both stages work through approximately the same pressure ratios.
a) What is the ratio of the mass rates of circulation of the two working fluids?
b) What is the overall theoretical coefficient of performance as a refrigerator?

9. Suppose that an inventor claims to have an absorption refrigeration machine that can freeze 1 ton of water per hour with an input to the generator of 50 lb of steam per hour, saturated at 25 psia. Would you think this possible? Support your answer.

10. For the lithium bromide/water absorption refrigeration system in Fig. 4.21, use steam tables and Li Br/water chart values to calculate the following:
a) Coefficient of performance.
b) Steam required per hour, saturated at 5 psig, per ton of refrigeration.
c) Heat removed by cooling water in the lower shell and in the upper shell per ton of refrigeration.
d) Rate of cooling water supply required per ton at 60 F with a temperature difference in the upper heat exchanger of 5 degrees F.
e) Rate of circulation of Li Br/water solution past point ② in the system per ton of refrigeration.
Note that water that is evaporated and condensed is pure water at saturation conditions governed by the existing vapor pressures.

11. Can a liquid ammonia and water solution exist at 30 psia and 300 F?

12. Two streams of ammonia-water solution, both saturated liquid at 50 psia, are mixed. One stream has an ammonia concentration of 25% and is supplied at 30 lb/min. The other stream, at an ammonia concentration of 75%, is supplied at 20 lb/min. At what rate must heat energy be supplied to convert all the mixture to saturated vapor at the same pressure?

13. Two streams of ammonia-water solution, both at 100 F and 30 psia, are mixed together with no change of pressure and no heat transfer. Stream *A*, 50 lb/min, has a concentration of 0.2 lb of ammonia per pound of solution. Stream *B*, 25 lb/min, has a concentration of 0.8 lb of ammonia per pound of solution.

    a) What are the concentration values for the resulting liquid and vapor?

    b) How many pounds per minute of vapor, and how many of liquid will result?

14. A liquid ammonia/water solution is supplied to the generator of a two-component separation system (Fig. 4.26) at 100 psia and 80 F, at a rate of 10 lb/min. The mixture is 40% ammonia by mass. The whole system operates at 100 psia. It is desired to produce a vapor with no more than 3% water content. Heat energy is supplied to the generator at a rate of 3200 Btu/min. At what rate must the dephlegmator be cooled?

15. An open cycle (Brayton) aircraft cabin cooler expands air at 80 F through a turbine that is 30% efficient from 30 to 15 psia. The cabin temperature is not to exceed 75 F. Estimate the pounds of air per minute necessary for each ton of cooling required.

16. A Brayton cycle with a regenerator can cool to very low temperatures. Draw such a cycle on *T-s* coordinates. Show what would happen to the temperature of refrigeration if the cooling load were doubled while the temperature at the start of compression, the pressure ratio, and the rate of circulation remained fixed.

17. An ideal closed Brayton cycle with regeneration uses air as the working fluid to produce refrigeration. Assume a constant specific heat of 0.24 Btu/lb-F and that $\gamma = 1.4$. All heat exchangers are "perfect." Given a pressure ratio of 2:1 and a minimum cycle temperature of -150 F, what is the maximum refrigeration rate per pound of air circulated?

18. Assume an air-to-air heat pump for building heating with a 50% relative efficiency. What would its coefficient of performance be in your city in January? See Table 6.5 for some assumptions and make others as required.

19. Draw a clear schematic apparatus diagram for a heat-pump arrangement, where ambient air at 20 F is the source for building heating, and hot water is delivered at 120 F. Assume a 10-degree temperature difference in the air-to-vapor heat exchanger and a 5-degree temperature difference between delivered water and condensing vapor. Also assume a compressor efficiency of 80% for a Refrigerant-12 system, and calculate the system's coefficient of performance.

REFERENCES

A. *Articles*

1. ALLCUT, E. A. and F. HOOPER, "The Possibilities of the Heat Pump in Canada," *Engineering Journal* (Canada) Vol. 35, No. 6, 1952, pp. 610–614.

2. ANDERSON, J. H., "Centrifugal Refrigeration Theory," *Industrial Refrigeration*, Sept., 1960, pp. 10–12, 23–24.

3. BRENNAN, P. J., "Fresh Water from Vapor-Compression Evaporation," *Chemical Engineering*, Vol. 70, Oct. 14, 1963, pp. 170–172.

4. CARR-HARRIS, "The Heat Pump for Large Scale Heating," *Canadian Refrigeration Journal*, Vol. 18, No. 9, 1952, pp. 19–22, 49–52.

5. CHINNAPPA, J. C. V., "Experimental Study of the Intermittent Vapour Absorption Refrigeration Cycle Employing the Refrigeration-Absorbent Systems of Ammonia Water and Ammonia Lithium Nitrate," *Solar Energy*, Vol. 5, No. 1, 1961.

6. CHINNAPPA, J. C. V., "Performance of an Intermittent Refrigerator Operated by a Flat-Plate Collector," *Solar Energy*, Vol. 6, No. 4, 1962.

7. CHUN, U., "Combination Centrifugal Absorption Cooling May Save Dollars," *Heating Piping and Air Conditioning*, Vol. 33, No. 3, 1961, pp. 108–109.

8. DAVIS, J. C. and P. R. ACHENBACK, "Energy Use and Power Demands in All-Electric House Equipped with Air-to-Air Pumps," *ASHRAE Trans.*, Vol. 68, 1962, pp. 390–411.

9. ENGLUND, J. S., "Air-to-Air Heat Pump," *ASHRAE Journal*, Vol. 2, No. 3, 1960, pp. 56–59, 70.

10. GOSNEY, W. B., "Calorimetric Determination of Refrigerating Capacity," *Modern Refrigeration*, Vol. 65, No. 770, 1962, p. 468.

11. HARNISH, J. R., Sr., "Capacity Control for Reciprocating Compressors," *Air Conditioning Heating and Ventilating*, Vol. 58, No. 2, 1961, pp. 102–112.

12. HEALY, C. T. and T. I. WETHERINGTON, JR., "Water Heating by Recovery of Rejected Heat from Heat Pumps," *ASHRAE Journal*, Vol. 7, No. 4, 1965, pp. 68–74.

13. HILSCH, R., "The Use of the Expansion of Gases in a Centrifugal Field as a Cooling Process," *Review of Scientific Instruments*, Vol. 18, Feb., 1947, p. 108.

14. JOHNSON, S. E., "Look at Today's Absorption Refrigeration," *ASHRAE Journal*, Vol. 2, No. 9, 1960, pp. 55–59.

15. LaFLEUR, J. K., "Cryogenics via the Gas Turbine," *Mechanical Engineering*, Vol. 87, No. 11, pp. 38–40.

16. McHARNESS, R. C. and D. D. CHAPMAN, "Refrigerating Capacity and Performance Data for Various Refrigerants, Azeotropes and Mixtures," *ASHRAE Journal*, Vol. 4, No. 1, 1962, pp. 49–58.

17. MELBY, E. E., "Newest Refrigerant for Transport Cooling," *SAE Journal*, Vol. 72, No. 4, 1964, p. 39.

18. MISSIMER, D. J. and W. L. HOLLADAY, "Cascade Refrigerating Systems—State of the Art," *ASHRAE Journal*, Vol. 9, No. 4, 1967, pp. 70–73.

19. NOLCKEN, W. G., "Babylke's Theory of Thermodynamic Similarity," *Journal of Refrigeration*, Vol. 5, No. 1, 1962, pp. 12–13.

20. OVERMYER, E. J., "Successive Developments That Led to Vapor Cycle System for Boeing 707," *ASHRAE Journal*, Vol. 3, No. 12, 1961, pp. 39–43.

21. PAVLOVA, I. A., "New Electronic Measuring Instruments for Investigation of Refrigerating Compressors," *Journal of Refrigeration*, Vol. 3, No. 6, 1960, pp. 141–143.

22. PERRY, E. J., and P. D. LAING, "Positive Displacement Rotary Compressor as Applied to Refrigeration," *Journal of Refrigeration*, Vol. 4, No. 1, 1961, pp. 2–5; see also *Modern Refrigeration*, Vol. 63, No. 753, 1960, pp. 1258–9.

23. PICHEL, W., "Development of Large Capacity Lithium Bromide Absorption Refrigeration Machines in the U.S.S.R.," *ASHRAE Journal*, Vol. 8, No. 8, 1966, pp. 85–88.

24. RANQUE, G., "Experiments on Expansion in a Vortex with Simultaneous Exhaust of Hot Air and Cold Air," *Le Journal de Physique et le Radium*, Vol. 4, June, 1933, p. 1125.

25. SCATCHARD, et al., "Thermodynamic Properties of the Saturated Liquid and Vapor of Ammonia-Water Mixtures," *Refrigerating Engineering*, Vol. 53, No. 5, 1947.

26. SOUMERAI, H., "Large Screw Compressors for Refrigeration," *ASHRAE Journal*, Vol. 9, No. 3, 1967, pp. 38–46.

27. SPENCER, E., "Steam Vacuum Refrigeration," *ASHRAE Journal*, Vol. 3, No. 11, 1961, pp. 59–65.

28. STEINHAGEN, W. K., and D. C. UNGER, "Preliminary Design and Development of 1962 GM Automotive Air Conditioning Compressor," *SAE Journal*, Vol. 70, No. 3, 1962, pp. 77–78.

29. STEVENSON, F. F., "Energy for Air Conditioning: Gas or Electricity," *Heating, Piping and Air Conditioning*, Vol. 34, No. 6, 1962, pp. 135–140.

30. STEVENSON, F. F., "Gas for Air Conditioning: How to Estimate Operating Costs," *Heating, Piping and Air Conditioning*, Vol. 34, No. 7, 1962, pp. 135–140.

31. TABOR, H., "Use of Solar Energy for Cooling Purposes," *Solar Energy*, Vol. 6, No. 4, 1962.

32. WEIBEL, J., JR., and R. N. MANTEY, "The Engineering Development of a Compressor for Automotive Air Conditioning Systems," *General Motors Engineering Journal*, Vol. 10, No. 4, 1963, pp. 19–24.

33. WHITLOW, E. P., "Trends of Efficiencies in Absorption Refrigeration Machines," *ASHRAE Jl.*, Vol. 8, No. 12, 1966, pp. 44–48.

B. *Books*

1. AMERICAN SOCIETY OF HEATING, REFRIGERATING AND AIR CONDITIONING ENGINEERS, *Handbook of Fundamentals*. New York: ASHRAE, Inc., 1967.

2. AMERICAN SOCIETY OF HEATING, REFRIGERATING AND AIR CONDITIONING ENGINEERS, *Guide and Data Books: Applications*, 1966, *Fundamentals and Equipment*, 1965. New York: ASHRAE, Inc.

3. BOSNJAKOVIC, F., and P. L. BLACKSHEAR, JR., *Technical Thermodynamics*. New York: Holt, Rinehart and Winston, 1965.

4. BROWN, G. G., et al., *Unit Operations*. New York: Wiley, 1950.

5. HOUGEN, O. A., K. M. WATSON, and R. A. RAGATZ, *Chemical Process Principles, Part I*. New York: Wiley, 1954.

6. JORDAN, R. C. and G. B. PRIESTER, *Refrigeration and Air Conditioning* (second edition). Englewood Cliffs: Prentice-Hall, 1956.

7. KEMLER, E. N. and S. OGLESBY, *Heat Pump Applications*. New York: McGraw-Hill, 1950.

8. MACINTIRE, H. J., and F. W. HUTCHINSON, *Refrigeration Engineering*. New York: Wiley, 1950.

9. RABER, B. F. and F. W. HUTCHINSON, *Panel Heating and Cooling Analysis*. New York: Wiley, 1947.

10. SHARPE, N., *Refrigerating Principles and Practices*. New York: McGraw-Hill, 1949.

11. SPORN, P., E. R. AMBROSE, and T. BAUMEISTER, *Heat Pumps*. New York: Wiley, 1947.

12. THRELKELD, J. L., *Thermal Environmental Engineering*. Englewood Cliffs: Prentice-Hall, 1962.

CHAPTER 5

CRYOGENICS

5.1 INTRODUCTION

In the preceding chapter, the thermodynamics of systems producing refrigeration or "pumping heat" at what might be called conventional temperatures was discussed. In this chapter, systems for producing very low temperatures will be dealt with. These very low temperatures have, in recent years, been referred to as **cryogenic temperatures**, and the study of the means for producing such low temperatures and of the properties of substances at these temperatures has become known as **cryogenics**. There is no clear-cut boundary between "conventional temperatures" and "cryogenic temperatures"; it is likely that experience will change the engineer's limits on what he thinks of as commonly encountered temperatures. Scott suggests [B.13] that current usage would designate temperatures below −150 C as cryogenic. (To the nearest degree, this is 123 K, 222 R, or −238 F.) Sittig and Kidd suggest −100 C (−148 F) as the dividing line [B.14]. Either limit is below −109 F, the temperature of atmospheric sublimation of solid carbon dioxide (dry ice) discussed in Section 4.4.

In scientific laboratories in the last quarter of the nineteenth century and industrially in the early part of the twentieth century, two of the main goals in the production of very low temperatures were the liquefaction and the separation of the so-called "permanent" gases. These remain important aspects of cryogenics. In addition, the investigation and utilization of the unusual properties of various substances at extremely low temperatures have become increasingly relevant to engineering work.

Important as the industrial applications of cryogenics and the physical properties of substances at low temperatures are, they will not be discussed here

241

in detail except where they explain the reason for or the function of some thermodynamic operation. To be consistent with the aims of the book, this chapter will be primarily an examination of the means for producing cryogenic temperatures.

It will become obvious to the reader that extremely effective heat transfer, particularly in regenerative heat exchangers, is essential to the operation of most of the systems under investigation. Detailed examination of the heat-transfer process and of the fabrication of the exchangers for cryogenic systems will be left to other sources. These topics are dealt with in some detail by Scott, by Daunt, and by Collins [B.13, A.2, and A.3]. Similarly, the special attention given to design of moving parts, such as pistons and bearings, and the extreme care required to avoid heat leaks in system operation and in cryogenic fluid storage are merely mentioned here to emphasize the number of factors which must be considered by the designer in this field.

Throughout this chapter, temperatures will frequently be given in degrees Kelvin and pressures in atmospheres or in millimeters of mercury, simply because those are the units usually employed by the sources of physical-property data and the texts and papers discussing cryogenics. Although degrees Rankine and pounds per square foot or per square inch have been used more frequently in earlier chapters, the practice of working in these other units will be worth developing. Conversions are noted as a reminder here:

$$
\begin{aligned}
1 \text{ Kelvin degree} &= 1.8 \text{ Rankine degrees (exactly)} \\
0.0 \text{ Celsius (C)} &= 273.15 \text{ Kelvin (K)} \\
&= 32.00 \text{ Fahrenheit (F)} \\
&= 491.67 \text{ Rankine (R)} \\
0.0 \text{ Fahrenheit (F)} &= 459.67 \text{ Rankine (R)} \\
1 \text{ I. T. calorie/gram-K} &= 1 \text{ Btu/lb-R} \\
1 \text{ kilogram} &= 2.2046 \text{ lb (mass)} \\
1 \text{ Btu} &= 1055.04 \text{ abs. joule} \\
1 \text{ abs. joule/gram} &= 0.42993 \text{ Btu/lb} \\
1 \text{ watt hour/gram} &= 1 \text{ kilowatt hr/kilogram} \\
&= 3600 \text{ joule/gram} \\
&= 1547.74 \text{ Btu/lb} \\
1 \text{ standard atmosphere} &= 760 \text{ mm Hg} \\
&= 14.696 \text{ psi} \\
1 \text{ micron} &= 10^{-6} \text{ meters Hg} \\
&= 10^{-3} \text{ mm Hg} \\
&= 1.933 \times 10^{-5} \text{ psi} \\
1 \text{ lb/ft}^3 &= 0.016018 \text{ gram/cm}^3 \\
1 \text{ liter} &= 61.025 \text{ cu in} \\
&= 0.26418 \text{ gal (U. S.)}
\end{aligned}
$$

See also, the conversion factors in Appendix I.

Table 5.1

TEMPERATURE AND PRESSURE POINTS FOR SELECTED CRYOGENIC FLUIDS[a]

	Molecular weight	Triple point		Atm B. P.	Critical point		Max. inversion[c] temp
		K	Atm	K at 760 mm Hg	K	Atm	K
Oxygen (O_2)	32	54.363	0.00150	90.19	154.7_8	50.1	893
Argon (A)	39.944	83.78	0.6800	87.29	$150._{65}$	48.0	723
Nitrogen (N_2)	28.016	63.156	0.1237	77.395	126.1_{35}	33.49	621
Air[b]	28.96	—	—	78.8[e]	132.52(max) (Contact point) / 132.42 (Plait point)	37.17 / 37.25(max)	603
e-Hydrogen (H_2)[f]	2.016	13.83_3	0.0695	20.27_8	32.9_{94}	12.7_{70}	—
n-Hydrogen (H_2)[g]	2.016	13.95_7	0.07105	20.39_0	33.19	12.98	204.6
Helium	4.003	None		4.214[h]	$5.20^{(i)}$	$2.26^{(i)}$	$40^{(d)}$

a) All data from NBS Circular 564, 1955, [B. 7] except as noted. Figures beyond verifiable limits are lowered half a line, but are the most likely values.
b) Data for air from F. Din [B. 4].
c) Maximum inversion temperatures from J. G. Daunt, 1956 [A. 3] except for helium.
d) From R. B. Scott [B. 13]. Note also Fig. 5.7.
e) Atmospheric condensation point for air: 81.8K.
f) Equilibrium hydrogen: 0.21% *orthohydrogen*, 99.79% *parahydrogen* at 20.4K.
g) Normal hydrogen: 75% *orthohydrogen*, 25% *parahydrogen* [B. 13].
h) From NBS Technical Note 154, 1962 [A. 19].
i) From W. H. Keesom, 1942 [B. 9].

5.2 PROPERTIES OF AIR AND OTHER "GASES" AT CRYOGENIC TEMPERATURES

Various systems use what used to be called permanent gases as the working fluids for refrigeration cycles to produce extremely low temperatures or to accomplish the liquefaction of the gases themselves. To understand the systems, it will be necessary to become familiar with the thermodynamic properties of these fluids at cryogenic temperatures. One must realize that these fluids will be operating near or below their critical temperatures. Therefore their properties will be more nearly like those of the vapors studied in connection with vapor power cycles and vapor-refrigeration cycles than like perfect gas properties. Also, it will be seen that the utilization of the Joule-Thomson effect below the inversion point of this effect is an essential part of the thermodynamics of some of the systems to be described.

The triple-point temperature and pressure, the atmospheric boiling point, the critical temperature and pressure, and the maximum inversion temperature for the most important cryogenic fluids are presented in **Table 5.1**. Unique values of temperature and pressure for each point are given for all fluids except air, which is a mixture of several gases.

The seven figures from Fig. II.1.2 to Fig. II.1.8 inclusive in Appendix II present the thermodynamic properties of oxygen, nitrogen, air, parahydrogen, and helium on temperature-entropy coordinates down to very low temperatures. Parahydrogen is the major component of equilibrium hydrogen at low temperatures (see Section 5.2.2). The data are the most reliable to date according to the United States National Bureau of Standards, Cryogenic Data Center. These charts, along with enthalpy-entropy charts for parahydrogen and helium, are published also in Reference [B.1].

5.2.1 Air

Atmospheric air is not a single substance; dry air (without water vapor) is a mixture of approximately 78% nitrogen, 21% oxygen, and 1% argon on a mole basis, with traces of other substances (see also Chapter 6, Table 6.1). Because the boiling and critical points of argon are between those of oxygen and nitrogen, and because argon is chemically inactive, it is often sufficiently accurate to consider air as a binary mixture of 79% nitrogen and 21% oxygen on a mole basis, or 77% nitrogen and 23% oxygen on a mass or weight basis.

The existence of an atmospheric-pressure boiling point for air slightly different from the atmospheric-pressure condensation point or dew point is to be expected from knowledge of evaporation and condensation of mixtures in general (see Section 4.6.3). The existence of two "critical" points for air requires further explanation. For simplicity, air will be considered as a binary fluid. The characteristics of such a fluid for the region near critical temperature and critical pressure were not discussed previously, because they were not of interest for ordinary absorption-refrigeration systems. A temperature-concentration (T-x) diagram for nitrogen and oxygen is shown in **Fig. 5.1**. This is similar to the T-x diagram for ammonia and water, Fig. 4.22, but it is extended to the supercritical region in which both constituents must be vapor at all temperatures and pressures.

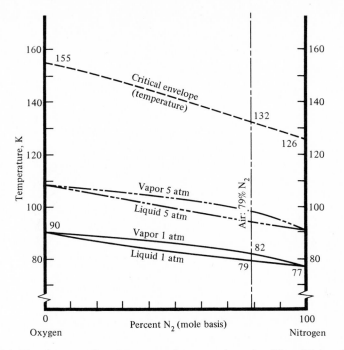

FIG. 5.1 T-x diagrams for nitrogen and oxygen (see also Figs. 5.14 and 5.15).

A small portion of a pressure-concentration (p-x) diagram is shown with a greatly enlarged scale in **Fig. 5.2**. Note that the saturated vapor line is below the saturated liquid line rather than above, as it was for the T-x diagram. A peculiarity of binary mixtures not mentioned before can be seen. For the constant temperature line T_1, the vapor line folds back to meet the liquid line at point A_1. At a slightly higher temperature, $T_1 + \Delta T$, the vapor and liquid lines meet at A_2. The temperature and pressure of A_1 are not the highest values at which vapor and liquid can exist in equilibrium for the concentration value of point A_1. Therefore, this cannot be considered a critical point with all that is implied by that term for a single substance.

If an envelope is drawn tangent to all the isotherms on p-x coordinates, the line will represent the maximum liquid-vapor equilibrium pressure for any concentration, but not the maximum concentration for the equilibrium temperature. A vertical line tangent to each isotherm will represent the maximum equilibrium concentration for each temperature. Thus two envelopes are developed. The vertical line for a particular concentration (say 79% N_2 and 21% O_2, mole basis) will be cut by the maximum-temperature line at a point called the **contact point**; this is slightly below the maximum pressure for liquid-vapor equilibrium at that concentration. The maximum-pressure line will cut the same concentration line at a point called the **plait point**; the temperature will be slightly below the temperature of the isotherm that would be tangent to the same concentration line at the

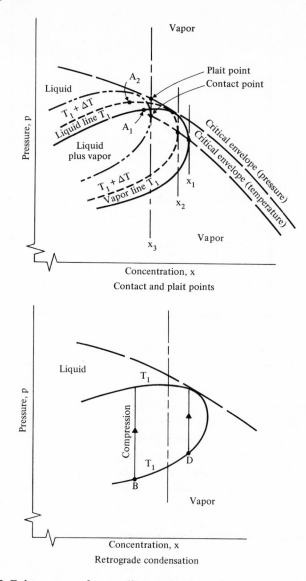

FIG. 5.2 Enlargement of a *p-x* diagram in the region of critical envelopes.

contact point. Table 5.1 shows that the differences between the contact point and the plait point are about 0.1 K degree and 0.1 atm [B.4].

The phenomenon known as **retrograde condensation** can also be explained with the aid of Fig. 5.2 [B.3]. At point *B* on the isotherm T_1, an increase of pressure at fixed concentration will create condensation as would be expected. At point *D*, which is at a concentration to the right of point A_1, compression will cause partial condensation initially, but will finally cause reevaporation if compression is continued to a point on the same isotherm where the vapor-equilibrium line is folded back.

These peculiarities of the properties of mixtures in the critical temperature and pressure range must be appreciated for establishment of exact values here, but are not important in general to the problems of liquefaction and separation to be discussed in the next section.

5.2.2 Hydrogen

Properties for two differently designated hydrogens are listed in Table 5.1. Hydrogen is unusual in that the diatomic molecule, composed of two protons and two electrons, may have both nuclei (the protons) spinning in the same direction (*ortho-***hydrogen**), or it may have the two nuclei spinning in opposite directions (*para-***hydrogen**). At normal ambient temperatures and at a pressure of one atmosphere, the equilibrium concentration of hydrogen gas is 75% *ortho*-hydrogen and 25% *para*-hydrogen. This composition is designated as **normal** or ***n*-hydrogen**. At any set of conditions, the equilibrium concentration is designated as ***e*-hydrogen**. Near the atmospheric-pressure boiling temperature of hydrogen, the equilibrium concentration is 0.21% *ortho*-hydrogen and 99.79% *para*-hydrogen [B.7]. **Deuterium**, a relatively rare isotope of hydrogen, at. wt. 2.0142, almost twice that of hydrogen, also is mostly *ortho* at normal room temperature and *para* at liquefaction temperature.

In the liquefaction of hydrogen, normal hydrogen may be brought very quickly to the extremely low temperatures of the liquid state. It is then possible to produce liquid with the *ortho/para* composition of normal hydrogen. The subsequent transition from predominantly *ortho-* to predominantly *para*-hydrogen is exothermic since the molecular energy level of *p*-hydrogen is lower. The energy of conversion is greater than the energy of vaporization, so that stored liquid *n*-hydrogen is difficult to keep. For this reason, catalysts are used during the liquefaction process to bring the *n*-hydrogen to a low-temperature equilibrium state before actual liquefaction and storage. The properties of the various forms of hydrogen and the conversion to equilibrium concentration are discussed in some detail by Scott [B.13] and by Barron [B.2].

5.2.3 Helium

A peculiarity of **helium**, noted in Table 5.1, is the absence of a solid–liquid–vapor triple point. Helium has been solidified under external pressure, but not by the reduction of its own vapor pressure above liquid. **Figure 5.3** shows a plot of the melting or fusion pressure of helium vs. temperature. The critical point is located for reference to show how far above critical pressure and below critical temperature the fusion conditions are found. These and other data from the extensive researches into the low-temperature properties of helium are reported in *Helium*, by W. H. Keesom. Some of Keesom's data and some later work are summarized by Scott and by Barron.

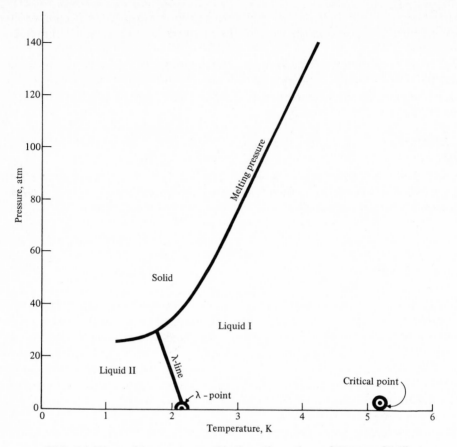

FIG. 5.3 The melting pressure of helium (from data of Keesom [B.9]).

Below a particular temperature (influenced by the pressure), liquid helium I undergoes a transformation to liquid helium II, which has markedly different properties. It is largely because of the extreme values of certain properties of liquid helium II that such a large amount of research has been carried out on this substance. The transition from I to II at a given pressure is at the λ (lambda) line shown in **Fig. 5.3**. At 0.50 atm and 2.18 K, the λ-line intersects the liquid-vapor equilibrium line at what is called the **λ-point**. (Except near the critical point, the liquid-vapor line is too close to the zero-pressure line to be shown on the same diagram, for which the pressure scale was chosen to show the melting pressure line.)

Some unusual properties will be mentioned here, but will not be treated in detail. The value of viscosity for liquid helium II, as determined by flow through a narrow passage, appears to be practically zero; as temperature is lowered, a container with extremely fine pores through which liquid I would not escape will suddenly allow the liquid to pass when the transition temperature is reached. On the other hand, the viscosity determined by a method such as torsion on a disc in rotating fluid is closer to the values found for liquid helium I. This "superfluidity" can cause a sudden "superleak" into a space that was previously at high vacuum. The extremely high rate at which liquid helium II will wet a surface, such as the walls of its container above the main liquid surface level, results in the phenomenon known as "creep." The wetting action overcomes gravity in the film and can continue in an open container until all the liquid has crept out.

The thermal conductivity of liquid helium II is another "super" property. While that for liquid helium I is nearly the same as for air at room temperature and pressure, the conductivity of liquid helium II is greater than that for copper. Furthermore, a temperature impulse will pass through the liquid as a sound wave passes through a compressible medium, leaving the region behind it unchanged. This heat transfer phenomenon has been referred to as "second sound." The electric properties of liquid helium are not so remarkable; the dielectric constant for liquid helium II is close to unity, which is the same order of magnitude as for liquid helium I.

5.2.4 The Joule-Thomson Effect

If an ideal gas were to expand from a higher to a lower pressure adiabatically and without doing external work, the internal energy and therefore the temperature would not change. For any real gas, such isenthalpic expansion does result in an increase or a decrease in temperature except at certain conditions. The ratio of temperature change to pressure change at constant enthalpy is called the **Joule-Thomson** or **Joule-Kelvin coefficient** (μ), which is defined as

$$\mu = \left(\frac{\partial T}{\partial p}\right)_h .$$

A positive value indicates a temperature fall on expansion.

The cooling effect as a gas is throttled is familiar, because μ is positive for almost all gases at ordinary temperatures and pressures; only hydrogen, helium, and neon exhibit a heating effect with throttling (negative μ) at normal ambient conditions. Nevertheless, all gases at sufficiently high temperature have negative values of the Joule-Thomson coefficient. This is shown by the plot of constant enthalpy lines on *T-s* coordinates for a typical gas in **Fig. 5.4**. The temperature and pressure at which μ changes from positive to negative, and is therefore zero, is the **inversion temperature**. As the typical plot shows, there is a maximum inversion temperature for each gas, and above that temperature μ is always negative producing a temperature rise with an isenthalpic pressure drop. That maximum inversion temperature is included in Table 5.1 for the gases listed.

Some insight into the Joule-Thomson effect is obtained from an investigation of thermodynamic relationships for a constant enthalpy operation. Various authors take different approaches [B.11, B.15, and B.18]. A functional relationship

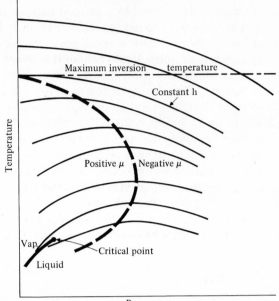

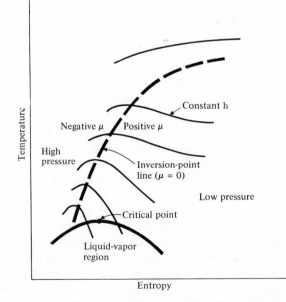

FIG. 5.4 The Joule-Thomson effect for a typical gas.

for temperature, pressure, entropy, and enthalpy is given by the following thermodynamic equation:

$$dh = T\,ds + v\,dp.$$

Because entropy (s) is a function of temperature (T) and pressure (p),

$$ds = \left(\frac{\partial s}{\partial T}\right)_p dT + \left(\frac{\partial s}{\partial p}\right)_T dp.$$

Then, for constant enthalpy,

$$T\left(\frac{\partial s}{\partial T}\right)_p dT + \left[T\left(\frac{\partial s}{\partial p}\right)_T + v\right] dp = 0.$$

Now

$$T\left(\frac{\partial s}{\partial T}\right)_p = \left(\frac{\partial h}{\partial T}\right)_p = c_p.$$

Therefore the Joule-Thomson coefficient μ is

$$\left(\frac{\partial T}{\partial p}\right)_h = -\frac{1}{c_p}\left[T\left(\frac{\partial s}{\partial p}\right)_T + v\right]. \tag{5.1}$$

To eliminate entropy from Eq. (5.1), the Gibbs function (g) can be used to provide an exact differential (dg). By definition,

$$g = h - Ts$$
$$dg = T\,ds + v\,dp - T\,ds - s\,dT = v\,dp - s\,dT.$$

Because this is an exact differential, we find that

$$\left(\frac{\partial v}{\partial T}\right)_p = -\left(\frac{\partial s}{\partial p}\right)_T.$$

(Note that this is one of the Maxwell thermodynamic relationships.) Substituting, we have

$$\mu = \frac{1}{c_p}\left[T\left(\frac{\partial v}{\partial T}\right)_p - v\right]. \tag{5.2}$$

This form of the equation for μ is useful in that it shows that μ must be zero for a gas following the perfect gas law $pv = RT$, from which

$$\left(\frac{\partial v}{\partial T}\right)_p = \frac{R}{p} = \frac{v}{T}.$$

While Eq. (5.2) shows that a gas exhibiting a value of μ other than zero cannot be a perfect gas, there is no indication that a zero value of the Joule-Thomson coefficient is sufficient to identify the gas as "perfect." The curves of Fig. 5.4 show that real gases can have zero μ at certain conditions. Equation (5.2) is not a completely satisfactory form, because it does not show clearly the two components to be found in the Joule-Thomson effect. An alternative expression for μ can be derived from the following relationship:

$$dh = T\,ds + v\,dp = T\Big(\frac{\partial s}{\partial T}\Big)_p dT + T\Big(\frac{\partial s}{\partial p}\Big)_T dp + v\,dp.$$

Thus from the partial derivative of h with respect to p at constant temperature, we have

$$\Big(\frac{\partial h}{\partial p}\Big)_T = T\Big(\frac{\partial s}{\partial p}\Big)_T + v.$$

This can be substituted in Eq. (5.1):

$$\mu = -\frac{1}{c_p}\Big(\frac{\partial h}{\partial p}\Big)_T.$$

But again,

$$h = u + pv,$$

so that

$$\mu = \frac{1}{c_p}\Big[-\Big(\frac{\partial u}{\partial p}\Big)_T - \Big(\frac{\partial pv}{\partial p}\Big)_T\Big]. \tag{5.3}$$

The first component within the square bracket indicates the deviation from **Joule's law** which states that the internal energy (u) is constant at constant temperature for a perfect gas. On expansion, the separation of the molecules of a real gas increases the potential energy of their mutual forces (decreasing the magnitude of their mutual attraction) so that $(\partial u/\partial p)_T$ is negative. In other words, this must contribute to a temperature drop with decreasing pressure at constant enthalpy, as indicated by a positive value of μ.

The second term indicates the departure of the gas from **Boyle's law** which states that volume varies inversely with pressure at constant temperature. For a real gas, the product of pressure and volume may increase or decrease along an isotherm. As a general rule, $(\partial pv/\partial p)_T$ is negative at low temperatures and pressures, so that the second term adds to the cooling effect of the first term on throttling. However, it becomes positive at higher temperatures and pressures.

When the two terms in the square bracket of Eq. (5.3) are equal but opposite in sign, the **inversion temperature** is reached. When the deviation from Boyle's law is of opposite sign and outweighs the deviation from Joule's law, throttling produces a heating effect.

As mentioned at the beginning of this discussion of the Joule-Thomson effect, only hydrogen, helium, and neon have maximum inversion temperatures below normal ambient temperature. Therefore, if the Joule-Thomson effect is to be used for cooling with these gases, precooling is necessary. Table 5.1 shows that this would not be difficult for the maximum inversion temperature of hydrogen, but for helium, that temperature is below the atmospheric boiling point of nitrogen.

The inversion temperature for a particular constant enthalpy line is not difficult to locate approximately on the T-s diagram for the gases, Figs. II.1.2 through II.1.8. Of this group, only Fig. II.1.6 for hydrogen and Fig. II.1.8 for helium show isenthalps above a maximum inversion temperature.

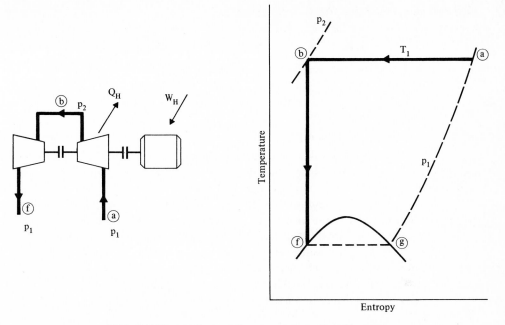

FIG. 5.5 Theoretical minimum work to liquefy a gas.

5.3 LIQUEFACTION OF GASES

Since the time of early cryogenic laboratories, many successful systems have been built to liquefy gases. Only a few will be presented here schematically in an attempt to show the basic differences. Details are to be found in references [A.2, A.3, B.13] and in *Experimental Cryophysics* [B. 8], which has a particularly interesting review of some historical developments.

The low temperatures necessary for liquefaction of gases are achieved in four basically different ways: (a) by external cooling by means of a separate refrigeration system, (b) by the Joule-Thomson effect in the throttling of the gas to be liquefied, (c) by the nonflow isentropic expansion of the gas, and (d) by the expansion of the gas in an engine or turbine. Examples will be given for each.

5.3.1 Minimum Work to Liquefy a Gas

The work required to change a gas from conditions p_1 and T_1 to a liquid at the same pressure should be the net work of reversible processes connecting the two state points. **Figure 5.5** gives a hypothetical system which, though impractical, allows a simple calculation of the minimum work required.

The gas at **a** is compressed isothermally to point **b** at p_2. The work of compression must be

$$w_c = (h_b - h_a) + q_H \text{ Btu/lb.}$$

In isentropic expansion from b to saturated liquid at f, which is also at p_1, the work recovered would theoretically be

$$w_E = (h_b - h_f) \text{ Btu/lb.}$$

The heat removed at the top temperature on isothermal compression is

$$q_H = T_1(s_a - s_b) = T_1(s_a - s_f).$$

Thus the net work required is

$$w_N = T_1(s_a - s_f) - (h_a - h_f) \text{ Btu/lb.} \tag{5.4}$$

The enthalpy difference term is represented by the area under the constant-pressure line from f to a. Inspection of the T-s diagram shows that the net work is represented by the enclosed area *abfga*.

Any one of the T-s diagrams for the gases already discussed shows that p_2 would be impractically high. However, the conditions at b do not appear in the final equation for net work. Although the system of Fig. 5.5 is impractical because of the very large pressure range required, it is a theoretical reversible cycle between points a and f, and therefore represents an ideal against which other systems may be compared. It is interesting to note that Eq. (5.4) can be derived by an application of the principle of increase of entropy. See, e.g., Reference [B.19], p. 164.

5.3.2 Separate Refrigeration System to Cool Gas Below Its Critical Point

With reference to Fig. 5.5, a gas at ambient conditions a could be liquefied if it were cooled at constant pressure p_1 to the liquid saturation line at f. This can be accomplished, except for hydrogen and helium liquefaction, by means of a separate refrigeration system that does not use the gas to be liquefied as part of the working fluid.

One refrigeration system that is being used for the liquefaction of air is the **Stirling-cycle** machine, which will be described in detail in Section 5.5.2. The working fluid for the cycle is helium gas for which the critical temperature is well below the atmospheric condensation temperatures of oxygen and nitrogen. Because of extremely efficient regeneration within the cycle, the temperature range from ambient (about 300 K) to less than 80 K can be spanned with a compression ratio of only about 2:1.

A more complex system, used for many years before being replaced by the Stirling-cycle system and more direct methods, is **multiple-fluid cascade refrigeration**. This was described in Section 4.2.5, where it was noted that cascading is useful where unusually large temperature ranges are encountered. A system for the liquefaction of nitrogen is shown in **Fig. 5.6**. The four working fluids are ammonia (NH_3), ethylene (C_2H_4), methane (CH_4), and nitrogen (N_2) itself. These substances were chosen so that each could evaporate at normal atmospheric pressure. Their respective atmospheric boiling points are approximately 240 K (NH_3), 169 K (C_2H_4), 111 K (CH_4), and 77 K (N_2).

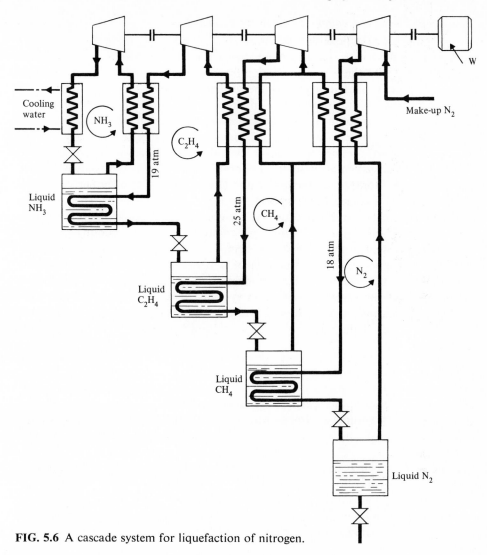

FIG. 5.6 A cascade system for liquefaction of nitrogen.

The efficiency has been greatly improved compared to the simpler cascade system of Fig. 4.10 by the use of the three additional counter-flow heat exchangers. The coefficient of performance of the whole system is good compared with other methods of liquefaction. The power consumption is reported [A.3] as only 0.54 kwh/kg of liquid nitrogen (about 0.7 kwh/liter) which is only 2.5 times the theoretical minimum. However, the arrangement is much more cumbersome than later systems.

5.3.3 *Joule-Thomson Cooling of Gas to Be Liquefied*

Because all gases at sufficiently low temperatures exhibit a positive Joule-Thomson coefficient (cooling with isenthalpic pressure drop), it is possible to liquefy a gas by throttling it into the wet vapor region from a pressure and temperature above its critical point. The feasibility of accomplishing this, at least with a simple system, does not depend on the Joule-Thomson effect at the throttling conditions alone; the values of enthalpy of the gas before and after compression at the upper tempera- ture of the system are also significant. External cooling may be necessary to bring the gas to a temperature below the inversion value for the pressures involved.

The simplest system utilizing the Joule-Thomson effect is that shown schema- tically in **Fig. 5.7** for the liquefaction of air. This is referred to as either the **Hamp- son** system or as a **simple Linde** system, since it was developed independently by both workers in the same year. Ideally, the compression would be isothermal as shown on the *T-s* diagram; a two-stage compressor with intercooling and after- cooling is indicated in the schematic system. A highly efficient counter-flow heat exchanger and good thermal insulation of that exchanger and of the expansion chamber below are essential to the attainment of liquefaction.

In this and the several liquefaction systems to be described, there are two important indications of effectiveness and efficiency. The first is the **yield** (Y) of the system, which is the ratio of the mass of liquid produced to the mass of gas compressed,

$$Y = \frac{\dot{m}_f}{\dot{m}_c}.$$

The second criterion is the **specific work consumption** (W_Y), or the energy required per unit mass or volume of liquid produced. This is expressed in various units, such as Btu per pound or kilowatt-hours per kilogram or kilowatt-hours per liter.

The system below points c and i can be referred to as a **Joule-Thomson refrigeration system** [A.4]. It will be found as the critical component of some other, more complicated arrangements for gas liquefaction.

The theoretical yield, assuming perfect insulation, can be determined by a mass and energy balance on the system below points c and i:

$$\dot{m}_c = \dot{m}_f + \dot{m}_i; \qquad \dot{m}_c h_c = \dot{m}_f h_f + \dot{m}_i h_i.$$

Combining these equations to eliminate $\dot{m}_i$, we have

$$\dot{m}_c(h_i - h_c) = \dot{m}_f(h_i - h_f).$$

The yield is, by definition,

$$Y = \frac{\dot{m}_f}{\dot{m}_c} = \frac{h_i - h_c}{h_i - h_f}. \tag{5.5}$$

The first point of interest in Eq. (5.5) is that no yield is possible unless h_i is greater than h_c, and yet T_i can only approach T_c as an upper limit for a perfect heat exchanger. This explains the necessity for a positive Joule-Thomson effect in

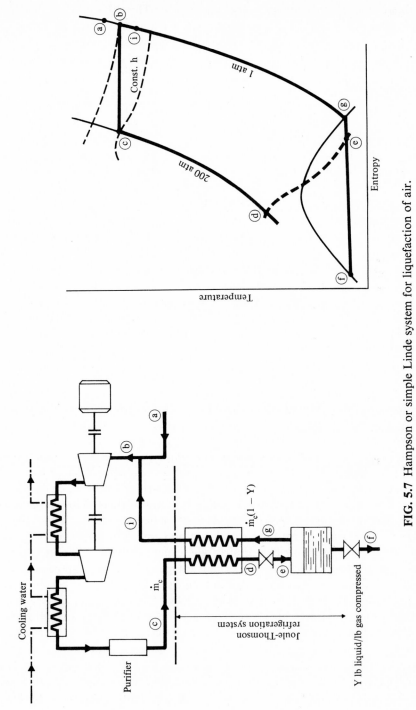

FIG. 5.7 Hampson or simple Linde system for liquefaction of air.

the ambient temperature range for this simple cycle. The value of h_i can be greater than the value of h_c, while T_i is less than T_c, only when p_i is below the inversion pressure for this temperature range. The maximum yield would occur if point c were at the inversion-point line. (See Fig. 5.4.) Compression beyond the inversion pressure would not only increase the necessary work of compression, but would decrease the numerator of Eq. (5.5).

The second point of interest in Eq. (5.5) is that conditions at d and e do not appear. For equilibrium operation, the enthalpy value h_d (which is equal to h_e) is established by the thermodynamic properties of the fluid and by the mass and energy balances already considered; once the system is in operation, T_d will reduce further and further until the equilibrium yield is established, so long as the system is operating in the region of positive Joule-Thomson coefficient.

The value of h_d can be determined from a mass and energy balance on the counter-flow heat exchanger:

$$\dot{m}_c h_c + \dot{m}_c(1 - Y)h_g = \dot{m}_c h_d + \dot{m}_c(1 - Y)h_i,$$

$$\frac{h_c - h_d}{h_i - h_g} = (1 - Y). \tag{5.6}$$

The work of an isothermal compressor was given in Chapter 3, Eq. (3.10a), which can be combined with Eq. (5.5) to determine the ideal specific work consumption. Alternatively, the development of Eq. (3.8a) can be used, from which, for isothermal compression, we have

$$w_C = T_c(s_b - s_c) - (h_b - h_c). \tag{5.7}$$

The combination of Eq. (5.7) with Eq. (5.5) allows us to calculate the ideal specific work consumption entirely from tabulated or graph values of temperature, enthalpy, and entropy. For instance, for air liquefaction, reasonable accuracy for property values can be obtained from the T-s diagram, Fig. II.1.4.

Both the yield and the specific work consumption of the Hampson or simple Linde cycle can be improved in two ways: (1) precooling of the high-pressure stream by a separate refrigeration system, and (2) throttling to an intermediate pressure, from which only a fraction of the fluid is throttled again to the lowest pressure. The advantage of precooling is that part of the refrigeration effect is accomplished with a conventional vapor-compression cycle which usually has a higher coefficient of performance than a gas cycle for reasons discussed at the beginning of Section 2, Chapter 4. The advantage of partial evaporation at an intermediate pressure results from the fact that a substantial part of the cooling of the high-pressure supply stream is accomplished by means of a gas refrigeration system operating over a pressure range substantially lower than the overall pressure range of the whole liquefaction system; the reduction in pressure and temperature range implies a higher potential coefficient of performance. Either refinement alone can be added to the simple system of Fig. 5.7. A **dual-pressure Linde system** for air liquefaction is shown in **Fig. 5.8**, including the optional feature of precooling by means of a separate ammonia vapor-compression system; state points are shown

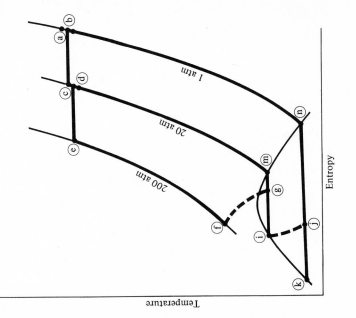

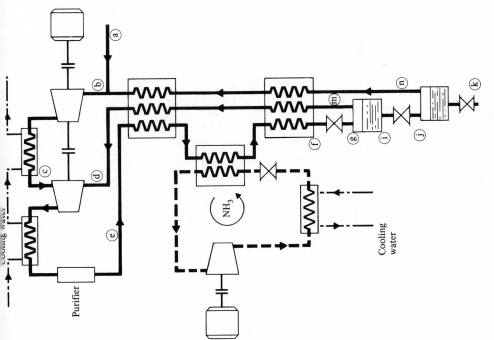

FIG. 5.8 Dual-pressure Linde system for air liquefaction with NH$_3$ precooling.

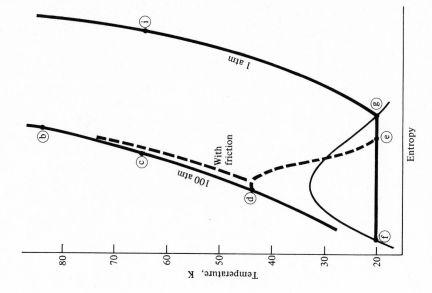

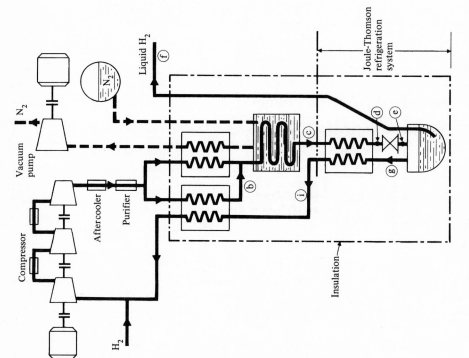

FIG. 5.9 National Bureau of Standards hydrogen liquefier. Precooling with liquid nitrogen.

on the accompanying idealized *T-s* diagram. It must be remembered that the mass rate of flow is not the same at all points. The intermediate pressure must be somewhat below the critical pressure for air (37 atm) so that liquefaction will occur in the intermediate-pressure chamber. The use of either the dual-pressure cycle or the ammonia system precooling has been found to decrease the specific work requirement by nearly half compared to the simple cycle of Fig. 5.7. Use of both additions together, as shown in Fig. 5.8, reduces the specific work by nearly two thirds. The yield of a simple system can be about 8%. It is increased to 20% by precooling, but it is reduced to 7% by the dual cycle without precooling. The combination gives a yield of about 17% [B.8].

To liquefy either hydrogen or helium by means of the Joule-Thomson effect of the gas itself requires that the gas be precooled to a temperature below the inversion point for the pressure from which throttling is to take place. **Figure 5.9** is a schematic representation of the United States **National Bureau of Standards** hydrogen liquefier. In it, the evaporation of liquid nitrogen at a partial vacuum is used to precool the hydrogen high-pressure stream. Approximate temperature and pressure values are taken from Scott. Essentially the same design is used for helium liquefaction, in which liquid hydrogen is substituted for the precooling medium. The system below points *c* and *i* is identical to the Joule-Thomson refrigeration system noted in Fig. 5.7. Therefore the yield (Y) can be calculated from mass and energy balances identical in form to those used in the development of Eq. (5.5), and the same equation can be used for the system of Fig. 5.9. The idealized yield would be computed with the assumption that T_i is equal to T_c, which is the limiting case. The constant pressure line *bcd* of Fig. 5.9 is idealized. It does not allow for the pressure drop between points *b* and *d* because of fluid friction. Point *d* is therefore moved to the right to coincide approximately with the inversion point of the Joule-Thomson effect for that temperature.

5.3.4 Nonflow Isentropic Expansion of Gas to be Liquefied

When the gas or vapor in a container is allowed to escape through a throttling valve, the flowing gas and the remaining gas undergo two entirely different processes. If zero heat transfer is assumed, the gas flowing through the valve is expanded to a final state at which the enthalpy is the same as it was before expansion; the gas remaining in the container does work of expansion against the progressively decreasing container pressure, and expands practically isentropically, since the nonflow process involves very little fluid friction.

This process of achieving helium liquefaction is used in the **Simon** nonflow or batch liquefier illustrated in **Fig. 5.10**. Chamber *D* contains liquid nitrogen or liquid hydrogen. Chamber *B* contains hydrogen vapor, and chamber *C* initially contains helium vapor to aid heat transfer from *A* to the liquid of chamber *D*. (Provision for filling *B* is not shown.) The helium vapor to be liquefied is admitted through valve V_1 to chamber *A* at a pressure of about 150 atm. During admission it is precooled in a coil immersed in the liquid bath of container *D*. Before the final expansion of the vapor in *A*, chamber *C* is evacuated to provide thermal insulation, and the

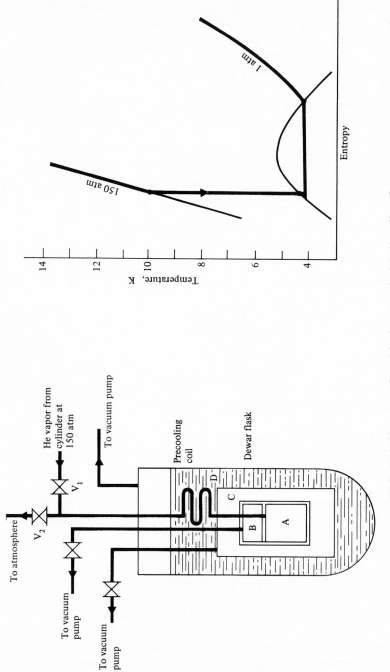

FIG. 5.10 Simon nonflow (single expansion) helium liquefier.

pressure of the hydrogen vapor in *B* is reduced to a very low value so that the hydrogen will solidify at about 10 K. From this temperature and about 150 atm, the helium in *A* is allowed to expand through valve V_2 to about atmospheric pressure. The nearly isentropic expansion results in liquefaction as shown on the *T-s* diagram.

The fraction of the volume of *A* that is finally filled with liquid is larger than might be expected. The density of the vapor at 10 K and 150 atm is more than one and one-half times the density of the liquid formed. One additional cryogenic property is fortuitous: although the high pressure requires that *A* be a thick-walled metal vessel, the specific heat of the metal at such low temperatures is so low that the heat capacity of the vessel is almost negligible. The basic design shown schematically in Fig. 5.10 has undergone numerous variations. Details for draining the liquid to other chambers or other apparatus for experimental investigation are given in Reference [B.8], p. 101.

5.3.5 Engine or Turbine Expansion of Gas to Be Liquefied

The nonflow isentropic expansion of a gas for liquefaction can produce only a single, relatively small batch at one time, and the whole process must be repeated for more liquid to be formed. The continuous liquefaction of some fraction of a stream of gas by isentropic expansion requires the use of an engine or turbine to do external work during the expansion process.

The earliest successful liquefaction system using an expansion engine was devised by **Claude** and is illustrated in **Fig. 5.11**. The temperatures and pressures noted are approximately the values used. Expansion in the engine from some point at the higher pressure could produce liquid directly, but this would introduce mechanical difficulties. The major purpose of the engine is to produce effective cooling in the upper heat exchangers; the resultant work may be worth recovering or may simply be dissipated through a brake. Again, it is apparent that below points *c* and *i*, the system is the Joule-Thomson refrigeration system noted in Fig. 5.7 for the Hampson or simple Linde system. The effective cooling of isentropic expansion has made it possible to precool the high-pressure supply stream with a lower overall compression ratio.

In a system similar to that devised by Claude, **Heyland** increased the upper pressure to about 200 atm, and was thereby able to eliminate the upper heat exchanger. Points *b* and *j* were then near ambient temperature. The expansion engine replaced the external cooling used in some Linde systems to increase yield and decrease specific power.

There are two important differences between the Claude system of air liquefaction and that of **Kapitza**, shown in **Fig. 5.12**. First, the expansion in the Kapitza system takes place in a turbine rather than in a reciprocating machine. Second, the heat transfer between high-pressure and low-pressure streams is accomplished in packed-bed regenerators rather than in counter-flow heat exchangers. The efficiency of the regeneration allows cooling to a very low temperature and operation with a

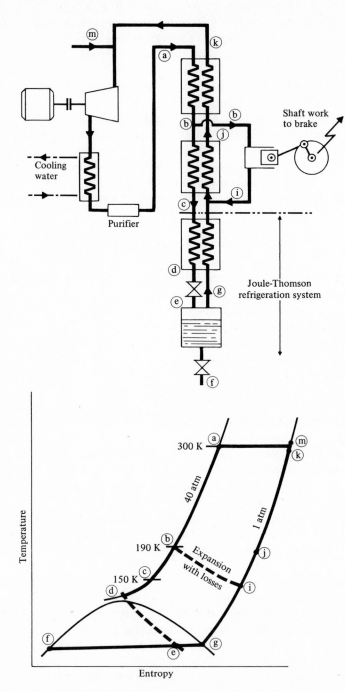

FIG. 5.11 Claude system for air liquefaction.

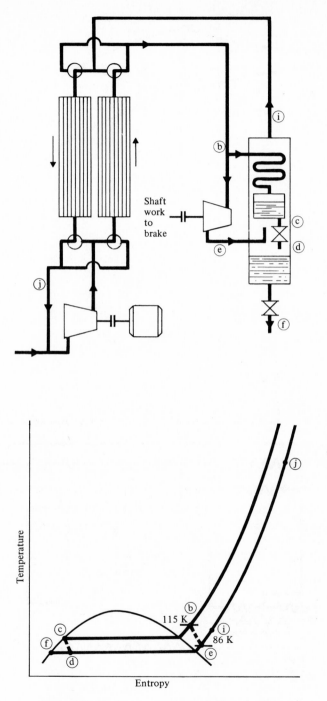

FIG. 5.12 Kapitza system for air liquefaction.

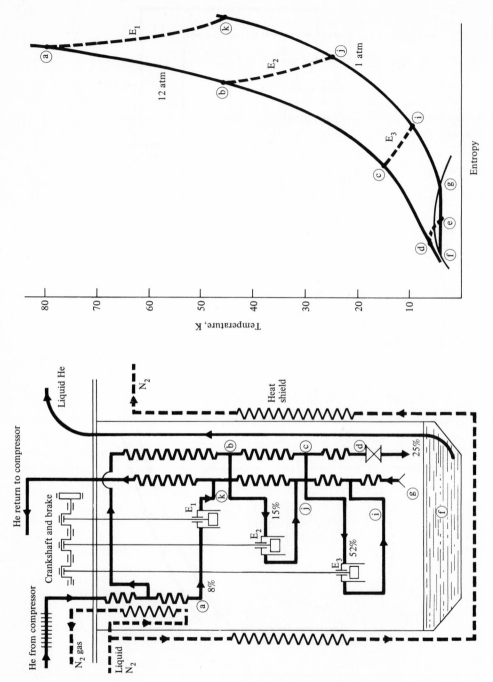

FIG. 5.13 Collins helium liquefier.

smaller pressure ratio. Two regenerator columns are required to allow heating of the return stream in one and cooling of the supply stream in the other until the packing of one is cooled and that of the other is heated. The valves (shown here schematically) are then reversed. The reversal is necessary for continuous heat transfer, and it also aids in clearing the colder regions of condensed or solidified impurities.

The **Collins** helium liquefier shown in **Fig. 5.13** has made an important contribution in allowing relatively small laboratories to produce their own liquid helium continuously rather than in batches. It employs three reciprocating expansion engines operating through three different temperature ranges, as noted. All three, through long tension connecting rods, are linked to the same crankshaft. The work of expansion is dissipated externally by a brake. The schematic diagram was adapted (and somewhat simplified) from Collins [A.2]. Theoretically, precooling is not an absolute necessity for attainment of liquefaction, because isentropic expansion rather than the Joule-Thomson effect is used. However, as in other systems, precooling does increase the yield.

5.4 SEPARATION OF GASES

Multicomponent gases such as air can be separated into their individual constituents by selective absorption and other chemical processes. However, the most important commercial methods depend on refrigeration to allow the components to vaporize and condense at varying equilibrium conditions in a separating column or tower. The processes depend on the same principles discussed in Chapter 4 in connection with the separation of ammonia from water for an absorption-refrigeration system. When the fluids to be separated are considered to be gases at normal temperatures and pressures, the refrigeration temperatures must be in the cryogenic range.

It will be recalled from the discussion in Section 4.6 that any desired degree of purity can be obtained in the separation process if heat addition in the generator and cooling in the dephlegmator, or reflux cooler, and in the rectifying column are properly adjusted. Full control is possible only with a relatively complex system. In the following paragraphs, two systems for air separation will be described. The first will be a simpler arrangement which will provide pure oxygen, but not oxygen-free nitrogen. The second will be more complex to make it theoretically possible to obtain complete separation of the two principal constituents.

For simplicity of presentation, it will be assumed that air is a binary mixture of nitrogen and oxygen. Through the addition of other heat exchange apparatus, the other constituents (principally argon) can be isolated as well. In fact, the principal source of argon and some of the more rare constituents of normal air is through liquefaction and separation procedures similar to those which will be illustrated, but with additional complexities. Some commercial systems employ expansion engines to provide part of the precooling of the supply gas, but those are a refinement of, or a variation in the second system to be presented here.

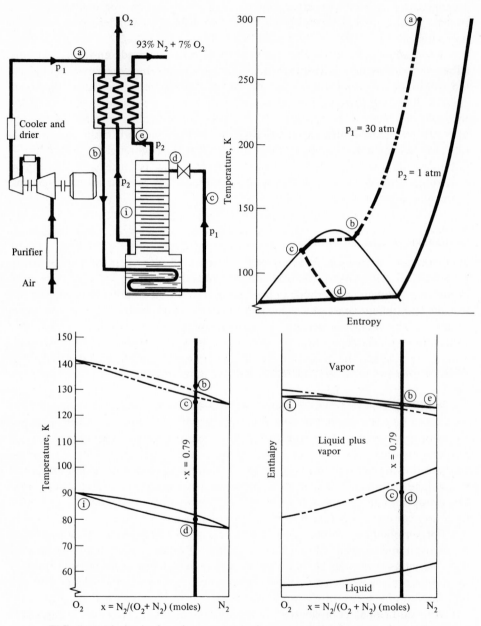

FIG. 5.14 Linde single-column process for separation of oxygen from air.

5.4.1 *Simple Single-Column Linde Air-Separation System*

An early system that is used for air separation, credited to Linde, is shown schematically in **Fig. 5.14**. The state points are shown on the accompanying temperature-entropy, temperature-concentration, and enthalpy-concentration diagrams. The concentration scale is expressed in moles of nitrogen per mole of mixture, and a binary mixture of concentration 0.79 is assumed for the entering air. Some references present concentration as a mass fraction which would give 0.77 lb nitrogen per pound of mixture as mentioned in Section 5.2.1.

As shown on the diagram, the air is purified and dried before delivery to the heat exchangers so that impurities, particularly water and carbon dioxide, will not condense or freeze, blocking passageways. After being precooled in the counter-flow heat exchanger, the compressed air is condensed in coils immersed in the liquid sump at the bottom of the column. The high pressure liquid at *c* is throttled to about atmospheric pressure at *d*. The temperature of the resulting mixture of liquid and vapor is lower than at *c*, but the enthalpy is the same. The isotherm in the wet vapor region on the *T-x* diagram shows that the vapor is closer to pure nitrogen, but still contains some oxygen. The liquid flowing downward through the baffles of the vertical column exchanges heat with rising vapor, preferentially vaporizing nitrogen (atmospheric boiling point 77 K) and condensing oxygen (atmospheric boiling point 90 K). Thus the vapor leaving at *e* is somewhat richer in nitrogen than the equilibrium vapor for point *d*. The liquid and the vapor in equilibrium at the bottom of a column will be pure oxygen if a sufficient number of stages or baffle trays is provided. An attempt to increase the yield of a given system by variation of the flow rates will usually decrease the purity of the product. As a general rule, high yield and high oxygen purity are incompatible. Because of the separation, state points beyond *d* cannot be represented on the *T-s* diagram for "air."

The equilibrium states shown are attained only after some time. State point *b* is important in determining states *e* and *i*. If perfect insulation of the column is assumed, the separation of stream *b* into the two streams *e* and *i* is the reverse of adiabatic mixing of two streams as discussed in Section 4.6. The same mass and energy equations govern, and it is found that points *e*, *b*, and *i* must be in a straight line on the *h-x* diagram. Close inspection of the constant enthalpy lines on the *T-s* diagram for air, Fig. II.1.4, will show that the enthalpy of saturated vapor at 30 atm is less than the enthalpy of saturated vapor at 1 atm. Thus it is possible for point *b* to be in the vapor region and still be on the straight line joining points *e* and *i*. Note the crossover of the saturation lines for 30 atm and for 1 atm on the enthalpy-concentration diagram. The enthalpy of vaporization for nitrogen is very small at 30 atm, because that pressure is close to the critical pressure for that gas.

The system just described can be compared with Fig. 4.27, in which the separation of ammonia from water was accomplished in a generator, rectifying column,

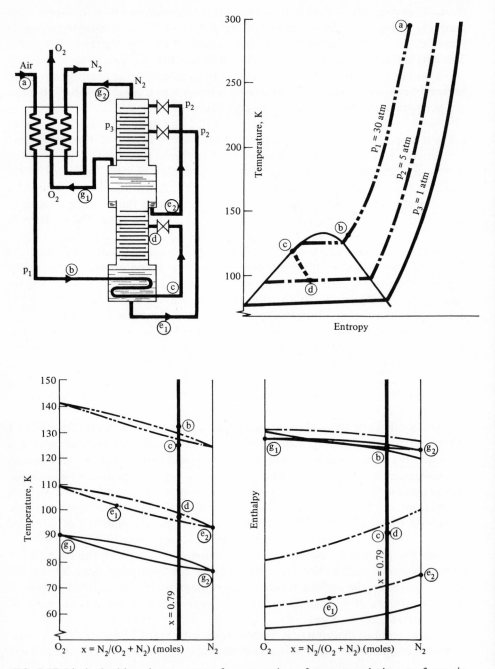

FIG. 5.15 Linde double-column process for separation of oxygen and nitrogen from air.

and dephlegmator. Cooling with ambient-temperature water or air is, of course, impossible for air separation. Furthermore, heat leakage to the system must be kept to a minimum. The generator heat source for air separation is therefore the supply air stream which must be cooled and condensed. The dephlegmator cooling must be accomplished entirely by counterflow of the liquid and vapor within the column. Because the two pressure levels of operation make the transfer of heat possible, the whole system can be thought of as an example of a heat pump as suggested by Bosnjakovic [B.3]. See Section 4.8.5 in the previous chapter.

5.4.2 Double Column Linde Air-Separation System

More than the simple system just described is necessary to obtain arbitrary purity in the nitrogen and oxygen yields. It must be possible to adjust the condition at which fluid is fed to the top of the column. This is accomplished in the double column of **Fig. 5.15**. The supply pressure, intermediate pressure (lower column), and final pressure (upper column) can all be adjusted so that vaporization of the liquid oxygen in the upper column results in the condensation of practically pure nitrogen at the top of the lower column. (The heat-transfer surface between the two is extended by vertical tubes not shown.) Note that the temperature at point g_1 is lower than the temperature at point e_2, as shown on the temperature-concentration diagram.

While the system of Fig. 5.15 is more complicated than that of Fig. 5.14, its explanation follows logically from that of the simpler system. Note again that point b must fall on a straight line joining points g_1 and g_2 on the h-x diagram, because the single supply stream and the two leaving streams ideally represent an example of adiabatic separation. Again also, it is noted that points beyond d cannot be represented on the T-s diagram for "air."

5.5 VARIOUS GAS CYCLES

Some gas cycles used for refrigeration were examined in Chapter 4, Section 4.7. It was noted then that some of these cycles could be used to produce very low temperatures, and that they would be introduced again in this chapter. Other cycles, used almost exclusively for very low-temperature work, will be presented here for the first time.

5.5.1 The Brayton Cycle with Regeneration

The simple Brayton cycle, without regeneration, was described in Section 4.7.1. As either an open or a closed cycle, it could be used to produce refrigeration in the conventional temperature range. In Section 4.7.3, it was shown that the Brayton cycle with regenerative heat transfer could pump heat from temperatures in the cryogenic range, even with relatively low values of pressure ratio. As a closed system, an improved COP was possible with monatomic gases, such as helium, which have higher values of specific heat ratio (γ).

There is no need to repeat the thermodynamic analysis. The cycle is mentioned again here because of its importance in cryogenic refrigeration. Note, however, that the attainable temperatures are not as low as in some other systems; there is no change of phase of the working fluid.

5.5.2 The Stirling Refrigeration Cycle

Although the Stirling power cycle was introduced in Chapter 1 and was described and analyzed in some detail in Section 2.8 of Chapter 3, the thermodynamic reversal of this cycle for the purpose of refrigeration was only briefly mentioned in Chapter 4, Section 7.4. It will be examined more closely here because its primary application in heat pumping is to produce cryogenic temperatures.

The basic idealized cycle is shown in **Fig. 5.16**. This is identical to the idealized power cycle of Fig. 3.27, but it is reversed to achieve refrigeration. The thermal efficiency for the power cycle was shown to be the same as that of a Carnot cycle between the same temperature limits in Chapter 1, Section 3.4. A different approach will be taken here to derive the expression for the coefficient of performance of the reversed cycle, although a corollary of the second law of thermodynamics tells us that it must be the same as for the Carnot cycle.

While there is heat exchange within the system from one constant-volume line to the other, all heat transfer from the surroundings takes place along the isothermal line *ab* and all heat is rejected to the surroundings along the isothermal line *cd*. The heat quantities can be determined indirectly from the work of isothermal expansion or compression because, with no change in internal energy, the heat transfer is equal to the work done. From the first law of thermodynamics, we have

$$Q = \Delta U + W = W.$$

For a perfect gas, we have

$$pv = RT.$$

The work of isothermal expansion from ① to ② is

$$w_t = \int p \, dv$$

$$= \int RT \, (dv/v)$$

$$= RT \ln (v_2/v_1) \tag{5.8a}$$

$$= -RT \ln (p_2/p_1). \tag{5.8b}$$

Because the expansion and compression in this cycle are between the same volumes, the work and therefore the heat transfer quantities are proportional only to the absolute temperatures. The sign is different for expansion or compression, of course. Thus the coefficient of performance of the ideal Stirling cycle as a refrigerator is

$$\text{COP}_R = \frac{\text{Heat received}}{\text{Heat rejected} - \text{Heat received}}$$

$$= \frac{T_L}{T_H - T_L}. \tag{5.9}$$

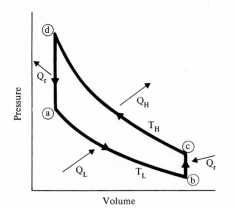

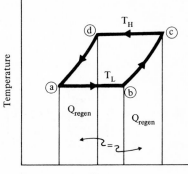

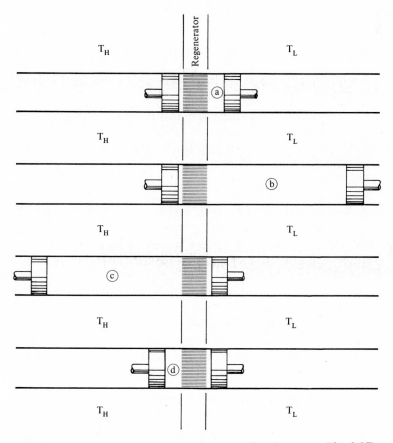

FIG. 5.16 Idealized Stirling cycle refrigeration (compare Fig. 3.27).

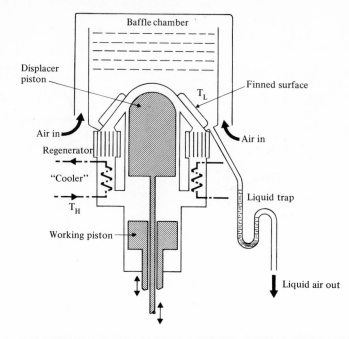

FIG. 5.17 Stirling cycle refrigeration machine for liquefaction of air.

As was noted for the power cycle, a practical machine operates on a distorted Stirling cycle. Without intermittent motion of the two pistons, and without sufficient time for heat transfer, flow of the gas through the regenerator in either direction is not at constant volume, and the heat addition and rejection are not isothermal. Nevertheless, the COP values are quite good for the large temperature ranges spanned.

A machine similar to the engine of Fig. 3.28 has been built by the Philips Research Laboratories for refrigeration [A.15]. In one version, the two pistons are driven by two out-of-phase cranks on a common shaft. **Figure 5.17** shows a schematic representation of the cylinder and two pistons in an arrangement used to produce liquefaction of air on the finned surface of the upper chamber. The heat is rejected to cooling water near ambient temperature. The working fluid is pressurized hydrogen or helium at several atmospheres with a pressure ratio of about $2:1$. A specific power consumption as low as 0.88 kwh/liter of liquefied air has been reported. This achievement in a single-stage unit indicates the usefulness of a Stirling-cycle machine for a wide range of applications where refrigeration below conventional temperatures is required with a small plant.

5.5.3 The Gifford-McMahon Refrigeration System

A displacer piston and a regenerator, but with a separate gas compressor, are combined in a different refrigeration system, which has become known as the Gifford-McMahon cycle. It is well suited to miniaturization for localized cooling, and will produce temperatures well into the cryogenic range [A.11].

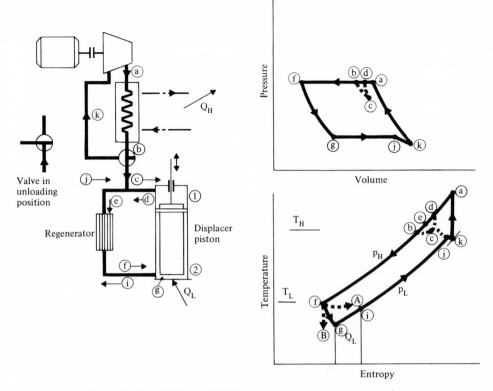

FIG. 5.18 The Gifford-McMahon refrigeration system, single-stage.

A schematic system diagram for a single-stage unit and thermodynamic representations of the cycle are shown in **Fig. 5.18**. It must be stated immediately that the *p-v* and *T-s* diagrams are hypothetical; they are an attempt to demonstrate what happens to the working fluid on the average. Actually, as will be seen, not all the fluid follows one cycle as in, say, a Brayton-cycle system. Not only is a stream separated into parts which follow different paths, but, also, fluid passing a given point at a different time will be in a different thermodynamic state. Despite these difficulties in describing all the thermodynamic processes, a meaningful expression can be written for the system's coefficient of performance.

Gas is compressed to p_H at point *a* and cooled to approximately ambient temperature at *b*. With the three-way valve in the loading position and the displacer piston in its lowest position as shown, high-pressure gas is admitted to space ①. Gas in that space is at a lower pressure p_L at condition *j*. Some incoming gas is initially throttled to a lower pressure, but as more gas follows, the pressure in ① is raised to p_H at condition *d*. The displacer piston is then moved upward so that gas can flow into space ②. The gas flowing through the regenerator is a mixture of gas at *d* from ① and more gas at *b* from the compressor. The mixture at *e* is cooled in the regenerator to *f*, at which condition it fills space ② until V_1 is essentially zero.

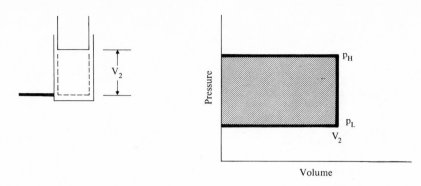

FIG. 5.19 The work done as fluid leaves space 2 below the displacer piston.

At this time, the three-way valve is indexed to the unloading position. The pressure in space ② begins to fall, and gas flows out through the regenerator, where it is heated up again. The first gas to leave space ② follows the path fA on T-s coordinates. Gas remaining in ② expands isentropically along path fB. On the average, the gas follows the path fg and is cooled well below T_f.

The refrigeration effect (Q_L) is accomplished as the gas is heated to T_i. Gas at i, flowing through the regenerator, is brought back to a temperature close to T_e. All the gas in space ② is displaced by the piston which is now lowered. Part of the returning gas flows into space ① at condition j while the rest, after a slight pressure drop through the valve, is taken into the compressor at k.

The cycle has been described with the assumption that it has been operating for some time to allow the temperatures in space ② to reach the lower levels. The repeated expansion of gas from f to g will continue to lower the temperatures in that region until an equilibrium energy balance is achieved.

To obtain an expression for the coefficient of performance in terms of pressures and temperatures, the refrigeration effect (Q_L) must be found. This can be done indirectly by a consideration of the work done in space ②. Let us consider the gas as it flows out of space ② from the time it fills the whole volume V_2 at pressure p_H until the displacer is fully lowered and the pressure has fallen to p_L. This is shown on the p-v diagram of **Fig. 5.19**. The work done is represented by the area of the diagram which is

$$W = V_2(p_H - p_L). \tag{5.11}$$

Now, for this flow process, the work, from Eq. (1.34b), is the change of enthalpy:

$$W = m_2\Delta h. \tag{5.12}$$

For other refrigeration processes, it has been usual to calculate the refrigeration effect as

$$Q_L = m_2 c_p \Delta T = m_2 \Delta h. \tag{5.13}$$

Therefore, for this type of expansion machine, the refrigeration effect is equal to the work of the flow process that we imagined:

$$Q_L = V_2(p_H - p_L). \tag{5.14}$$

The ideal work input to the system is the work of isentropic compression and delivery as given by Eq. (3.9b):

$$W_c = m_k c_p T_k [(P_H/P_L)^{(\gamma-1)/\gamma} - 1]. \tag{5.15}$$

Since part of the gas flowing out of ② goes to the compressor, and part goes to space ①,

$$m_a = m_2 \left[\frac{p_H V_2}{RT_f} - \frac{p_L V_1}{RT_j} \right] = m_2 \frac{V}{R} \left[\frac{p_H}{T_L} - \frac{p_L}{T_H} \right], \tag{5.16}$$

since V_2 and V_1 are the same, and T_f and T_j are logically designated as T_L and T_H, respectively.

Combining Eqs. (5.14), (5.15), and (5.16), we find that

$$\mathrm{COP}_R = \frac{Q_L}{W}$$

$$= \frac{R(p_H - p_L)}{c_p T_H [(p_H/T_L) - (p_L/T_H)][(p_H/p_L)^{(\gamma-1)/\gamma} - 1]} \tag{5.17}$$

$$= \frac{R[(p_H/p_L) - 1]}{c_p [(p_H/p_L)(T_H/T_L) - 1][p_H/p_L)^{(\gamma-1)/\gamma} - 1]}. \tag{5.18}$$

Although there were many simplifying assumptions in this development, the same is true for the COP expressions for the Stirling and for the regenerative Brayton cycles with which this might be compared.

The loading and unloading valve shown in Fig. 5.18 is purely schematic for ease of illustration. One can be designed easily to perform the same actions while rotating continuously rather than by oscillating through 90°. The mechanism to move the displacer piston is not shown and can be any one of several different linkage systems that will correlate the valve and the piston motions.

The effectiveness or efficiency of the regenerator in both the Stirling cycle and the Gifford-McMahon cycle is extremely important. Very high values (close to 1.0) are required if extremely low temperatures are to be reached. In any system which depends on a reversed-flow regenerator, the small value of heat capacity of the thermal storage material at low absolute temperatures puts a minimum limit on the temperatures that are obtainable in practice.

An advantage of the Gifford-McMahon cycle is that it can readily be staged to achieve lower temperatures than would be obtainable in a single stage. **Figure 5.20** shows a three-stage system followed by a Joule-Thomson refrigeration system, which has been suggested for helium liquefaction [A.5]. The working fluid throughout would be the gas to be liquefied, as it was in some other systems described in Section 3 of this chapter.

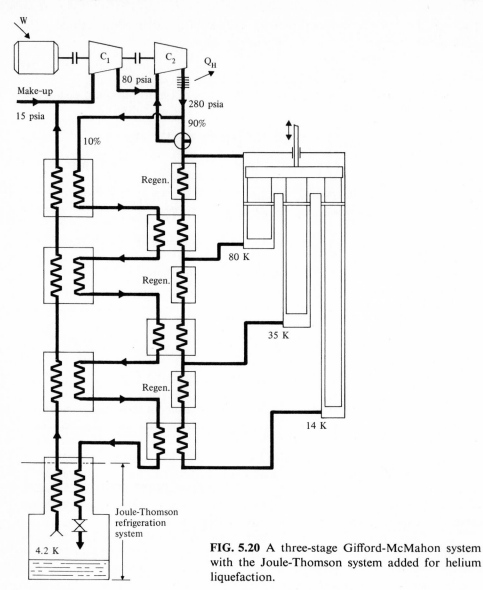

FIG. 5.20 A three-stage Gifford-McMahon system with the Joule-Thomson system added for helium liquefaction.

5.5.4 The Pulse-Tube System

One of the simplest devices for producing a cooling effect is the **pulse-tube**, which, with a regenerator, can pump heat from a lower to a higher temperature when provided with a compressed gas supply. This device, developed by Gifford, has no moving parts outside the conventional compressor except a control valve [A.9, and A.10]. **Figure 5.21** illustrates the heat pumping action within the tube. If the

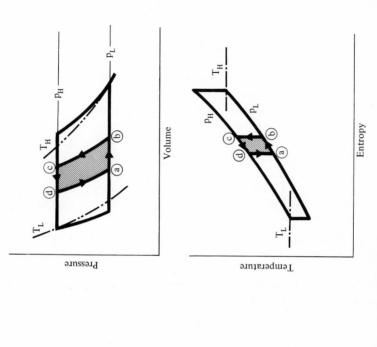

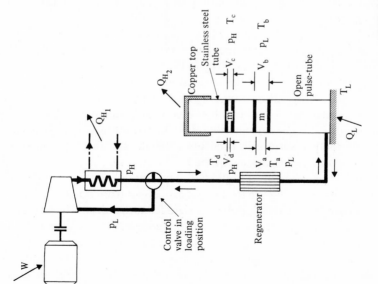

FIG. 5.21 Pulse-tube refrigeration.

pressure ratio p_H/p_L were high enough to force gas all the way through the regenerator and lower heat exchanger and bring it in contact with the upper heat exchanger, it is clear that the temperature rise resulting from compression would allow heat transfer at the upper temperature T_H. Reexpansion to p_L would provide cooling at the lower temperature T_L in a manner similar to that of the Gifford-McMahon cycle.

It is interesting to find, however, that heat pumping from T_L to T_H can be effected at relatively low-pressure ratios (about 2:1), at which only a small fraction of the gas in the tube will be brought in contact with the upper heat exchanger. Again, it is difficult to represent the cycle of the whole system on thermodynamic coordinates. In a steady-flow system, conditions of the working fluid are constant at each point in the system. In a simple reciprocating engine or compressor, conditions at any instant are assumed constant throughout the system for an idealized cycle. Neither situation is found here; gas properties vary at any instant throughout the gas column from the valve to the top heat exchanger, and they vary at any point with time.

The heat-pumping effect can be explained if we examine the cycle executed by a small slug of gas marked m on the diagram. We assume that the gas, at pressure p_L, is in approximate thermal equilibrium with the tube wall at T_b. When the control valve is moved to the loading position shown and the pressure in the pulse tube is raised rapidly to p_H, we find that not only is the slug m compressed from V_b to V_c, but also it is moved to the higher position shown and its temperature is raised to T_c. The temperature gradient in the tube wall is such that the wall temperature in the new position is below T_c and the slug of gas is cooled to T_d, while still at pressure p_H. The volume of m is decreased to V_d. When the control valve is moved to the unloading position, the gas pressure drops rapidly to p_L doing work against the regenerator and valve friction. The slug m moves back approximately to the original lower position, but on expansion cools to T_a and occupies volume V_a. In returning to thermal equilibrium with the wall at T_b, the gas expands at constant pressure p_L to the original volume V_b.

The small slug of gas is therefore seen to execute a Brayton cycle between p_L and p_H with a temperature range T_a to T_c. The gas above and below slug m will execute similar Brayton cycles between p_L and p_H, but through different temperature ranges. The gas within the tube can therefore operate between T_L and T_H, although the pressure ratio need not be high; the effect is that of an infinite number of Brayton-cycle stages with the tube wall acting as a continuous regenerator. An external regenerator of high efficiency is necessary to ensure a temperature near T_L for all gas entering the lower heat exchanger from outside the tube.

The actual cycle must, of course, deviate from the ideal. The work of reexpansion is dissipated to friction loss. Also, the compression and expansion are not instantaneous and therefore not adiabatic. Because of friction loss, particularly in the regenerator, the processes are not reversible as shown in the idealized p-v and T-s diagrams. The friction loss at the wall distorts the slug m from the flat shape shown, because the center will move more rapidly than the boundary layer

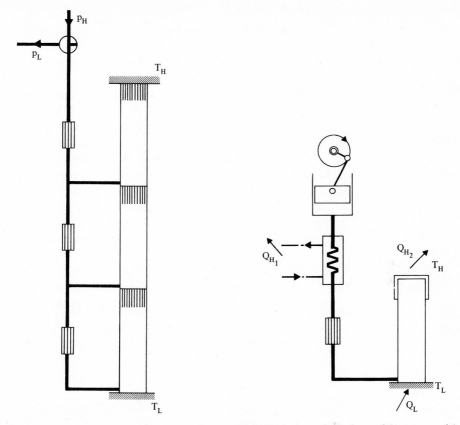

FIG. 5.22 Three-stage pulse-tube refrigeration.

FIG. 5.23 A pulse-tube refrigerator with a direct-acting compressor.

of gas. The boundary layer will impede heat transfer at the wall. The axial heat transfer, by means of the thin metal wall and through the gas column itself, will be against the desired direction.

Cascading to increase the possible temperature range, as shown in **Fig. 5.22**, is accomplished very simply. No moving parts are added to the system. The refrigeration effect at the bottom of each tube removes the heat rejection from the top of the stage below.

In the alternative arrangement shown in **Fig. 5.23,** the control valve is done away with and the change in total volume of the gas is controlled by the position of the compressor piston [A.8]. Deviation from isentropic compression within the pulse tube is likely to be greater than in the first arrangement described.

Over a given temperature range, the capacity is increased in any of the pulse-tube systems by the operation of several tubes in parallel from the same compressed air source. One tube cannot be enlarged beyond an optimum surface-to-volume ratio, which depends on cycle frequency and the desired temperature range [A.18].

5.6 ADIABATIC DEMAGNETIZATION

There is a limit below which temperatures cannot be achieved by any of the systems previously described, since they all use fluids as the thermodynamic working substance. Below 1 K, only helium is a liquid, and at that temperature, the vapor pressure that would have to be maintained is very low. However, certain solids, susceptible to a magnetic field, can be used as the "working substance" to reach temperatures within a very small fraction of a degree of absolute zero.

Like ferromagnetic substances, **paramagnetic** substances exhibit positive magnetic susceptibility, though the relationship between susceptibility and temperature is different. In the absence of a field they are nonmagnetic. (Diamagnetic substances exhibit negative susceptibility or are repelled by a magnet.) Certain paramagnetic salts such as iron ammonium alum and chromium potassium alum at very low temperatures have been subjected to strong magnetic fields isothermally, and then demagnetized adiabatically to produce a reduction in temperature. The two processes are analogous to the isothermal compression and subsequent adiabatic expansion of a gas.

A schematic arrangement for cryogenic magnetization and demagnetization is shown in **Fig. 5.24.** (The arrangement is somewhat similar to that for nonflow gas liquefaction; see Fig. 5.10.) The salt is precooled in a bath of liquid helium,

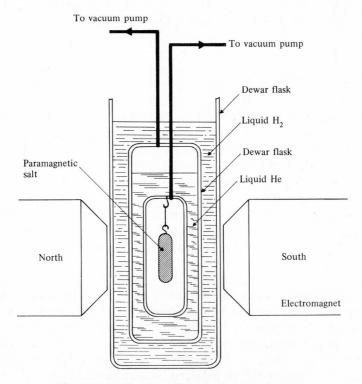

FIG. 5.24 Cooling by adiabatic demagnetization.

insulated from the surroundings as shown. Helium vapor in the chamber immediately surrounding the salt provides a heat-transfer medium between the salt and the liquid bath. The pressure above the liquid helium is lowered until 1 K or less is reached. When the electromagnet is turned on, the solid substance is magnetized, which is to say that its magnetic dipoles are oriented into the direction of the magnetic field lines. This operation would normally heat the substance, but the energy is transferred to the liquid helium, which is kept at a constant vapor pressure. The operation is isothermal, or at least the material returns gradually to its original equilibrium temperature. The chamber in which the paramagnetic salt is suspended can then be evacuated to insulate the salt from its surroundings. When the electromagnet is turned off, removing the field, the dipoles can then assume a more random orientation. This, however, takes place adiabatically so that the process lowers the temperature of the solid itself. A magnetic refrigerator, in which a working salt repeatedly cools a reservoir salt, has been developed [A.12].

Adiabatic demagnetization becomes less and less effective as the temperature is lowered to within 0.01 degree of absolute zero. Below that temperature, another phenomenon becomes useful. Here it is possible to orient the nuclear spin of a substance, such as copper, by the application of a magnetic field. Adiabatic demagnetization again results in temperature reduction. By this method, temperatures of the order of 20 millionths of a degree absolute have been achieved [A.16].

To measure temperatures within a fraction of a degree of absolute zero, conventional thermometry is not at all practical: magnetic properties must be used. The simple Curie relationship [B.18] between magnetic susceptibility and temperature is valid to about 1 K. Below that, deviations can be accounted for in the calibration of a magnetic thermometer. The calibration must consist of isothermal magnetization at say 1 K of a paramagnetic specimen with measurement of heat removal, subsequent adiabatic demagnetization of the specimen, and finally measurement of the heat input to bring it back to its original temperature. The temperature is then computed from thermodynamic relationships [B.2 and B.13].

PROBLEMS

1. When helium is drawn from a high-pressure tank which is initially at room temperature, the stream of gas beyond the throttling valve is found to be at a temperature higher than room temperature, while the tank becomes progressively colder. Explain both results.

2. Is the value of the Joule-Thomson coefficient any indication of the deviation of real gas properties from those of a "perfect gas"?

3 Air is initially at 300 K and 70 atm. Use chart values to find the final temperature at 1 atm after
   a) throttling,
   b) isentropic expansion, and
   c) expansion with an isentropic efficiency of 70%.

4. What is the minimum net work theoretically required to condense a gram of oxygen initially at 1 atm and 300 K?

5. a) What would be the theoretical yield of liquid air per pound of dry gas compressed for an ideal, simple Linde (Hampson) system operating between 1 and 200 atm with isothermal compression at 300 K? For a first approach, assume a very small yield and make-up rate, so that the enthalpy of returning gas is practically that of compressor intake gas.

 b) How much work must be supplied per pound of liquid air produced?

 c) Compare your answer in part (b) with the minimum work theoretically required.

6. Using the diagram in Fig. 5.13, make a rough estimate of the efficiency of expansion in each of the three cylinders in the Collins helium liquefier.

7. In Fig. 5.15 for the Linde double-column process for air separation, several facts are apparently contradictory. Explain the following:

 a) Point b is in the vapor region on the T-s and T-x diagrams, but is between g_1 and g_2 and apparently in the wet vapor region of the h-x diagram.

 b) Points g_1 and g_2 have lower temperatures but higher enthalpies than points e_1 and e_2.

 c) Points c and d are coincident on one diagram, but not on the other two.

8. It was noted in the text that a specific power consumption of 0.88 kwh/liter of liquefied air has been reported for a Stirling cycle machine.

 a) What coefficient of performance does this imply?

 b) What is the relative efficiency of the cycle?

 c) How does the reported value compare with the minimum net work theoretically required for liquefaction? (See Problem 4.)

 Use chart values and make assumptions where necessary.

9. What would the theoretical coefficient of performance be for a Gifford-McMahon refrigeration system operating between 300 K and 100 K with a pressure ratio of 2:1? The working fluid is helium for which the specific heat at constant pressure can be assumed constant at 1.25 Btu/lb-degree R (or 1.25 calorie/g-degree K).

REFERENCES

A. *Articles and Separate Publications*

1. BIRMINGHAM, B. W., H. SIXSMITH, and W. A. WILSON, "Applications of Gas-Lubricated Bearings to Miniature Helium Expansion Turbine," *Inst. Int. du Froid Annexe au Bul*, Commission 1, London, Sept. 20–22, 1961, pp. 85–103.

2. COLLINS, S. C., "Helium Liquefiers and Carriers," Reference B. 6, pp. 112–136.

3. DAUNT, J. G., "The Production of Low Temperature down to Hydrogen Temperature," Reference B. 6, pp. 1–111.

4. DEAN, J. W. and D. B. MANN, "The Joule-Thomson Process in Cryogenic Refrigeration Systems," NBS TN 227, *National Bureau of Standards*, Washington, 1965, 39 pp.

5. GIFFORD, W. E., "Novel Refrigeration Cycles and Devices," *Progress in Cryogenics*, Vol. III, K. Mendelssohn (ed), Heywood, London, 1961, pp. 51–73.

6. GIFFORD, W. E., "The Gifford-McMahon Cycle," paper for Cryogenic Engineering Conference, Rice University, 1965.

7. GIFFORD W. E. and T. E. HOFFMAN, "A New Refrigeration System for 4.2 K," *Advances in Cryogenic Engineering*, Vol. 6, K. D. Timmerhaus (ed), Plenum Press, New York, 1961, pp. 82–94.

8. GIFFORD, W. E. and G. H. KYANKA, "Reversible Pulse Tube Refrigeration," paper for Cryogenic Engineering Conference, U. of Colorado, June, 1966.

9. GIFFORD, W. E. and R. C. LONGSWORTH, "Pulse Tube Refrigeration," *Trans. ASME*, Jl. of Engineering for Industry, Aug. 1964, p. 264.

10. GIFFORD, W. E. and R. C. LONGSWORTH, "Pulse Tube Refrigeration Progress," *Advances in Cryogenic Engineering*, Plenum Press, New York, 1965.

11. GIFFORD, W. E. and H. O. McMAHON, "A New Refrigeration Process," *Proc. Tenth Int. Cong. of Refrig.*, Copenhagen, August, 1959.

12. HEER, C. V., G. B. BARNES, and J. G. DAUNT, *Rev. Sci. Inst.*, Vol. 25, 1954, p. 1088.

13. JACOBS, R. B., "The Efficiency of an Ideal Refrigerator," *Advances in Cryogenic Engineering*, Vol. 7, K. D. Timmerhause (ed), Plenum Press, New York, 1962.

14. JOHNSON, V. J. (ed), "The Role Cryogenics is Playing in Expanding Mechanical Engineering," *ASHRAE Symposium*, New York, Feb., 1963, 83 pp.

15. KOHLER, J. W. L., and C. O. JONKERS, "Fundamentals of the Gas Refrigerating Machine," Vol. 16, No. 3, p. 69, and "Construction of the Gas Refrigerating Machine," Vol. 16, No. 4, p. 105, *Philips Tech. Rev.*, 1954.

16. KURTI, N., F. N. H. ROBINSON, F. SIMON, and D. A. SPOHR, *Nature*, Vol. 178, p. 450, 1956.

17. LaFLEUR, J. K., "Cryogenics via the Gas Turbine," *Mechanical Engineering*, Nov. 1965, pp. 38–40.

18. LONGSWORTH, R. C., "An Experimental Investigation of Pulse Tube Refrigeration Heat Pumping Rates," paper for Cryogenic Engineering Conference, U. of Colorado, June, 1966.

19. MANN, D. B., "The Thermodynamic Properties of Helium from 3 to 300 K between 0.5 and 100 Atmospheres," NBS TN 154, *National Bureau of Standards*, Washington, 1962, 95 pp.

20. McCORMICK, J. E., "Cryogenic Refrigeration Efficiency Nomograph," *ASHRAE Jl.*, Vol. 8, No. 11, 1966, p. 60.

21. OLCOTT, T. M., and H. A. BLUM, "Thermodynamic Investigation of Refrigerant Expansion Engine," *ASHRAE Jl.*, Vol. 3, No. 8, 1961, pp. 75–82.

22. RODER, H. M., R. D. GOODWIN, and L. A. WEBER, "Thermodynamic and Related Properties of Parahydrogen from the Triple Point to 100 K at Pressures to 340 Atmospheres," NBS Monograph 94, *National Bureau of Standards*, Washington, 1965.

23. STROBRIDGE, T. R., "The Thermodynamic Properties of Nitrogen from 64 to 300 K between 0.1 and 200 Atmospheres," NBS TN 129, *National Bureau of Standards*, Washington, 1962.

24. STROBRIDGE, T. R. and D. B. CHELTON, "Size and Power Requirements of 4.2 K Refrigerators," Paper No. U-3, Cryogenic Engineering Conference, June, 1966, *National Bureau of Standards*, Boulder, Colorado, 1966, 15 pp.

B. *Books*

1. AMERICAN SOCIETY OF HEATING, REFRIGERATING AND AIR CONDITIONING ENGINEERS, *Handbook of Fundamentals*. New York: ASHRAE Inc., 1967.

2. BARRON, R., *Cryogenic Systems*. New York: McGraw-Hill, 1966.

3. BOSNJAKOVIC, F., and P. L. BLACKSHEAR, JR., *Technical Thermodynamics*. New York: Holt, Rinehart, and Winston, 1965.

4. DIN, F. (ed), *Thermodynamic Functions of Gases—Vol. 2*. London: Butterworths, 1962.

5. DIN, F. (ed), *Thermodynamic Functions of Gases—Vol. 3*. London: Butterworths, 1961.

6. FLUGGE, S. (ed), *Encyclopedia of Physics—Vol. XIV, Low Temperature Physics I*. Berlin: Springer-Verlag, 1956.

7. HILSENRATH, J., et al, *Tables of Thermal Properties of Gases*, (NBS Circular 564), U.S. Government Printing Office, Washington, 1955.

8. HOARE, F. E., L. C. JACKSON, and N. KURTI (eds), *Experimental Cryophysics*. London: Butterworths, 1961.

9. KEESON, W. H., *Helium*. New York: Elsevier, 1942.

10. MacDONALD, D. K. C., *Near Zero*. Garden City, N. Y.: Doubleday, 1961.

11. ROBERTS, J. K., *Heat and Thermodynamics*. London: Blackie and Sons, 1944.

12. RUHEMANN, H., *The Separation of Gases* (second edition). Fair Lawn, N. J.: Oxford University Press, 1949.

13. SCOTT, R. B., *Cryogenic Engineering*. Princeton, N.J.: D. Van Nostrand, 1960.

14. SITTIG, M., *Cryogenics—Research and Applications*. Princeton, N.J.: D. Van Nostrand, 1963.

15. THRELKELD, J. L., *Thermal Environmental Engineering*. Englewood Cliffs: Prentice-Hall, 1962.

16. VANCE, R. W. and W. M. DUKE, *Applied Cryogenic Engineering*. New York: Wiley, 1962.

17. VANCE, R. W., *Cryogenic Technology*. New York: Wiley, 1963.

18. ZEMANSKY, M. W., *Heat and Thermodynamics* (fourth edition). New York: McGraw Hill, 1957.

19. ZEMANSKY, M. W. and H. C. VAN NESS, *Basic Engineering Thermodynamics*. New York: McGraw-Hill, 1966.

CHAPTER 6

AIR CONDITIONING

6.1 INTRODUCTION

To the layman, the term air conditioning seems to mean, exclusively, cooling for comfort in the summertime. Actually, much more is encompassed in this broad field of engineering. The term really should be interpreted literally, because any process by which atmospheric air is made more suitable for a particular use is a conditioning of that air. Cooling plus dehumidification for human comfort in the summer, and heating plus humidification in the winter are special applications. The control of environment for livestock, for preservation of stored material, for manufacture and weaving of various materials are only a few other familiar examples. The removal of contaminants by filtering or other methods is part of air conditioning. In many industries, large quantities of air are used in various processes and the conditions of its supply must often be carefully controlled.

There are many aspects of air conditioning that are not thermodynamic in nature. Some of these will be mentioned, but only insofar as they are relevant to a discussion of heat-transfer quantities, energy sources, and power requirements. For instance, an extended discussion of physiological factors in air conditioning for comfort or for survival is included in this chapter because the engineer must appreciate the reasons for various requirements and the means used to evaluate the effectiveness of attempts to provide suitable conditions. The section on heating and cooling loads stops short of a detailed examination of solar heat gains, heat-transfer coefficients, wind effects, etc., but does attempt to provide a brief survey of the type of factors affecting design loads and an appreciation of the complexity of the problem.

Table 6.1

COMPOSITION OF "DRY AIR"

Substance	Molecular weight (definition)	Molecular fraction in dry air (definition)	Partial molecular weight in dry air (calculated)	Weight fraction in dry air (calculated)
Oxygen (O₂)	32.000	0.2095	6.704	0.2315
Nitrogen (N₂)	28.016	0.7809	21.878	0.7552
Argon (A)	39.944	0.0093	0.371	0.0128
Carbon Dioxide (CO₂)	44.01	0.0003	0.013	0.0005
Dry Air		1.0000	28.966	1.0000

6.2 PSYCHROMETRICS

Basic to the solution of any air conditioning problem is the understanding of **psychrometrics**, which is the study of the thermodynamic properties of **moist air**. In engineering, the term "moist air" means the combination of water vapor in a binary mixture with dry air. This may seem to be a contradiction of terms since atmospheric air is known to consist of a dozen or so gases, predominantly nitrogen (N_2), oxygen (O_2), argon (A), and carbon dioxide (CO_2). However, the composition of atmospheric air excluding the variable water vapor is very nearly constant. If the molecular ratios are assumed to be fixed, the constituents of moist air other than the water vapor can be assumed to act as a single gas.

The International Joint Committee on Psychrometric Data recommended in 1949 that the arbitrary definition of **dry air** be that given in **Table 6.1**. In the final report of the committee, the variation in composition of actual atmospheric air was pointed out, particularly in terms of the proportions of carbon dioxide and the traces of other gases. Still, it was shown that the disagreements among various authorities affected only the fourth significant figure. The arbitrary nature of the exact definition should therefore not be overstressed; all the figures are consistent with the most reliable current knowledge. The molecular weights listed constitute part of the definition. (The molecular weight for water vapor in moist air is taken by the committee to be 18.016).

A first approach to the properties of a mixture of gases can be obtained by the application of **Dalton's law** of partial pressures. This, however, is a highly idealized relationship, which applies only to perfect gases and is inconsistent with the best current empirical data. The statement of Dalton's law that the total pressure of a mixture of gases would be the sum of the pressures of its constituents, each occupying the same volume separately, disregards the fact that molecules of different gases exert forces on one another; the idea that one gas exerts a pressure exactly proportional to the mole fraction present disregards the intermolecular forces among identical molecules. The **Gibbs-Dalton** formulations, which allow the summation of individual enthalpies and entropies as well as specific weights and pressures, are similarly in error. The error is greatest for the entropy term, which ignores the significant entropy of mixing.

Despite these shortcomings, the Gibbs-Dalton laws will be used to help explain the basic terms in psychrometry. The corrections to these laws that are necessary to bring their results in accord with the best empirical and theoretical data will be given, and the basis of accepted tabulated and chart data for the thermodynamic properties of moist air will be mentioned briefly. The definitions which follow are correct and do not depend on the idealized relationships used for elucidation.

We will assume that the **perfect gas law** is applicable in the first approach, acknowledging a slight error. Although water vapor is part of the mixtures considered, the partial pressure of water vapor in atmospheric air is so low that the application of gas laws to this vapor has approximately the same degree of validity as it has with regard to the other components. Thus we will assume that

$$pv = RT. \tag{6.1}$$

For the combined constituents of dry air, the value of R_a is 53.35 ft-lb/(lb)(degree F abs). For water vapor the value of R_w is 85.76.

Dalton's law, applied to moist air, would be

$$p = p_a + p_w. \tag{6.2}$$

Consequently,

$$p_a V = m_a R_a T,$$
$$p_w V = m_w R_w T. \tag{6.3}$$

The **humidity ratio** (W) of moist air is simply the weight (mass) of water vapor in the mixture per pound of dry air. (This has also been called the "humidity mixing ratio" or the "mixing ratio," but humidity ratio is preferred to avoid ambiguity when several streams of moist air are mixed. An older synonymous term is "specific humidity." Some references still use the units *grains of moisture per pound of dry air*. There are 7000 grains in a pound).

By definition,

$$W = m_w/m_a. \tag{6.5}$$

Applying perfect gas laws, we find that

$$\begin{aligned} W &= (p_w V/R_w T)(R_a T/p_a V) \\ &= (53.35/85.76)p_w/(p - p_w) \\ &= 0.622\, p_w/(p - p_w) \end{aligned} \tag{6.6}$$

Note that since the individual gas constants are obtained by dividing the universal gas constant by the molecular weights, the figure 0.622 could be obtained alternatively by the ratio of molecular weights (18.016/28.966), at least to three significant figures.

The **degree of saturation** (μ) is simply the ratio of the actual humidity ratio of a moist air to the humidity ratio that would exist at saturation for the same temperature and pressure. Thus

$$\mu = W/W_s. \tag{6.7}$$

The term **saturation** itself should be explained. It is defined as the condition of moist air that can exist in neutral equilibrium with a condensed phase (liquid or solid), presenting a flat face to it. Actually, the effect of curvature of the surface on the equilibrium condition is very small unless the condensed phase droplets have an extremely small radius, which of course they could have in a very fine mist. The condensed phase is not strictly pure water since some gas will be dissolved in it, but even in the development of the tabulated data for real properties, a consideration of the mixture as pure water and dry air introduces no appreciable error. An attempt to explain saturation in some form acceptable to the layman, such as "the air or the volume being saturated with water vapor," should be avoided; the condition exists when the temperature, pressure, and chemical potentials of the vapor phase have the same values as they do for the associated condensed phase.

The **relative humidity** (ϕ) is perhaps the most familiar term in a description of moist air conditions in both air conditioning and meteorology. Thermodynamically, it is not as directly related to the basic properties of the mixture as is the degree of saturation, and would probably be abandoned as a redundant term if it were not for its entrenched position in the vocabulary. It is defined by

$$\phi = \frac{\text{Mole fraction of water vapor in moist air}}{\substack{\text{Mole fraction of water vapor saturated at} \\ \text{the same temperature and total pressure}}}.$$

The ratio of mole fractions for the same substance becomes a mass ratio, so that

$$\phi = m_w/m_{ws}. \tag{6.8}$$

By using the perfect gas law, we find that

$$\phi = (p_w V/R_w T)(R_w T/p_{ws} V)$$
$$= p_w/p_{ws} \tag{6.9a}$$
$$= (W/0.622)(p - p_w)/p_{ws}$$
$$= (W/0.622)(p_a/p_{ws}). \tag{6.9b}$$

If the relative humidity ϕ, the ambient temperature, and the barometric pressure are known, the humidity ratio W can be calculated when the saturation pressure is determined from the vapor tables:

$$W = 0.622\, p_w/(p - p_w)$$
$$= 0.622\, \phi p_{ws}/(p - \phi p_{ws}). \tag{6.9c}$$

The relative humidity ϕ is different numerically from the degree of saturation μ because the former is concerned only with the water vapor present, while the latter takes into account the ratio of the vapor to the dry air present. Consequently,

$$\phi = m_w/m_{ws} = (Wm_a)/(W_s m_{as}) = \mu(m_a/m_{as}) = \mu(p_a/p_{as}). \qquad (6.9d)$$

From the definitions of relative humidity and degree of saturation, the last expression can be reduced to

$$\phi = \frac{\mu}{1 - (1 - \mu)(p_{ws}/p)}. \qquad (6.10)$$

At low temperatures, where the vapor pressure of water is small compared to atmospheric pressure, the numerical difference between ϕ and μ is not great, but at temperatures over about 60 F, it becomes appreciable. Of course, from their definitions, the values of the two are identical for saturated air (1.0) and for dry air (zero), regardless of temperature.

To continue with the assumptions of the perfect gas law and the Gibbs-Dalton rule, the specific volume, enthalpy, and entropy of moist air can be easily approximated. From Eqs. (6.1) and (6.2), the **specific volume** would be

$$v = \frac{R_a T}{p_a}$$

$$= \frac{R_a T}{p - p_w} \text{ cu ft/lb d.a.} \qquad (6.11)$$

Alternatively, the effect of the total pressure on the mixture could be considered so that

$$v = \frac{(R_a + R_w W)T}{p}$$

$$= \frac{R_a T}{p}\left(1 + \frac{W}{0.622}\right) \text{ cu ft/lb d.a.} \qquad (6.12)$$

Equations 6.11 and 6.12 should give the same values within the accuracy of these assumptions. Note that both give the volume per pound of dry air rather than per pound of mixture. This is preferred in most air-conditioning calculations where moisture may be added to or removed from a constant mass rate of dry air in a steady-flow operation. The volume per pound of dry air applies to the water vapor as well as to the dry air or to the mixture; the two components occupy the same volume.

The **enthalpy** of the mixture would ideally be the sum of the enthalpies of the components. That is,

$$h = h_a + W h_w \text{ Btu/lb d.a.} \qquad (6.13)$$

It happens that for water vapor at the partial pressures existing in atmospheric air, the enthalpy of the vapor changes very little with temperature and the value of h_w can be approximated by the value h_g (enthalpy of saturation) taken from the steam tables at the same temperature. A better approximation is obtained from the empirical relationship

$$h_w = 1061 + 0.444\, t \qquad (6.14)$$

above zero enthalpy for liquid at 32 F.

The convention for both tabulated and chart values of enthalpy is to introduce factors making the enthalpy of dry air equal to zero at zero F and, following steam-table practice, making the enthalpy of water equal to zero at 32 F. Thus

$$h_a = 0.240\, t,$$

and from Eq. (6.13), we find that

$$h = 0.240\, t + W(1061 + 0.444\, t)\ \text{Btu/lb d.a.} \qquad (6.15)$$

It cannot be overemphasized that the above relationships are approximations only, though often satisfactory for air-conditioning work near normal ambient temperatures. The specific heat for both components, of course, increases with temperature.

The values of **entropy** should also be additive according to the Gibbs-Dalton rules, so that

$$s = s_a + Ws_w\ \text{Btu/(lb d.a.)--}R. \qquad (6.16)$$

In this instance, it is important to note that the entropy values of the components must be calculated with the assumption that each had filled the whole volume. The components could not be imagined as being separated into their "partial volumes" by an imaginary barrier, both at the total pressure, because an increase in entropy would accompany an expansion to the total volume and partial pressures.

6.2.1 Real Properties of Moist Air

To a high degree of accuracy, the actual properties of moist air are now available. Goff and Gratch [A.21] have developed tables of the thermodynamic properties of moist air, which are generally accepted and the most reliable to date. They are based on the principles of statistical mechanics applied through "semirational" equations fitting the best empirical data. The consideration of moist air as a binary mixture of water vapor with dry air of fixed composition allows the formulations to be based on two chemical potentials only: that for water vapor and that for dry air.

It is worth noting that in developing the properties of moist air, Goff and Gratch found it necessary to reject the generally accepted steam tables of Keenan and Keyes in the range below 212 F as being of insufficient accuracy, and to present their own tables. These were found to be in close agreement with the later work of Keyes.

The Goff and Gratch tables for moist air are presented (in the ASHRAE Guide and Date Books. [B.3]) for saturation conditions only, but can be applied to any conditions through the use of the degree of saturation (μ) and, at higher temperatures, correction coefficients. The arrangement is similar to that for the

Table 6.2

COEFFICIENTS A, B, C, APPEARING IN EQUATIONS FOR $\bar{v}, \bar{h}$, AND $\bar{s}$. MAXIMUM VALUES OF $\bar{v}, \bar{h}, \bar{s}$, AND $\bar{\bar{s}}$. VALUES OF μ AT WHICH MAXIMA OCCUR AT STANDARD ATMOSPHERIC PRESSURE.*

t, F	A, ft³/lb$_a$	B, Btu/lb$_a$	C, Btu/lb$_a$	$\bar{v}_{max}$, ft³/lb$_a$	$\bar{h}_{max}$, Btu/lb$_a$	$\bar{s}_{max}$, Btu/F/lb$_a$	$\bar{\mu}_m$	s_{max}, Btu/F/lb$_a$	$\bar{\bar{\mu}}_m$
96	0.0018	0.0268	0.00004	0.0004	0.0069	0.00001	0.4925	0.0015	0.3650
112	0.0042	0.0650	0.00009	0.0010	0.0155	0.00002	0.4878	0.0025	0.3632
128	0.0096	0.1439	0.00020	0.0022	0.0332	0.00005	0.4805	0.0040	0.3602
144	0.0215	0.3149	0.00042	0.0047	0.0693	0.00009	0.4691	0.0065	0.3557
160	0.0487	0.6969	0.00091	0.0099	0.1418	0.00019	0.4511	0.0106	0.3485
176	0.1169	1.636	0.00207	0.0207	0.2903	0.00037	0.4213	0.0179	0.3363
192	0.3363	4.608	0.00567	0.0451	0.6180	0.00076	0.3662	0.0333	0.3129

* Reprinted by permission from ASHRAE Guide and Data Book, 1965–66.

familiar steam tables, so that for saturation,

$$v_s = v_a + v_{as}, \tag{6.17}$$
$$h_s = h_a + h_{as}, \tag{6.18}$$
$$s_s = s_a + s_{as}. \tag{6.19}$$

Up to about 150 F, the specific volume and the enthalpy for conditions other than saturation can easily be obtained by the two following equations:

$$v = v_a + \mu v_{as}, \tag{6.20}$$
$$h = h_a + \mu h_{as}. \tag{6.21}$$

To retain the accuracy of the basic tables above 150 F, the corrections $\bar{v}$ and $\bar{h}$ must be added, so that

$$v = v_a + \mu \, v_{as} + \bar{v}, \tag{6.22}$$
$$h = h_a + \mu \, h_{as} + \bar{h}, \tag{6.23}$$

where

$$\bar{v} = \frac{\mu \, (1 - \mu) \, A}{1 + 1.6078 \, \mu \, W_s}, \qquad \bar{h} = \frac{\mu \, (1 - \mu) \, B}{1 + 1.6078 \, \mu \, W_s}.$$

(The constant 1.6078 is simply the inverse of 0.622).

Even below 150 F, mixing entropy introduces an appreciable error, so that to retain table accuracy, two corrections must be added:

$$s = s_a + \mu \, s_{as} + \bar{s} + \tilde{s}, \tag{6.24}$$

where

$$\bar{s} = \frac{\mu \, (1 + \mu) \, C}{1 + 1.6078 \, \mu \, W_s},$$

and where $\tilde{s}$ is a more complicated function of μ and W_s, which is given in the references.

Table 6.2 is reproduced from the ASHRAE Guide and Data Book [B.3], and gives numerical values of the coefficients A, B, and C from 96 F to 192 F. Below 96 F, they are negligible. Also tabulated are the maximum values at each temperature for $\bar{v}$, $\bar{h}$, $\bar{s}$, and $\tilde{s}$, and the values of μ at which these maxima occur. These values can show quickly whether or not the refinement of adding the corrections is warranted for a particular calculation.

The **dew-point temperature** (t_d) of moist air might be loosely defined as the temperature at which condensation of moisture would begin if that air were cooled at constant pressure. This statement implies a process, and introduces the problem of instrumentation to determine the dew-point temperature. Thus a rigid mathematical definition is given as the solution of $t_d(p, W)$ in the equation $W_s(p, t_d) = W$. In other words, t_d, which is a function of pressure and humidity ratio, must have the numerical value to bring W to its saturation value in a valid functional relationship involving the pressure and temperature. For a given pressure, t_d depends only on the humidity ratio, and one is as much a property of the mixture as the other.

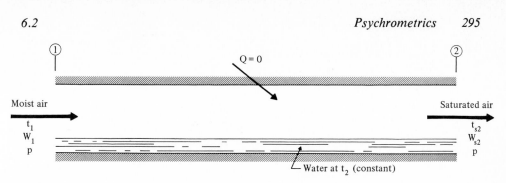

FIG. 6.1 The adiabatic saturation process.

The **thermodynamic wet-bulb temperature** (t^*) is another, slightly less obvious property of moist air. This is the temperature that would be achieved if sufficient water (liquid or solid) were added to the moist air to bring it to saturation adiabatically, the water being added at the temperature of the final equilibrium. The rigorous mathematical definition is the solution of $t^*(p, t, W)$ in the equation

$$h(p, t, W) = W h_w^*(p, t^*)$$
$$= h_s(p, t^*) - W_s(p, t^*) h_w^*(p, t^*).$$

That is, the enthalpy of the moist air must be that for adiabatic saturation less the enthalpy introduced with the water required to saturate.

Because this temperature is defined, by implication, as the endpoint of a process, it is not immediately obvious that its value depends on the original condition, and that it is therefore a property of the original moist air. An examination of the hypothetical process will verify that it is. **Figure 6.1** shows such a process. It must be imagined that the duct is so long that the residence time of the moist air in the duct allows equilibrium to be reached with the water surface; it will be recalled that this defines saturation. There must also be sufficient water to accomplish the saturation.

Consider a mass and energy balance between sections ① and ②.

In		Out
$h_1 + \Delta h$	$=$	h_2
$h_1 + (W_{s2} - W_1)h_{f2}$	$=$	$h_{s2}.$

$$h_1 + (W_{s2} - W_1)h_{f2} = h_{s2}. \tag{6.25}$$

Note that the only enthalpy added is that of the liquid "picked up" by the air. The latent heat required to evaporate that liquid has to come from the moist air, the temperature of which is lowered.

In Eq. (6.25), W_{s2}, h_{f2}, and h_{s2} are all determined by the saturation temperature t_{s2}. The only other factors in the single equation are h_1 and W_1. Thus a state point establishing h_1 and W_1 also establishes t_{s2}, and that temperature is a property of the original moist air condition. It is given the symbol t^*, and is called the thermodynamic wet-bulb temperature. Equation (6.25) can be rewritten for a general point without reference to the hypothetical process from section ① to section ②:

$$h + (W_s^* - W)h_f^* = h_s^*. \tag{6.26}$$

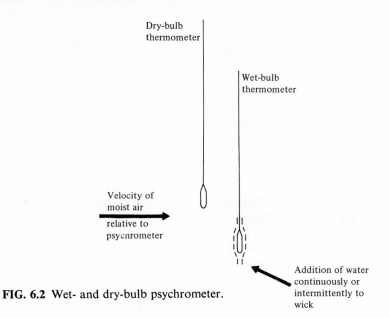

FIG. 6.2 Wet- and dry-bulb psychrometer.

Equation (6.26) is correct, depending only on the definition of thermodynamic wet-bulb temperature. It can be combined with an empirical equation for the value of h to yield a useful approximate expression for the determination of W. This should require only table data depending on a knowledge of t and t^*. From Eqs. (6.15) and (6.26), we find that

$$W = \frac{h_s^* - W_s^* h_f - 0.240\, t}{1061 + 0.444\, t - h_f^*}. \tag{6.27}$$

Or, to introduce Eq. (6.15) again for h^*,

$$W = \frac{W_s^*(1061 + 0.444\, t^* - h_f^*) - 0.240(t - t^*)}{1061 + 0.444\, t - h_f^*}. \tag{6.28}$$

An examination of the equations presented so far in this chapter shows that at a particular barometric pressure the state point of moist air can be established by the values of any two of the following properties: t, t_d, t^*, v, h, s, W, μ, ϕ, or p_w. Of all these, only temperature lends itself to direct, relatively simple determination. Thus most **psychrometers** or **hygrometers** attempt to measure t and either t_d or t^*. The wet- and dry-bulb psychrometer, which attempts to measure t and t^*, is the most familiar instrument. Many forms have been devised. Various other instruments are in use that are responsive to variation in the humidity ratio, but do not measure moist air properties directly. All these are discussed in the references.

The wet- and dry-bulb psychrometer illustrated in **Fig. 6.2** utilizes two identical temperature measuring elements (mercury in glass thermometers, thermocouples, etc.). One is dry and presumably measures the actual air temperature (dry-bulb temperature), while the other is kept wet by a cloth cover retaining liquid water on the sensing element, and this is cooled to a lower temperature (wet-bulb temperature) by evaporation, unless the ambient air is at saturation conditions. Whether or not this lower temperature is close to the thermodynamic wet-bulb temperature depends on many factors. Factors affecting the deviation between these two temperatures include the relative velocity between the air and the wet bulb, the diameter of the wet bulb, the presence or absence of radiant shielding, the actual difference between t and t^*, which is a function of t, and the humidity ratio. Surprisingly, an unshielded wet bulb gives better results under the usual circumstances than a shielded wet bulb if surrounding surfaces are at ambient air temperature. The reason is that the energy loss by evaporation (mass transfer) is larger than the gain by conduction heat transfer at the desired temperature. The additional gain by radiant heat transfer from the environment very nearly compensates by the correct amount to bring the wet-bulb temperature close to the desired thermodynamic wet-bulb temperature. This is discussed in some detail by Threlkeld [B.11], Chapter 10.

A wet- and dry-bulb psychrometer, properly used under normal conditions, will give wet-bulb readings within $\frac{1}{2}$ degree F of the thermodynamic wet-bulb temperature. Below 32 F, the "wet-" and dry-bulb psychrometer can still be used. Water frozen on the sensing element will sublimate to produce a "wet-bulb" reading. However, greater care is required in obtaining reliable readings.

6.2.2 The Psychrometric Chart

The real properties of moist air already discussed, which are determinable through the use of the tabulated properties, have also been presented in charts of various forms. These charts speed psychrometric or air-conditioning calculations when, very often, the assumptions and approximations necessary preclude the degree of accuracy ostensibly achievable from the tables. As an aid to the understanding of psychrometric processes, a chart presentation can be far superior to tabulated data and formulas. The various processes can be visualized and explained relatively easily once the chart form is understood.

Many engineers familiar with the use of a psychrometric chart are unaware of the efforts and judgements lying behind the selection of scales and terms incorporated in the final format. The American Society of Heating, Refrigerating and Air Conditioning Engineers has published a set of charts which avoids most of the objections raised to other forms. The first criterion to be met was that the chart, in general, and all its terms should be thermodynamically consistent with the published tabulated data. Second, it was generally agreed that the basic coordinates for the chart should be enthalpy of the mixture per pound of dry air (h) and humidity ratio (W). These coordinates were suggested many years ago by Mollier, after whom the h-s chart for steam was named.

An objection has been raised to certain features of the ASHRAE chart. The use of relative humidity seems inconsistent with the use of W as an ordinate, and degree of saturation would be preferable in referring to table data; nevertheless, relative humidity has been used for so long that it cannot easily be abandoned or redefined. The inclusion, although in a minor way, of the terms "sensible heat" and "total heat" would be better abandoned, since thermodynamics has for many years used enthalpy as a property and rejected "heat" as belonging to the old caloric theories.

Although W and h have been selected as coordinates, they have not been made mutually perpendicular on the chart. In older charts, dry-bulb temperature has been the abscissa, and this format can be approached by sloping the enthalpy lines properly to make the dry-bulb temperature approximately vertical. The degree of approximation must vary with temperature, since constant t is not quite a straight line on h-W coordinates, and two constant temperature lines will diverge slightly.

The basic geometric construction of such a psychrometric chart has been explained by Goodman [B.5], but the data available to him have since been improved. Threlkeld [B.11] has adapted Goodman's procedure to a new and more reliable set of charts, explaining in some detail the choice of scales.

A valuable term used in connection with psychrometric charts is **enthalpy moisture ratio** q'. It is defined by

$$q' = \frac{h_2 - h_1}{W_2 - W_1} \text{ Btu/lb moisture.} \tag{6.29}$$

This gives the slope of a straight-line process on h-W coordinates and can often be used to establish endpoints even when the process is not a straight line.

Figure 6.3 gives an outline of the ASHRAE Psychrometric Chart 1 for normal temperature. ASHRAE Charts 2 and 3 cover the low and high ranges from -40 to $+50$ F and $+65$ to $+250$ F, respectively, so that there is an overlapping with the normal temperature chart which covers $+32$ to $+120$ F; all are for standard atmospheric pressure only. Chart 4, covering the same temperature range as Chart 1, is drawn for a standard altitude of 5000 ft, at which the barometric pressure would be 24.89 in Hg. Copies of these charts are available from the Society. Other isobaric charts will probably be published in the future.

The basic lines of these charts will now be considered:

a)   Since the humidity ratio (W) is chosen as a vertical ordinate, lines of **constant** W are horizontal straight lines on the charts.

b)   Enthalpy, being one of the two coordinates, gives straight parallel lines of **constant** h which are not vertical, but arbitrarily sloped to the horizontal at an angle β.

c)   The **saturation line** is plotted directly from h and W values in the tables of moist air properties. This results in a curved line sloping upward to the right. Below and to the right of this line, the water vapor in the moist air is superheated, and the humidity ratio is less than unity. Above and to the left of the line, any water greater in amount than that required for saturation at the existing dry-bulb temperature must be in the liquid or solid phase, presumably in a fine suspension. This is commonly called the **fog region.**

d)   The characteristic of lines of **constant dry-bulb temperature** can be determined by the approximate relationship of Eq. (6.15):

$$h = 0.240t + W(1061 + 0.444 \, t).$$

At constant dry-bulb temperature, we find that

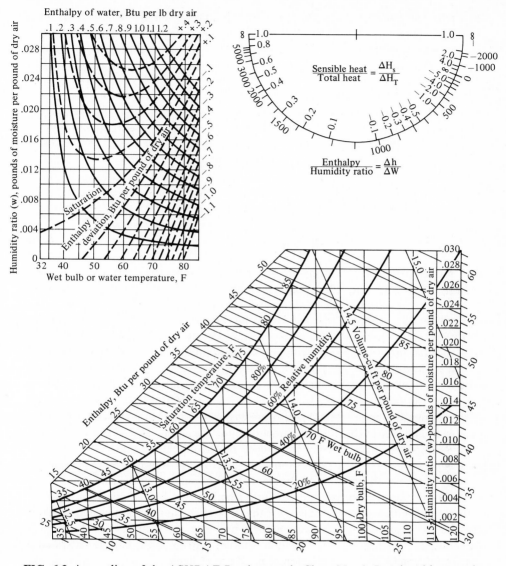

FIG. 6.3 An outline of the ASHRAE Psychrometric Chart No. 1. Reprinted by permission from the ASHRAE Handbook of Fundamentals, 1967.

$$h_2 - h_1 = (W_2 - W_1) \times \text{(a constant)},$$

or,

$$q' = \frac{h_2 - h_1}{W_2 - W_1} = \text{const}, \tag{6.30}$$

and lines of constant t are straight on h-W coordinates. The slope of each line will depend on the temperature, the lines diverging as W increases. Insofar as the real properties deviate from Eq. (6.15), and this deviation is small, the constant t lines will deviate from straight lines. The angle β for enthalpy lines is chosen for each chart to make constant t lines nearly vertical, but the dry-bulb temperature must not be mistakenly construed as the abscissa. Equation (6.15) applies only in the region below the saturation line, and constant t lines cannot be extrapolated into the fog region (see the following paragraph).

e)   Any two points with the same **thermodynamic wet-bulb temperature** can, by definition, be brought to saturation at that temperature by the hypothetical adiabatic saturation process. The equation for that process is Eq. (6.26):

$$h + (W_s^* - W)h_f^* = h_s^*,$$

in which (h, W) and (h_s^*, W^*) are two points on the line which contains all points of identical values of t^*. For that line, we find that

$$q' = \frac{h_s^* - h}{W_s^* - W} = h_f^*. \tag{6.31}$$

Now for a particular thermodynamic wet-bulb temperature, h_s^* (which is the enthalpy of saturated liquid at that temperature) is a known constant. Thus lines of constant t^* are straight with a slope dictated by h_f^* and therefore not quite parallel. For lines of constant enthalpy, q' must be zero; therefore the value of h_f^* shows the difference in slope between constant h and constant t^* lines. The numerical value of difference in slope is small, and so these lines provide a poor intersection. Since it is more common for a user of the psychrometric chart to start with a knowledge of t^* than of h, the constant h lines are not extended into the region below and to the right of the saturation line except for every 5 Btu/lb d.a. Since constant h lines are parallel, intermediate points are easily projected to scales at the edges of the chart. The wet-bulb lines are therefore not obscured. In the **fog region**, Eq. (6.26) still applies, since h_s^* may be reached by the adiabatic removal of moisture from any point above the saturation line. During the removal of the moisture, the temperature of the moist air will remain unchanged so that the line of adiabatic saturation or the thermodynamic wet-bulb line is also a line of constant dry-bulb temperature. With the slope unchanged, the lines of constant t^* can be extrapolated into the fog region. This simultaneously provides lines of identical dry-bulb temperature. The fog region, which is limited on the chart, can be extended if desired by straight line extrapolation of h, W and t^* lines.

f)   Lines of **constant specific volume** are found to be very nearly straight (within chart accuracy) and are so drawn. The slope is a weak function of W so that these lines diverge slightly. The units for convenience are cubic feet/pound dry air.

g)   The **degree of saturation** μ can easily be calculated for any value of W and the table value of W_s at the same dry-bulb temperature. The **relative humidity** ϕ is determined from μ by Eq. (6.10); alternatively, W and ϕ can be related by the use of Eq. (6.9a) with tabulated saturation pressure values at the dry-bulb temperature. Either μ or ϕ could be plotted, but to plot both would confuse the chart since they begin to diverge, as previously noted, only at about 60 F. While the use of μ would be more consistent with the ordinate W, convention has dictated the use of ϕ in the ASHRAE psychrometric charts.

Two appendages to the ASHRAE psychrometric charts are found on the top or left side of each. One of these is an **enthalpy/humidity ratio protractor**, which gives the slope of the line for any value of $q' = \Delta h/\Delta W$ for the particular chart. This is extremely convenient, and its use will be illustrated in examples that follow. The other appendage is a nomograph that can be used for two purposes: (a) as one means of determining enthalpy for a state point by applying a correction to the approximate enthalpy at wet-bulb saturation, and (b) as a means of determining the enthalpy of liquid water added or removed in a process.

Parenthetically, it is noted that the enthalpy/humidity ratio protractor has a second scale, which is called the ratio of **sensible heat** to **total heat**. The former implies heat transfer that results in a "sensible" change of temperature, while the latter includes moisture enthalpy. The inappropriateness of using the term heat on a thermodynamic property chart has already been mentioned. The terms "sensible heat" and "total heat" are not, in fact, clearly defined in terms of moist air properties. The importance of the enthalpy/humidity ratio protractor in determining heat transfer quantities will be demonstrated for several different situations later in this chapter.

It was noted that each of the current ASHRAE psychrometric charts is drawn for only one barometric pressure. Threlkeld [B.11] has developed charts which include saturation lines for pressures from 14.696 to 10.0 psia. These lines provide a means for correcting wet-bulb temperatures and relative humidity for the effect of change in pressure. An alternative would be to prepare more isobaric charts (each for one barometric pressure only) to cover the range most used in engineering work. The present charts for standard atmospheric pressure are not, however, useless at other pressures. Enthalpy and mixture ratios, being coordinates, remain applicable as does dry-bulb temperature which is unaffected. The specific volume values are useful if corrections are made simply by using the gas laws. According to Eq. (6.12),

$$v = \frac{R_a T}{p}\left(1 + \frac{W}{0.622}\right).$$

The temperature and mixture ratio lines on the chart will establish the state point; the specific volume at some pressure other than standard atmospheric pressure can be calculated from

$$v = (\text{chart } v)\frac{(14.696)}{p}. \tag{6.32}$$

6.2.3 Use of the Psychrometric Chart

The use of the ASHRAE psychrometric chart will be explained with examples. For convenience, these will all be taken in the "normal temperature" range, and we will use chart No. 1. The examples here will be similar to those of the ASHRAE Guide and Data Book, but more effort will be made to explain principles, and

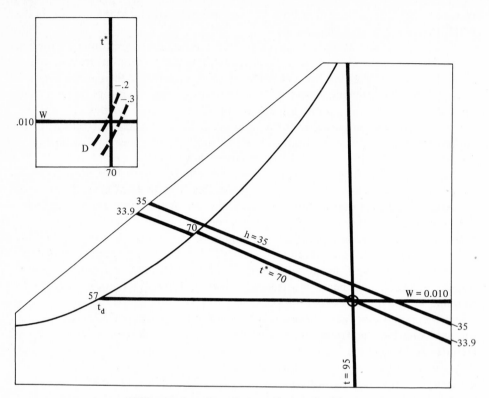

FIG. 6.4 Reading the psychrometric chart.

numerical values will not be used except in the first example. The charts will be used again in Section 6.6 to explain certain air-conditioning systems.

Example 6.1 As an example of how to determine the psychrometric properties of moist air for a given state point, **Fig. 6.4** shows one point chosen at random. Any two properties in addition to the total pressure, assumed here to be standard atmospheric pressure, will establish the point. The values will be read to the accuracy of the chart.

The point shown is at the intersection of $t = 95$ F, $t^* = 70$ F and $W = 0.010$ lb moisture per pound of dry air; these three may be read directly. By visual or more careful interpolation, the relative humidity ϕ is seen to be 28.7%. The dew point t_d, which is the temperature at which moist air would be saturated with $W = 0.010$, is 57 F.

Constant enthalpy lines are not shown in this region. A line drawn through the point parallel to the closest enthalpy line (35 Btu) will intersect the margins at a value of $h = 33.9$ Btu/lb of dry air. The value should be checked with a straight edge through the point at both margins to ensure accuracy of direction. An alternative procedure to determine h would be for us to read the enthalpy value at the

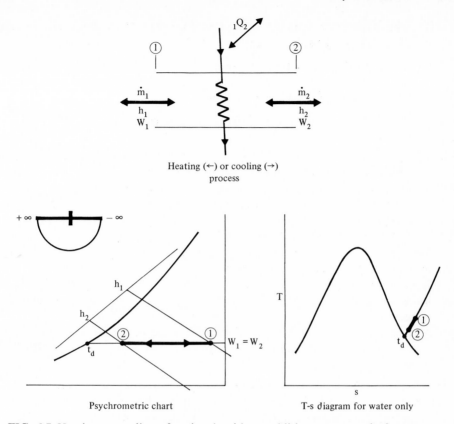

FIG. 6.5 Heating or cooling of moist air without addition or removal of water.

saturation line for $t* = 70$ F, and then apply the correction D from the nomograph on the left using

$$\begin{aligned} H &= h* - (W_s^* - W)h_w^* \\ &= h* + D \\ &= 34.1 - 0.23 \\ &= 33.9 \text{ Btu/lb d.a.} \end{aligned} \tag{6.33}$$

The specific volume can be determined, by linear interpolation between the bracketing values, to be 14.20 cu ft/lb dry air.

The degree of saturation can be obtained from Eq. (6.9d) using the tabulated value of p_{ws} at $t = 95$ F:

$$\begin{aligned} \mu = \phi\left(\frac{p_{as}}{p_a}\right) &= \phi\left(\frac{p - p_{ws}}{p - \phi p_{ws}}\right) \\ &= 28.7\left[\frac{29.921 - 1.6607}{29.921 - 0.287(1.6607)}\right] = 27.5\%. \end{aligned}$$

Had the barometric pressure been, say, 13.0 psia, the following numerical values would still apply to the state point:

$$t = 95 \text{ F}, \qquad W = 0.010 \text{ lb}_w/\text{lb d. a.}, \qquad h = 33.9 \text{ Btu/lb d. a.}$$

Also, the specific volume could be calculated:

$$v = (14.20)\left(\frac{14.696}{13.0}\right) = 16.05 \text{ cu ft/lb d.a.}$$

Example 6.2 Simple **heating or cooling** of moist air in the superheated vapor zone without addition or removal of moisture is illustrated in **Fig. 6.5**. Since the amount of moisture associated with the air is unchanged, the process from ① to ② on the psychrometric chart must be a horizontal line established by $W_1 = W_2$. Note that q' would be plus or minus infinity, and that the enthalpy/moisture-ratio protractor indicates a horizontal line. In the constant total-pressure process, the partial pressures remain unchanged since, according to the gas laws, both constituents will be equally affected by changes of temperature and pressure. Thus the line ①—② on the T-s diagram for water is a constant pressure line in the superheat region. If heat removal were to be carried to the saturation line, the dew-point temperature would be reached.

Because no moisture is removed and any change of enthalpy is reflected directly in a change of temperature, older nomenclature would call this "sensible" heating or cooling.

Example 6.3 When moist air is **cooled below its dew point**, some water vapor is condensed to liquid. An idealized situation will first be considered in which two somewhat unreasonable assumptions will be made: (a) that all the moist air is cooled to the same temperature before leaving the cooling coils, and (b) that all the liquid condensate is cooled to this temperature before leaving the system. These assumptions are used in **Fig. 6.6**. The implications are: (a) that the moist air leaving the system will be at saturation conditions, and (b) that the enthalpy of the liquid leaving is easily determined as the enthalpy of saturated liquid at t_2. The assumptions are best represented by the process shown on the psychrometric chart as heat removal from ① to point X and then liquid removal at constant wet-bulb temperature from X to ②. Note though that the enthalpy of the mixture at X is not that of the final condition ② because the enthalpy of the liquid has not yet been removed. On the T-s diagram for the water vapor alone, points X and ② are identical since the water vapor, whether removed or not, is saturated at t_2^*.

The following equations can be written for the process just described:

$$_1Q_2 = \dot{m}_a(h_1 - h_2) - \dot{m}_f h_{f2},$$
$$\dot{m}_f = \dot{m}_a(W_1 - W_2),$$
$$_1Q_2 = \dot{m}_a[(h_1 - h_2) - (W_1 - W_2)h_{f2}]. \tag{6.34}$$

The difficulty of the first assumption is shown if we try to visualize the alternative path ①— Y —②. Certainly some of the moisture will begin to condense as

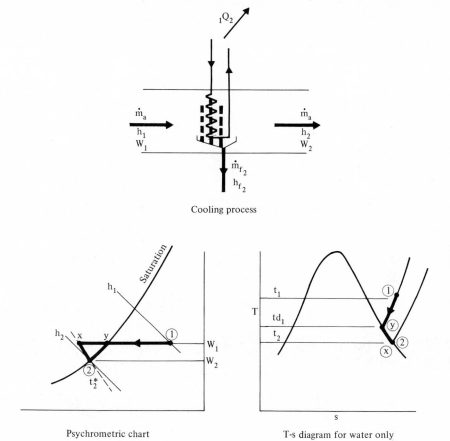

Cooling process

Psychrometric chart   T-s diagram for water only

FIG. 6.6 Cooling below the dew-point temperature.

soon as the dew-point temperature of the original condition is reached at **Y**. The moisture will continue to condense at varying temperature from **Y** to ②.

The second assumption is unrealistic simply because, for a real cooling coil, some of the air passes close to the cooling surfaces and reaches a saturation condition, while air even a little distance from the surface will not be cooled to the same temperature. Subsequent mixing will produce air with a temperature above the coil surface temperature and at a relative humidity less than 100 %. This is illustrated in **Fig. 6.7**, where the broken line from ① to ② is not intended to represent a process path; only the endpoints are significant. Values of W and h for these points can be used in Eq. (6.34). Even with the acknowledged simplification of assumption (b), the error in determining heat removal would be quite small.

The nomograph in the upper left corner can be used to determine the last term of Eq. (6.34) for the process of either Fig. 6.6 or 6.7. This avoids the necessity of using vapor tables to find h_f.

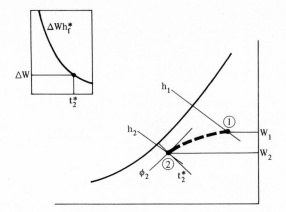

FIG. 6.7 Cooling and dehumidification with imperfect mixing.

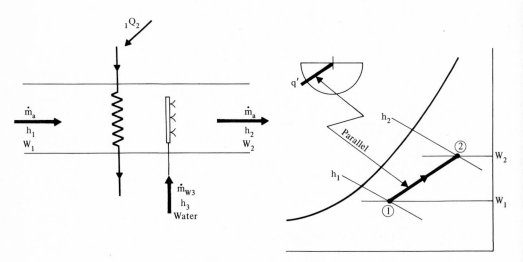

FIG. 6.8 Heating and humidification.

Example 6.4 The **addition of heat and moisture** to moist air is a common problem, particularly for winter air conditioning. **Figure 6.8** shows this process. The water added may be liquid or vapor. The following equations apply:

$$\dot{m}_{a1} = \dot{m}_{a2},$$

$$\dot{m}_a h_1 + \dot{m}_3 h_3 + {}_1Q_2 = \dot{m}_a h_2,$$

$$\dot{m}_3 = \dot{m}_a(W_2 - W_1),$$

$${}_1Q_2 = \dot{m}_a(h_2 - h_1) - \dot{m}_3 h_3,$$

$$q' = \frac{(h_2 - h_1)}{(W_2 - W_1)} \tag{6.35}$$

$$= \frac{{}_1Q_2 + \dot{m}_3 h_3}{\dot{m}_3}. \tag{6.36}$$

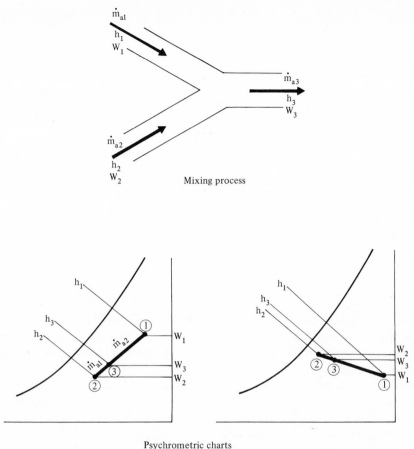

Mixing process

Psychrometric charts
Two possible combinations
with $\dot{m}_{a2} > \dot{m}_{a1}$

FIG. 6.9 Adiabatic mixing of two streams of moist air.

A problem can usually be solved either by the calculation of endpoints using Eq. (6.35) or semigraphically with Eq. (6.36) and the aid of the humidity/moisture ratio protractor to establish the slope of the line ①–②. The nomograph for water enthalpy would be useful only if the water added were liquid.

A special case of this example would be humidification without heating. The value of q' would then be numerically equal to the enthalpy h_3 of the water added. The slope may be upward to the left, to the right, or vertical, depending only on h_3.

Example 6.5 Adiabatic mixing of two streams of moist air provides an example similar to that presented for aqua/ammonia solutions in Chapter 4. In **Fig. 6.9**, the process is illustrated with two possible combinations of $(h_1 - h_2)$ and $(W_1 - W_2)$. Just as for the aqua/ammonia solution, the equations indicate a straight-line relationship for points ①, ②, and ③ on enthalpy-mixture ratio coordinates:

$$\dot{m}_{a1}h_1 + \dot{m}_{a2}h_2 = \dot{m}_{a3}h_3,$$
$$\dot{m}_{a1}W_1 + \dot{m}_{a2}W_2 = \dot{m}_{a3}W_3,$$
$$\dot{m}_{a1} + \dot{m}_{a2} = \dot{m}_{a3}.$$

Combining these equations, we have

$$\frac{\dot{m}_{a1}}{\dot{m}_{a2}} = \frac{h_3 - h_2}{h_1 - h_3} = \frac{W_3 - W_2}{W_1 - W_3}. \tag{6.37}$$

Not only do the points fall in a straight line, but the division of the line is inversely proportional to the ratio of the mass rates of flow.

6.3 PHYSIOLOGICAL FACTORS

The air-conditioning engineer must have some appreciation of the physiological responses of people to their thermal environment. Of course, there are many reasons for conditioning air, such as prevention of corrosion of metals, control of dimensions for gaging and manufacturing, control of moisture content of hygroscopic materials during manufacture and storage, optimization of environment for perishable substances, and prevention of overheating of electrical circuits. Still, the majority of air-conditioning applications must deal with the problem of providing comfortable conditions for people and sometimes animals, or when this is not possible, avoiding exposure to conditions beyond endurance limits. While always important in certain industries and the armed services, endurance limits have received additional attention because of the unprecedented demands of manned space flight.

In general, psychrometric conditions cannot economically be held within a fraction of a degree on the temperature scale or a very small percentage variation in humidity ratio. This usually is not called for because comfort conditions exist over a region rather than at a point on the psychrometric chart. Nevertheless, to establish the **comfort region** during the design stage, and to maintain it in operation is more complicated than might first be imagined. Age, sex, clothing, physical activity, previous conditioning, and state of health all combine to establish psychrometric limits for comfort for a particular individual. In addition, endurance limits vary widely among individuals.

A vast amount of research has been conducted in the last 40 or 50 years to provide data useful to the engineer, and many attempts have been made to reduce the findings to reliable mathematical formulations or single terms that combine many effects. Most of the information to date, however, must be used with caution and a full awareness of the limitations of the original sources. The greatest limitations have been the number of subjects tested and the problem of extending data to clothing and activity conditions other than those prevailing for the experiments. The whole field of physiological response to psychrometric conditions must be considered to be in a state of flux.

There are two extremely useful sources of information available. One is the *Handbook of Fundamentals* [B.2], particularly Chapter 7, Physiological Principles, and Chapter 8, Air Conditioning in the Prevention and Treatment of Disease. The other is the *Bioastronautics Data Book* [B.12], particularly Chapter 1, Atmosphere; Chapter 7, Temperature; Chapter 9, Combined Stresses; and Chapter 10, Energy. While there is considerable overlapping of the two sources, it might be said that, by their nature, the former is primarily concerned with comfort conditions and the latter with tolerance for extreme conditions. Both present extensive bibliographies with sources of specific data carefully noted. In what follows, most of the information will be taken from one or the other of these two references and their original sources. Some additional references will also be noted. It would be impractical and unnecessary for the present purpose to reproduce these bibliographies completely.

6.3.1 The Necessity of a Heat Balance

The human body is continuously producing heat through oxidation of carbohydrates, fats, and proteins. This **metabolic rate** of heat production for any individual depends primarily on the amount of his physical activity at the time. Metabolism has been measured directly by calorimetry and indirectly by the measurement of the oxygen consumed. **Table 6.3,** which is taken from the *Handbook of Fundamentals*

Table 6.3*

ESTIMATES OF ENERGY METABOLISM (M)
OF VARIOUS TYPES OF ACTIVITY*

Kind of work	Activity	M, Btu/hr
Light work	Sleeping	250
	Sitting quietly	400
	Sitting, moderate arm and trunk movements (e.g., desk work, typing)	450–650
	Sitting, moderate arm and leg movements (e.g., playing organ, driving car in traffic)	550–650
	Standing, light work at machine or bench, mostly arms	550–650
Moderate work	Sitting, heavy arm and leg movements	650–800
	Standing, light work at machine or bench, some walking about	650–750
	Standing, moderate work at machine or bench, some walking about	750–1000
	Walking about, with moderate lifting or pushing	1000–1400
Heavy work	Intermittent heavy lifting, pushing or pulling (e.g., pick and shovel work)	1500–2000
	Hardest sustained work	2000–2400

* Reprinted by permission from ASHRAE Handbook of Fundamentals (1967). The values apply for a 154 lb man, and do not include rest pauses.

Table 6.4

Extreme heat stress	↑	Disastrous if continued.
Inevitable body heating		Heat exhaustion, cramps, stroke.
Evaporative regulation		Operation of the sweat glands.
Vaso-motor regulation		Dilation of surface blood vessels. Increased blood flow through skin. Rise in skin temperature.
Neutral zone	◯	No body reaction. Comfort.
Vaso-motor regulation		Drop in skin temperature. Constriction of surface blood vessels. Decreased blood flow through skin.
Metabolic regulation		Increased muscular tension.
Inevitable body cooling		Shivering. Spontaneous activity.
Extreme cold stress	↓	Disastrous if continued.

[B.2], shows typical metabolic rates for an "average" man weighing 154 pounds. Note that the range covers a ratio of about 10 to 1. The body must lose this heat to the environment or suffer a temperature rise. Conversely, if the heat loss exceeds the metabolic production, the body temperature must fall. As will be seen, man can endure a wide range of environmental conditions if certain responses occur and precautions are taken, but his body temperature must remain within quite narrow limits to avoid tissue or brain damage.

The exchange of heat with the environment can be accomplished by **evaporation** (almost always a positive body loss), **radiation,** and **convection** (the last two being either positive or negative). Heat conduction to the air requires continuous removal from the body surfaces of warmed air, and is therefore classed as convection. Conduction to other contact surfaces is usually, but not necessarily, negligible. The difference between metabolism and the sum of the body heat losses is **stored energy**. This may be positive or negative at any moment, but over a long period of time, it must be zero. The **heat-balance equation** for the body and its environment therefore is

$$M = \pm S + E \pm R \pm C \text{ Btu/hr,} \tag{6.38}$$

where S is the rate of heat storage in the body, and the other symbols have meaning obvious from the preceding discussion.

Within limits, the body will respond to positive or negative heat storage in such a way as to increase or decrease one or more of the other terms in an attempt to achieve an equilibrium condition. This response above or below a null position can be interpreted as a **strain** in the body in response to a **stress,** which is a combination of the environmental conditions and the metabolic energy rate demanded by the activity. The stress-strain relationship however is not a simple one. The responses of the body are summarized very briefly in **Table 6.4**. The **neutral zone** or null point from which one should start reading is located in the middle of the table be-

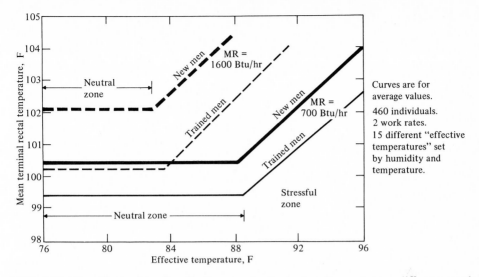

FIG. 6.10 The effect of acclimatization on body temperature at two different work rates (from Wyndham, *et al.* [B.12]).

tween extreme heat stress at the top and extreme cold stress at the bottom. Displacement up or down from the neutral zone is in the direction of increasing stress.

Slightly above or below the neutral zone, Table 6.4 indicates that the body heat balance is maintained by variation in the flow of blood to the skin. An increased flow near the surface improves heat transfer by blood convection to the surface with a resulting increase in skin temperature for greater heat transfer to the air. A decreased convection of heat by the blood to the skin allows a drop in skin temperature and thus better insulation between deep body temperature and the environment. The more powerful responses of operation of the sweat glands for increased heat stress and voluntary and involuntary activity to raise the metabolism rate in response to cooling are certainly indications of body strain precluding comfort. The limit of the effectiveness of either of these modes of heat balance regulation becomes the tolerable limit of stress in the hot or cold regions. When the body strain cannot be increased beyond these limits, stress must be limited to prevent disaster.

Evidence is plentiful that man can be **acclimatized** to increased heat stress. This acclimatization is primarily through increased ability to lose perspiration with a simultaneous reduction in the salt content of the fluids lost. Apparently, the ability to endure an increased heat stress for longer periods without undue body temperature rise can only be accomplished when lost fluids are regularly replaced by drinking water; thirst is not a sufficient guide to the optimum frequency of drinking; regular, controlled replacement has been found more effective than *ad lib* replacement in limiting body temperature rise. **Figure 6.10** shows the difference

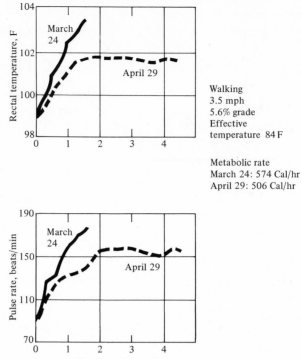

FIG. **6.11** An example of rapid acclimatization to heat stress (from Robinson, *et al.* [B.12]).

in final equilibrium rectal temperature for both trained (acclimatized) and untrained men working at two different rates in a range of **effective temperatures** above the comfort zone. (Effective temperature is intended to indicate the combined effect of dry-bulb and wet-bulb temperature and air movement, and will be discussed later). The so-called "neutral zone" here probably extends through the region of vaso-motor regulation. The stressful zone is indicated by environment driven body temperature. **Figure 6.11** illustrates the very rapid acclimatization to heat stress for one subject who, over a period of five weeks, was exposed to the conditions of test only eleven times in the first $2\frac{1}{2}$ weeks and once again at the end of the fourth week. Note incidentally that the efficiency of work has apparently increased since the metabolism has decreased for the same task. On the first day (March 24), the test was terminated at 90 min because the subject collapsed. On the last test, equilibrium was being maintained with ease after $4\frac{1}{2}$ hours. The results were similar for a second subject.

Acclimatization to cold does not seem to take place. While knowledge of how to prevent body damage by proper dress, diet, and other habits can be acquired, the apparent ability to tolerate discomfort or even a reduced body temperature seems to be psychological rather than physiological. Where differences have been noted between ethnic groups in the tolerance of cold during sleep, the differences have been attributed to acceptance of some discomfort by one group.

Pain limited tolerance of heat seems to be very clearly defined at a skin temperature of 113 F, one degree more being intolerable. (This does not apply to the elbow or knee which might receive second degree burns without pain). The exposure time to reach this value depends of course on many factors. Under conditions of extreme cold, various areas of the skin usually tolerate different low temperatures. The hands for instance are typically "painful and numb" at 50 F skin temperature and the feet at $55\frac{1}{2}$ F. The typical sensation for a weighted mean skin temperature of 84 F is "extremely cold."

6.3.2 Calculation of Individual Components of the Heat-Balance Equation

It was mentioned that the **metabolic rate** of heat generation could be determined by direct calorimetry with subjects placed in chambers with water cooled surfaces, or could be inferred from the consumption of oxygen. The results of these two techniques have been well correlated by various authorities. The relationship depends on the **respiratory quotient** (R.Q.) which is the volume ratio of carbon dioxide expired to the oxygen absorbed. This in turn depends on whether carbohydrates, fats, or protein are being consumed and on the activity level. Hawk, Oser, and Summerson [B.6] show a range of metabolism only from 4.7 to 5.0 Cal/liter of oxygen consumed over a range of R.Q. covering likely values. Thus average values cannot be greatly in error. Fahnestock and his coworkers [A.16] used a value of 4.825 Cal/liter for a wide range of work rates. The use of 1 liter/min as equal to 1158 Btu/hr is reported by Webb [B.12]. This is equal to 4.76 Cal/liter of oxygen, since

$$1 \text{ Calorie} = 1 \text{ Kilocalorie} = 3.966 \text{ Btu}.$$

For air conditioning design work, as opposed to research for determining these quantities, the metabolic rate is best estimated from a description of the activity level, as shown in Table 6.3.

The metabolism cannot easily be estimated from the actual power necessary to do a task, because the efficiency of the human body is widely variable. The ratio of work accomplished to metabolic energy is unfortunately called "**mechanical efficiency**" by physiologists, and is distinguished from the mechanical engineering term of the same name by the use of quotation marks here. The work of Fahnestock referred to in the previous paragraph shows a wide range of values for "mechanical efficiency" as summarized in **Fig. 6.12**. Four work rates (600, 1400, 2000, and 2400 ft lb/min) were used. Efficiencies based on the difference between total working metabolism and resting metabolism are of interest in indicating the efficiency of the extra energy required for work. The lower values, based on total working metabolism, allow a conversion from job requirement to heat release. Values from 5 to 35% are reported by others.

The **convective heat** loss from the body could be calculated in a conventional way as

$$C = K_c A_c (t_{skin} - t) \text{ Btu/hr}$$

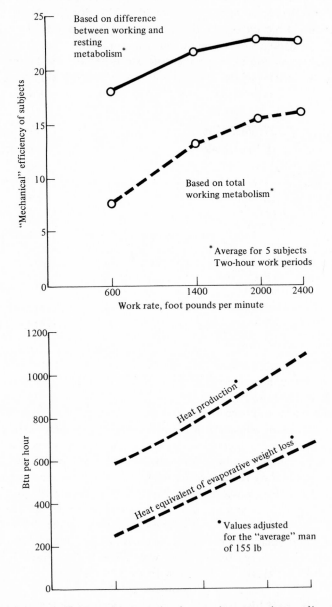

FIG. 6.12 Mechanical efficiency, heat production, and evaporative cooling for four work rates [A.16].

if the coefficient K_c could be well established. It must certainly depend on air velocity, body movement, and clothing. In calculating values of the **heat stress index**, to be discussed below, Belding and Hatch [A.5] used a relationship determined for nude men in a standing position, and this is one of the present limitations to the usefulness of their charts. Taking $K = 0.1 V^{0.5}$ and $A = 20$ sq ft, they used

$$C = 2 V^{0.5}(t_{skin} - t) \text{ Btu/hr.} \tag{6.39}$$

The **radiant heat** loss from the body to the room could be calculated from the basic radiation heat laws if all the terms were known:

$$R = \epsilon K_R A_R(T^4_{skin} - T^4_w).$$

The **emissivity** (ϵ) is between 0.7 and 0.9 for most surfaces and nearly 1.0 for the body or clothing at wavelengths beyond the visible range. The radiation constant (K_R) is 1.7×10^{-9} Btu/hr-ft²-R. The effective radiation surface of the human body is naturally less than the total skin surface and is taken by Haines and Hatch to be 15.5 sq ft. Skin temperature is in the range of 95 F. The last term (T_w), the mean effective radiant temperature of the surroundings, is the most difficult to evaluate. A **globe thermometer** can be used for this purpose. The globe thermometer is one for which the temperature-sensing element is in the center of a 6-in. diamater hollow metal sphere painted flat black. At equilibrium, which should be reached in 20 minutes, the radiant heat gain is balanced by the convective heat loss of the globe:

$$R_g = C_g,$$
$$(0.95)(1.7 \times 10^{-9})(T^4_w - T^4_g) = 0.17 V^{0.5}(t_g - t).$$

This approximate formula [B.2] permits the calculation of T_w when air velocity, globe temperature, and air temperature are measured. The equation

$$R = \epsilon K_R A_R(T^4_{skin} - T^4_w)$$

can be simplified to

$$R = K'_R 15.5(t_{skin} - t), \tag{6.40}$$

where

$$K'_R = K_R \frac{(T^4_{skin} - t^4_w)}{(T_{skin} - T_w)}$$
$$= K_R(T^3_s + T^2_s T_w + T_s T^2_w + T^3_w),$$

which is a function of T_w only, for an assumed constant skin temperature. (Note that where differences between temperatures to the first power are desired, either Fahrenheit or Rankine values may be used, but this is not true for other powers.)

Reasonable accuracy (within 50 Btu/hr) can be achieved, as noted by Belding and Hatch, for wall temperatures not higher than 175 F, by use of an average value for K'_R to reduce Eq. (6.40) to

$$R = 22(t_{skin} - t_w) \text{ Btu/hr.} \tag{6.41}$$

The **evaporative heat** loss from the body can be estimated in different ways. The most direct in laboratory work is to determine the weight loss over a period of time. Each pound of moisture removes about 1050 Btu (each gram a little less than 0.6 Cal, or 600 cal). But there is a limit to the application of this value: once the skin is completely wet and perspiration drips off, the heat of vaporization is not taken from the body. Whenever the evaporation is at some distance from the skin, as on hairy areas or clothing, part of the latent heat of vaporization may come from the air.

Again, Belding and Hatch suggest a formula for evaporative heat removal, but only for the maximum condition of a nude man with fully wetted skin:

$$E_{max} = 533 \, V^{0.4} \, (p_{ws,skin} - p_w) \text{ Btu/hr}, \qquad (6.42)$$

where $p_{ws,skin}$ is the saturation pressure of water vapor, psia, at skin temperature, and p_w is the ambient vapor pressure. At 95 F, $p_{ws,skin}$ would be 0.816 psia. Even below conditions of elevated heat stress, the body secretes a fairly stable minimum rate of perspiration equivalent to about 10% of the maximum. This constitutes a part of the body cooling process under comfort and even cold stress conditions.

6.3.3 The Evaluation of Comfort and Thermal Stress

Unfortunately there is no simple, reliable method at the present time for evaluation of relative comfort and relative thermal stress from known psychrometric and air movement conditions. Several systems of evaluation will be described, some fairly well established and others recently proposed and requiring considerable verification.

The **ASHRAE Comfort Chart** has long been used to show lines of equal comfort or discomfort in terms of psychrometric conditions. Its basis is the concept of **effective temperature** introduced by Houghten and Yaglou in 1923 [A.27]. The experimental method originally was to have observers pass from one test room to another, expressing each time whether one room was warmer or cooler than the other. The conditions in one room were held constant while those in the other were slowly changed to bracket a feeling of equal effective temperature. All lines of identical effective temperature were identified with the value of the temperature at saturation (equal dry-bulb and wet-bulb) falling on the line. The chart has been revised since the original work, but has been criticized as indicating equal "immediate-temperature perception" without an opportunity for equilibrium to be established between the body and the environment.

Figure 6.13 shows the comfort chart with effective-temperature lines and distribution curves of preferred conditions for summer and winter. Superimposed on the chart are lines of **equal comfort**, determined more recently by Koch, Jennings, and Humphreys. To establish these lines, observers wearing light indoor clothing remained in a constant condition test room moving about only slightly for three hours. They registered their subjective thermal sensations by means of a scale from 1 to 7 with 4 indicating comfort. These curves would indicate that, in the comfort

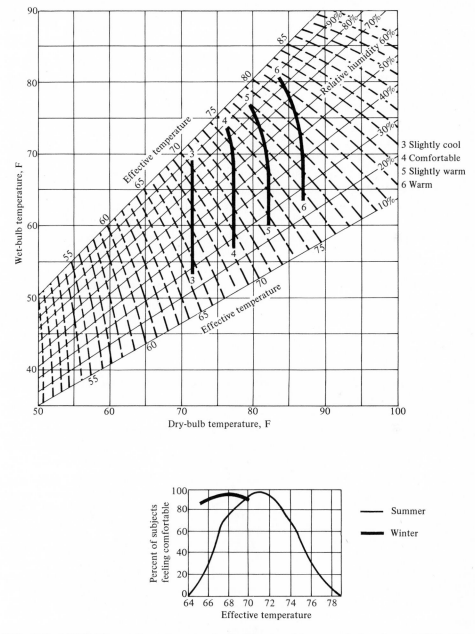

FIG. 6.13 ASHRAE comfort chart (revised). Reprinted by permission from ASHRAE Handbook of Fundamentals, 1967.

region, dry-bulb temperature alone is important except at very high relative humidity. For warmer conditions, where perspiration is above the normal minimum level, the wet-bulb temperature assumes more significance. These tests extended over summer and winter periods, and the lines shown are averages. Summer lines would fall only about 0.6 degree above, and winter lines fall only about 0.6 degree below.

It is interesting to note that the desired effective temperature for winter, found in tests since the early 1920's, has been gradually rising from about 66 F to the present value of 68 F (which incidentally coincides with the comfort No. 3 line at 60% relative humidity). This rise has been attributed to the fact that indoor winter clothing is lighter than before. A similar difference would be expected between countries where habits of dress are not the same.

The effective temperature is affected by **air movement** as would be expected. The values of Fig. 6.13 were obtained with air velocities of 20 ft/min or less. **Figure 6.14** allows adjustment for velocities up to 700 ft/min (about 8 mi/hr). This chart was also developed from large numbers of subjective responses to thermal environment. Note that the cooling effect of an increased air movement reverses to a heating effect for air temperatures over 100 F. It is advised that globe temperature should be used rather than dry-bulb temperature with Fig. 6.14, where the difference, because of radiant heat, is more than 2 F degrees. Endurance limits of effective temperature have been noted at 90 F, 85 F, and 80 F for continuous light, moderate, and heavy work, respectively, for heat acclimatized men.

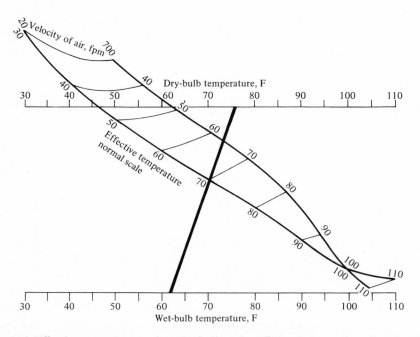

FIG. 6.14 Effective temperature chart including the effect of air motion. Reprinted by permission from ASHRAE Handbook of Fundamentals, 1967.

Usually it is very difficult to obtain a meaningful measurement of air velocity. To eliminate this necessity, the term **wet-bulb globe-temperature index** (WBGT) has been suggested [B.2]. Air movement will have an effect on globe temperature and should not have to be measured separately. A weighted average of the two temperatures (70 % wet-bulb and 30 % globe) is recommended for a single value to replace effective temperature when radiant heat is a significant factor.

A measurement of heat strain has been attempted [B.2] in the development of an **index of physiological effect** (Ep). The heart rate, skin temperature, rectal temperature, and sweat rate have been measured on a limited number of subjects in an attempt to produce a correlation between heat strain and psychrometric conditions.

Most authorities agree that heart rate and skin temperature are unreliable indications of heat stress. As a matter of fact, skin temperature falls when sweating becomes profuse. But the tendency to produce sweat is an indication of strain, whether it is heat strain or some other strain such as anxiety or pain. An excellent index of heat stress or strain, which is unfortunately difficult to apply outside the

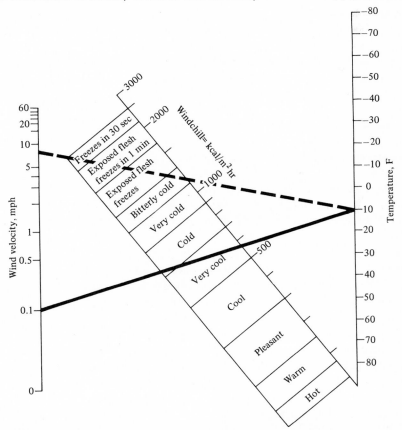

FIG. 6.15 Windchill index of heat loss (from Bioastronautics Data Book [B.12]).

laboratory, is the **predicted four-hour sweat rate** (P4SR). A sweat loss of 4.5 liters in four hours is considered an upper limit for acclimatized men [B.2].

A term that has been developed [B.2] from a detailed consideration of the body heat balance is the **heat stress index** (HSI) of Belding and Hatch, based on the empirical relationships for C, R, and E presented earlier. Charts have been devised for an evaluation of the HSI, but are not presented here because of limited verification to date. An index of 100 is the "maximum strain tolerated daily by fit, acclimatized, young men." When the HSI equals zero, it indicates "no thermal strain," and -20 indicates "mild cold strain." The considerations are similar to those for the P4SR since the calculated evaporation required for body heat balance and the ability of the environment to dissipate this are the factors responsible for establishing the HSI. Regardlesss of other factors, the limit of sweating allowed is 1 liter/hr for 8 hr, which is equivalent to $E = 2400$ Btu/hr.

At temperatures well below the comfort region, the wind velocity becomes a critical factor in establishing tolerance limits. **Figure 6.15** gives some indication of the wind effect. The **windchill index** of heat loss, which is a function of air temperature and wind velocity, was actually determined by the cooling effect on a container of water; it cannot be rigorously applied to the human body because physiological responses are partially additive [B.12].

For comfort, positive or negative heat stress, or endurance limits, clothing is extremely important. In hot, dry, desert conditions, clothing can increase the surface and improve evaporative heat loss while protecting against conduction and radiation to the body. In cold environments, it controls, to a large degree, the tolerable exposure time. The extremities, particularly the feet, are difficult to protect adequately against cold. The value of clothing insulation is given by the unit *clo*, where one *clo* is equal to an insulation value of 0.88 F-ft^2-hr/Btu. This is about the effect of a mechanic's light-weight coverall.

Other attempts to evaluate comfort, stress, and endurance limits have been and will be made. None is perfect. Where possible, evaluations should be made by different methods to seek verification, particularly where conditions might be critical.

Finally, the physiological factors involved in **hospital air conditioning** should be mentioned. The control of air-borne infection, the provision of satisfactory conditions for new-born children and particularly premature infants, the special conditions for post-operative recovery rooms, and the alleviation of certain allergic conditions all require special considerations. In addition, the creation of high or low body temperature and environmental fog conditions for therapeutic purposes requires the cooperation of engineers and physicians. A particularly dangerous condition that can be alleviated by air conditioning is that encountered in cases of extensive burned skin area; the damaged areas lose moisture at an extremely rapid rate unless very high humidity conditions are provided in the environment, and this rate of body cooling and moisture loss cannot be tolerated for long by a patient who may already be suffering from shock. Kranz [A.33] mentions this problem and also a system of complete environmental isolation of body areas during surgery by means of a plastic isolation canopy, where humidity and sterility can be controlled apart from the rest of the operating room. The difference in requirements between those of the patient who may be anesthetized and those of the operating staff can perhaps be met only by these isolators. The danger of explosion in the operating room because of the use of oxygen and certain anesthetics is an additional air-conditioning problem.

6.4 DEHUMIDIFICATION

The removal of moisture from the air is an essential part of air conditioning in a wide variety of applications. For small container storage or large warehouse storage of materials subject to rust or mildew damage, and in manufacturing processes and laboratory testing involving hygroscopic materials, control of the moisture content of ambient air is often more important than temperature control.

In the discussion of the psychrometric chart, Example 6.3 showed that moisture can be removed from the air by a **cooling process** extending below the dew point. Unless the endpoint temperature of that process is desired, cooling for dehumidification may be uneconomical. To produce extremely dry air requires refrigeration to extremely low dew-point temperature. If a higher final temperature is desired, reheat may be necessary. This heat is not necessarily from an outside source, since part or all of the reheating may be accomplished by the condenser of the refrigeration system. Still, power must be supplied to the refrigerator.

Moisture can be separated from the air by **compression**. If the volume of the mixture per pound of water is reduced below the specific volume for saturated water vapor at the temperature of the mixture, some vapor must condense to liquid. The presence of the air has a practically negligible effect on the volume considerations, although it usually contributes the major part of the total pressure. If the compressed moist air is cooled back to approximately its original temperature, and the liquid water is separated out, the saturated high-pressure air can expand to its original pressure at a reduced humidity (established by the new partial pressure of the water vapor and the final temperature) and a reduced mixture ratio (established by the pressure and temperature before reexpansion). This is a useful method of dehumidification where compression of the air is necessary in any case. It can be supplemented by sorbent dehumidification after compression if very dry compressed air is needed, as in shock-tube operation. The removal of much of the moisture by compression and aftercooling reduces the moisture load on the sorbent material.

The change in the dew-point temperature caused by the compression of moist air can be determined approximately from the pressure ratio. If the original saturation vapor pressure is multiplied by the pressure ratio, the dew-point saturation pressure after compression is obtained. The new dew-point temperature can be determined from vapor tables. If the final temperature of the compressed moist air is below the new dew point, condensation of moisture must occur. The calculation is approximate, because it assumes that Dalton's law of partial pressures applies, whereas actually the air and water vapor do not act completely independently.

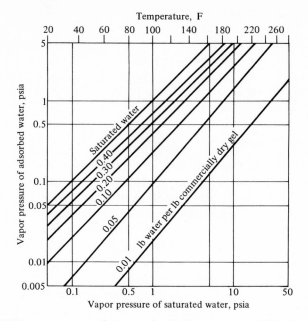

FIG. 6.16 Vapor pressure of water adsorbed on commercially dry silica gel. Reprinted by permission from *Unit Operations* by Brown, *et al.* [B.4].

More effective and economical dehumidification is often achieved by **sorption**. Various substances have a high affinity for water, and will remove water from the atmosphere whenever the ambient vapor pressure exceeds that of the water associated with the sorbent materials. **Absorption** and **adsorption** were both mentioned in the introductory paragraph for absorption refrigeration in Chapter 4. When there is a chemical or physical change in the sorbent material as the water content increases, the process is called **absorption**. This is the case for all liquid sorbents where the concentration of the solution is changed; it is the case for certain solids such as calcium chloride, where the addition of water forms a hydrate, $CaCl_2 \cdot 2H_2O$, and addition of water beyond this proportion results in a solution. The phenomenon called **adsorption** occurs only with certain solid sorbents. These do not change phase as water or other vapor is attracted to their surfaces because of unbalanced molecular forces which exist there. For both phenomena, the latent heat of condensation is given up by the change of phase of the water from vapor to liquid. For adsorption, the relatively small heat of **wetting** is also released. For absorption, the heat of **solution** and, for some substances, the heat of the **chemical reaction** will be added to or subtracted from the exothermic heat of condensation.

When water is adsorbed onto a solid sorbent, the vapor pressure of water is reduced much below the saturation pressure for the existing temperature. This **hygroscopic depression** can be expressed in terms of the difference between the temperature of the material and the saturation temperature for water that would produce the same vapor pressure. Alternatively, the effect can be shown on a **Cox chart** such as **Fig. 6.16**, which plots the vapor pressure of the adsorbed water

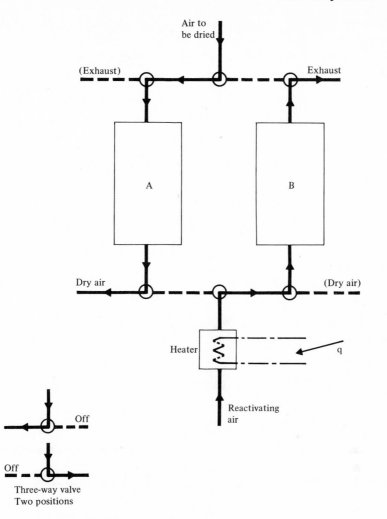

FIG. 6.17 Alternate dehumidification and reactivation in two beds of solid adsorbent.

vs. the vapor pressure of saturated water at the same temperature. In the example shown, the sloped lines are for constant ratios of adsorbed water per pound of commercially dry silica gel which retains about 5% moisture by weight. (Silica gel is a colloidal form of SiO_2). For a given temperature, the vapor pressure depends on the amount of moisture adsorbed. Moisture will continue to be adsorbed from the atmosphere as long as the atmospheric vapor pressure exceeds the adsorbed moisture vapor pressure. The process will be reversed and water will vaporize and leave the adsorbent if its vapor pressure exceeds the ambient value. Because of this, a nearly saturated adsorbent can be **reactivated** if its temperature is raised and there is ventilation for removal of the water vapor. In addition to silica gel, activated aluminas, bauxites, and alumino-silicates are practical commercial **dessicants**.

Figure 6.17 shows a typical system in which continuous dehumidification is achieved by the use of two beds of solid adsorbent. While one is drying air, the other is being reactivated by heated air. When the first becomes ineffective, valves are reversed to reactivate it, while the other dehumidifies. **Figure 6.18** shows a system with a bed of solid adsorbent rotating continuously and slowly (one revolution in about 10 min) so that a freshly reactivated portion of the bed is continuously being brought into the dehumidifying segment.

Dehumidification with liquid **absorbents** can often be incorporated more easily in the continuous flow of an air-conditioning system. As a general rule, the extreme dryness obtainable with solid sorbents cannot be reached, but precise control of outlet conditions is possible through control of both the temperature and the absorbent solution concentration. **Figure 6.19** is a typical performance chart for a

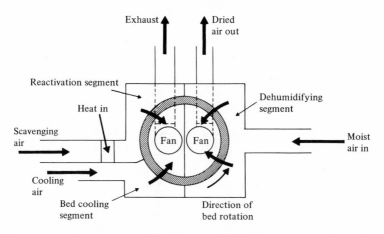

FIG. 6.18 Continous dehumidification with rotating bed of solid adsorbent.

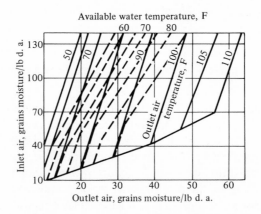

FIG. 6.19 A performance chart for liquid absorbent dehumidifier using lithium chloride spray and water cooling. Reprinted by permission from **ASHRAE Guide and Data Book,** 1967.

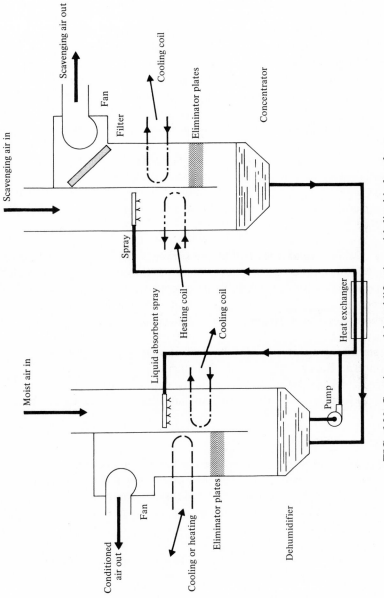

FIG. 6.20 Continuous dehumidification with liquid absorbent.

system in which air is simultaneously passed through a lithium chloride solution spray and over a coil cooled by water. Other salt solutions are used, but some are more corrosive. Triethylene glycol in water is another commercial absorbent. Because the solution of a liquid absorbent is continuously diluted by the moisture taken from the atmosphere, a portion of the pump supply to the spray is bled off to a separate section where moisture is revaporized.

Figure 6.20 is a schematic diagram of a dehumidifying unit with a separate solution concentrator. The moisture content of the conditioned air is controlled by the temperature of the cooling coils over which the solution and air pass, and by the concentration of the solution which itself is controlled by the temperature of the heating coils in the concentrator. Carry-over of solution in the conditioner is prevented by eliminator plates. The final temperature can be established by heating or cooling coils in the same unit. In the concentrator, the warmed air would carry some absorbent past the eliminator plates. To reclaim this, a cooling coil is added along with a filter which has a large surface, allowing the entrained solution to drain back to the pump. The heat exchanger between the concentrator and the air conditioner enables the warm returning liquid to be cooled while it preheats the concentrator spray.

There are additional advantages to the dehumidification methods described here. One of these advantages is that the sorbents are not completely selective in removing one or another vapor or contaminant. Consequently, while water vapor is being removed, odors are likely to be removed too. One of the most effective solid sorbents for odor removal is **activated charcoal** (charcoal from which all hydrocarbons have been driven off). The liquid sorbents provide air washing and quite effectively remove such solid irritants as pollen. Lithium chloride is quite an effective sterilizing agent and may be used in some applications where this is of prime concern. Because the liquid absorbents have low freezing temperatures, simultaneous dehumidification and cooling can be carried to well below 32 F without frost formation on the cooling coil surfaces which are washed by the spray.

6.5 HEATING AND COOLING LOADS

A short presentation of the methods used in calculating heating and cooling loads might be misleading. There is no quick formula or rule of thumb that can be applied on the basis of building type and size. Each installation presents a problem in heat transmission in which indoor and outdoor temperature conditions and all sources of heat must be considered. With this word of caution, an outline of the general methods will be presented. The reader who must undertake a load calculation is referred to the ASHRAE Guide and Data Books for specific information on design temperatures, heat conductivity values, solar heat loads, infiltration estimates, and so on. The heating load calculation is somewhat simpler than the calculation of cooling load, primarily because the contribution of the undependable solar heat must be ignored in designing for winter conditions, while it cannot be ignored for summer air conditioning.

Table 6.5

CLIMATIC CONDITIONS*

State	United States City	Elev. ft	Winter				Summer						
			Design winter temp.			Wind	Design DB			Daily range	Design WB		
			Extr.	99%	97½%		1%	2½%	5%		1%	2½%	5%
Ala.	Birmingham	610	14	19	22	L	97	94	93	21	79	78	77
Alaska	Fairbanks	436	−59	−53	−50	VL	82	78	75	24	64	63	61
Ariz.	Phoenix	1117	25	31	34	VL	108	106	104	27	77	76	75
Ark.	Little Rock	257	13	19	23	M	99	96	94	22	80	79	78
Calif.	Fresno	326	25	28	31	VL	101	99	97	36	73	72	71
	Los Angeles	99	36	41	43	VL	86	83	80	15	70	69	68
	San Francisco	8	32	35	37	VL	83	79	75	20	65	63	62
Colo.	Denver	5283	−9	−2	3	L	92	90	89	28	65	64	63
Conn.	Hartford, Brainerd Fld	15	−4	1	5	M	90	88	85	22	77	76	74
Del.	Wilmington	78	6	12	15	M	93	90	87	20	79	77	76
D. C.	Washington, National	14	12	16	19	M	94	92	90	18	78	77	76
Fla.	Miami	7	39	44	47	M	92	90	89	15	80	79	79
Ga.	Atlanta	1005	14	18	23	H	95	92	90	19	78	77	76
Hawaii	Honolulu	7	58	60	62	L	87	85	84	12	75	74	73
Idaho	Boise	2842	0	4	10	L	96	93	91	31	68	66	65
Ill.	Chicago, O'Hare	658	−9	−4	0	M	93	90	87	20	77	75	74

* Reprinted by permission from ASHRAE Guide and Data Book 1965.
All temperature values are given in degrees F for airport locations.
Extreme winter temperatures are medians of yearly extremes over many years.

Wind

VL = Very Light, 70% or more of cold extreme hours are ≤ 7 mph.
 L = Light, 50 to 69% cold extreme hours are ≤ 7 mph.
 M = Moderate, 50 to 74% cold extreme hours are > 7 mph.
 H = High, 75% or more cold extreme hours are > 7 mph, and 50% are > 12 mph.

(continued)

Table 6.5 (*continued*)
CLIMATIC CONDITIONS

State	United States City	Elev. ft	Winter — Design winter temp. Extr.	99%	97½%	Wind	Summer — Design DB 1%	2½%	5%	Daily range	Design WB 1%	2½%	5%
Ind.	Fort Wayne	791	−5	0	5	M	93	91	88	24	77	76	75
Iowa	Des Moines	948	−13	−7	−3	M	95	92	89	23	79	77	76
Kans.	Wichita	1321	−1	5	9	H	102	99	96	23	77	76	75
Ky.	Louisville	474	1	8	12	L	96	93	91	23	79	78	77
La.	Baton Rouge	64	22	25	30	L	96	94	92	19	81	80	79
Me.	Augusta	350	−13	−7	−3	M	88	86	83	22	74	73	71
Md.	Baltimore	146	8	12	15	H	94	91	89	21	79	78	77
Mass.	Boston	15	−1	6	10	M	91	88	85	16	76	74	73
Mich.	Detroit, Met.	633	0	4	8	L	92	88	85	20	76	75	74
	Sault Ste. Marie	721	−18	−12	−8	M	83	81	78	23	73	71	69
Minn.	Duluth	1426	−25	−19	−15	L	85	82	79	22	73	71	69
Miss.	Jackson	330	17	21	24	M	98	96	94	21	79	78	78
Mo.	St. Louis	535	−2	4	8	L	98	95	92	21	79	78	77
Mont.	Helena	3893	−27	−17	−13	M	90	87	84	32	65	63	61
Neb.	Omaha	978	−12	−5	−1	L	97	94	91	22	79	78	76
Nev.	Las Vegas	2162	18	23	26	VL	108	106	104	30	72	71	70
	Reno	4404	−2	2	7	VL	95	92	90	45	64	62	61
N.H.	Concord	339	−17	−11	−7	M	91	88	85	26	75	73	72
N.J.	Newark	11	6	11	15	M	94	91	88	20	77	76	75
N.M.	Albuquerque	5310	6	14	17	L	96	94	92	27	66	65	64
N.Y.	New York, Kennedy	16	12	17	21	H	91	87	84	16	77	76	75
	Syracuse	424	−10	−2	2	M	90	87	85	20	76	74	73
N.C.	Charlotte	735	13	18	22	L	96	94	92	20	78	77	76
N.D.	Bismarck	1647	−31	−24	−19	VL	95	91	88	27	74	72	70

Table 6.5 (*continued*)
CLIMATIC CONDITIONS

State	United States City	Elev. ft	Winter				Summer						
			Design winter temp.			Wind	Design DB			Daily range	Design WB		
			Extr.	99%	97½%		1%	2½%	5%		1%	2½%	5%
Ohio	Columbus	812	−1	2	7	M	92	88	86	24	77	76	75
Okla.	Oklahoma City	1280	4	11	15	H	100	97	95	23	78	77	76
Ore.	Portland	21	17	21	24	L	89	85	81	23	69	67	66
Pa.	Philadelphia	7	7	11	15	M	93	90	87	21	78	77	76
	Pittsburgh	1137	−1	5	9	M	90	87	85	22	75	74	73
R. I.	Providence	55	0	6	10	M	89	86	83	19	76	75	74
S. C.	Charleston	41	19	23	27	L	94	92	90	18	81	80	79
S. D.	Pierre	1718	−21	−13	−9	M	98	96	93	29	76	74	73
Tenn.	Nashville	577	6	12	16	L	97	95	92	21	79	78	77
Texas	Fort Worth	544	14	20	24	H	102	100	98	22	79	78	77
	Houston	50	23	28	32	M	96	94	92	18	80	80	79
Utah	Salt Lake City	4220	−2	5	9	L	97	94	92	32	67	66	65
Vt.	Burlington	331	−18	−12	−7	M	88	85	83	23	74	73	71
Va.	Roanoke	1174	9	15	18	L	94	91	89	23	76	75	74
Wash.	Seattle–Tacoma	386	14	20	24	L	85	81	77	22	66	64	63
W. Va.	Elkins	1970	−4	1	5	L	87	84	82	22	74	73	72
Wis.	Madison	858	−13	−9	−5	M	92	88	85	24	77	75	73
Wyo.	Casper	5319	−20	−11	−5	L	92	90	87	31	63	62	60
Province	Canada City												
Alta.	Edmonton	2219	−30	−29	−26	VL	82	79	76	25	67	65	63
B. C.	Vancouver	16	13	15	19	L	77	74	72	20	67	65	64

(*continued*)

Table 6.5 (*continued*)
CLIMATIC CONDITIONS

Province	City	Elev. ft.	Winter				Summer						
			Design winter temp.			Wind	Design DB			Daily range	Design WB		
			Extr.	99%	97½%		1%	2½%	5%		1%	2½%	5%
Man.	Churchill	115	−43	−40	−38	H	79	74	69	18	67	64	61
	Winnipeg	786	−31	−28	−25	M	89	85	82	24	74	72	70
N.B.	Fredericton	74	−19	−16	−10	L	76	73	70	17	69	67	65
Nfld.	St. John's	463	1	2	6	H	77	74	72	17	64	63	61
N.W.T.	Yellowknife	682	−51	−49	−47	VL	78	75	73	18	70	68	67
N.S.	Halifax	136	−4	0	4	H	89	86	83	23	75	74	72
Ont.	Ottawa	413	−21	−17	−13	M	91	87	84	20	76	74	73
	Toronto	578	−10	−3	1	M	79	76	74	15	71	69	68
P.E.I.	Charlottetown	186	−11	−6	−3	H	88	85	82	17	76	74	72
Que.	Montreal	98	−20	−16	−10	M	87	83	80	25	69	66	64
Sask.	Saskatoon	1645	−37	−34	−30	M	79	75	72	22	61	59	58
Y.T.	Whitehorse	2289	−45	−45	−42	VL	79	75	72	22	61	59	58

The **outdoor design conditions** are determined for each locality from records kept for many years at hundreds of weather stations (usually airport readings) throughout the continent. A few cities are listed in **Table 6.5** with extreme summer and winter conditions as presented in the ASHRAE Guide and Data Book, 1965–66. It is usually uneconomical and unnecessary to design a heating or cooling system for the most extreme conditions on record or even for a median value of most extreme conditions over several years. For this reason, winter temperatures are given for the 99% and $97\frac{1}{2}$% low values, and summer temperatures are given for the 1%, $2\frac{1}{2}$% and 5% high values. These percentage figures indicate the percentage of the total number of hours in the season during which the temperature has been equal to or greater than the listed value. The winter season is taken as December, January, and February (2160 hours), and summer is taken as June through September (2928 hours). For Canadian cities, winter conditions are considered only for January, the coldest month.

It has been recommended by Crow [A.13] that corrections can be applied to these tabulated outdoor temperatures to prevent overdesign. The reason is that the extreme temperature is seldom held for more than two or three hours. These times are approximately 5 to 7 AM for winter extremes and 2 to 4 PM for the summer. Conditions ameliorate rather rapidly after these times, so that even for a small, poorly insulated building, and particularly for a large, well insulated building, inside wall temperatures would never reach the equilibrium values predicted for extreme conditions. Thus installed heating or cooling capacities can be safely reduced. In the article referred to, Crow suggests values by which extreme outdoor temperatures can be raised or lowered according to the rate of temperature change following the extreme for various regions of the continental United States.

Indoor design conditions vary widely for industrial applications. For comfort air conditioning, they are set largely by the physiological considerations of Section 3 of this chapter. The summer comfort value of about 77 F dry-bulb temperature from 30 to 70% relative humidity, determined with wall temperatures equal to air temperature, can be lowered to about 75 F at 50% relative humidity, where inside surfaces of outside walls are bound to be at higher temperatures. Where people continually pass from unconditioned to conditioned spaces, temperature differences should not exceed about 15 F degrees, although comfort for continued occupancy may be sacrificed.

The concept of a **degree day** (DD) applies only to winter air conditioning, i.e., heating calculations. It is useful in estimating the energy consumption over a long period of time (not less than one month), but not in sizing the heating system in terms of Btu per hour. Experience has shown that fuel consumption is proportional to the difference between 65 F and the average outdoor temperature for buildings with indoor temperatures from 68 F to 72 F. The lower value of 65 F apparently compensates, over a long period of time, for solar heat gains, heat from occupants, and other minor sources. Thus for a 31-day month, the total degree days would be

$$\sum DD = 31(65 - t_{o,\,av}), \tag{6.43}$$

Table 6.6

AVERAGE MONTHLY AND YEARLY DEGREE DAYS FOR CITIES IN THE UNITED STATES AND CANADA (BASE 65 F)*

United States

State	City	July	Aug.	Sept.	Oct.	Nov.	Dec.	Jan.	Feb.	Mar.	Apr.	May	June	Yearly Total
Ala.	Birmingham	0	0	13	123	396	598	623	491	378	128	30	0	2780
Ark.	Little Rock	0	0	10	110	405	654	719	543	401	122	18	0	2982
Calif.	Los Angeles	0	0	17	41	140	253	328	244	212	129	68	19	1451
Colo.	Denver	5	11	120	425	771	1032	1125	924	843	525	286	65	6132
D. C.	Washington	0	0	37	237	519	837	893	781	619	323	87	0	4333

* Reprinted by Permission from ASHRAE Guide and Data Book, 1964. Values are given in degrees F for airport locations.

Table 6.6 *(continued)*

AVERAGE MONTHLY AND YEARLY DEGREE DAYS FOR CITIES IN THE UNITED STATES AND CANADA (BASE 65 F)

State	City	July	Aug.	Sept.	Oct.	Nov.	Dec.	Jan.	Feb.	Mar.	Apr.	May	June	Yearly Total
Fla.	Miami	0	0	0	0	8	52	58	48	12	0	0	0	178
Ga.	Atlanta	0	0	8	110	393	614	632	512	404	133	20	0	2826
Ill.	Chicago	0	0	90	350	765	1147	1243	1053	868	507	229	58	6310
Kan.	Wichita	0	0	32	219	597	915	1023	778	619	280	101	7	4571
La.	New Orleans	0	0	0	7	169	308	364	248	190	31	0	0	1317
Md.	Baltimore	0	0	50	278	582	908	955	840	676	378	115	5	4787
Mass.	Boston	0	7	77	315	618	998	1113	1002	849	543	236	42	5791
Mich.	Detroit City	0	8	96	381	747	1101	1203	1072	927	558	251	60	6404
Minn.	Duluth	56	91	298	651	1140	1606	1758	1512	1327	846	474	178	9937
Mo.	Saint Louis	0	0	45	233	600	927	1017	820	648	297	101	11	4699
Neb.	Omaha	0	5	88	331	783	1166	1302	1058	831	389	175	32	6160
Nev.	Reno	27	61	165	443	744	986	1048	804	756	519	318	165	6036
N.Y.	New York	0	0	28	250	546	908	992	907	760	447	141	10	4989
	Syracuse	0	29	117	396	714	1113	1225	1117	955	570	247	37	6520
N.C.	Charlotte	0	0	7	147	438	682	704	577	449	172	29	0	3205
N.D.	Bismarck	29	37	227	598	1098	1535	1730	1464	1187	657	355	116	9033
Ohio	Columbus	0	8	69	337	693	1032	1094	946	781	444	180	31	5615
Pa.	Philadelphia	0	0	47	269	573	902	986	879	704	402	104	0	4866
Tenn.	Nashville	0	0	22	154	471	725	778	636	498	186	43	0	3515
Texas	Fort Worth	0	0	0	58	299	533	622	446	308	90	5	0	2361
	Houston	0	0	0	77	181	321	394	265	184	36	0	0	1388
Wash.	Seattle-Tacoma	75	70	192	412	633	781	862	675	636	477	307	155	5275
Wis.	Madison	10	30	137	419	864	1287	1417	1207	1011	573	266	79	7300

(continued)

Table 6.6 (*continued*)

AVERAGE MONTHLY AND YEARLY DEGREE DAYS FOR CITIES IN THE UNITED STATES AND CANADA (BASE 65 F)

Canada

Province	City	July	Aug.	Sept.	Oct.	Nov.	Dec.	Jan.	Feb.	Mar.	Apr.	May	June	Yearly Total
B.C.	Vancouver	60	60	200	410	620	790	850	710	650	460	290	130	5230
Man.	Winnipeg	40	70	300	690	1250	1770	2000	1710	1440	810	400	150	10630
N.S.	Halifax	50	70	180	470	740	1120	1260	1180	1050	760	480	210	7570
Ont.	Toronto	20	30	140	460	770	1130	1260	1160	1020	640	320	70	7020
Que.	Montreal	10	40	180	530	890	1370	1540	1370	1150	700	300	50	8130
Y.T.	Dawson	170	320	660	1170	1890	2410	2510	2160	1830	1100	570	250	15050

where $t_{o,av}$ is the average outdoor temperature over the 31-day period, all temperatures above 65 F being ignored. Negative values of degree days have no significance since heat gains cannot usually be stored effectively for later use. A value of degree days is useful in the economic considerations of building insulation. The monthly degree days for a few cities are listed in **Table 6.6.**

The various positive and negative **components of a heating or cooling load** are grouped in **Fig. 6.21.** Some may contribute either a cooling or a heating effect, while others are always positive heat gains to the space. Some contribute only heat, while others may add moisture or require that additional moisture be supplied. The algebraic sum of these constitutes the load, a positive gain indicating a cooling load.

Normal **transmission** through walls, ceiling, floor, and glass are gains by heat conduction. The conductivity of a particular type of construction is determined as the inverse of the sum of all the heat resistances in series as in any calculation of heat transfer by conduction. Thus the overall rate of heat conduction U for a wall with an air space would be

$$U = \frac{1}{1/f_i + x_1/k_1 + 1/a + x_2/k_2 + 1/f_o} \text{ Btu/ft}^2\text{-hr-F}, \qquad (6.44)$$

where f_i and f_o are surface conductances, k_1 and k_2 are conductivities of wall materials, x_1 and x_2 are thicknesses of wall materials, and a is the air-space conductance.

Composite values of U for different combinations of building materials are given in tables in the ASHRAE Guide and Data Book. These are calculated values

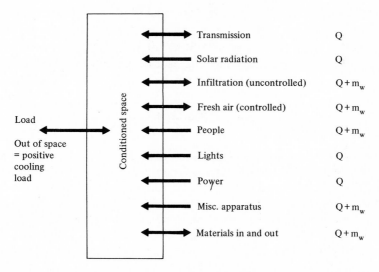

FIG. 6.21 Components of heating and cooling loads.

from Eq. (6.44) using test values of k for various materials, and values of f_i and a depending on surface orientation, direction of heat flow, air space thickness, and surface type; values of f_o are taken as 6.0 for a 15 mph wind for winter and 4.0 for a 7.5 mph wind for summer. Charts and tables allow correction for framing and insulation of various types in the air space and for different wind velocities.

The **solar radiation** contribution to the heat gain through walls, ceilings, and glass depends on the time of day, time of year, geographical longitude, surface orientation, atmospheric conditions, and surface emissivity, in addition to the materials of the barrier. For summer air conditioning, a very good approximation of the sum of the transmission and the solar heat gains is obtained by using the **total equivalent temperature difference** t_e for opaque walls and roof areas. Values of t_e have been worked out for different types of construction at specific times of day for roofs and for different wall orientations. These are presented in tables in the ASHRAE Guides and are to be used with overall heat transfer coefficients U for the corresponding construction. Corrections can be made for daily temperature ranges known to be different from the 20 degrees assumed in developing the tables. Transparent and transluscent areas are handled differently. For these, **solar heat gain factors**, in Btu per square foot, have been worked out for various dates, latitudes, and times of day, and these, too, are presented in the ASHRAE tables. They must be multiplied by coefficients for the type of shading used.

Infiltration, which must be balanced at some part of the building by **exfiltration**, brings with it outside air conditions. If a rate can be established, then the amount of heat addition or removal to bring the air to the desired temperature can easily be calculated:

$$\dot{Q}_a = c_p \frac{\dot{V}}{v}(t_i - t_o) \text{ Btu/hr.} \tag{6.45}$$

The specific heat can be taken as 0.24 Btu/lb of air and $1/v$ as 0.075 lb/cu ft; $\dot{V}$ must be measured in cubic feet per hour. The heat to be added or removed to make up for the difference in the humidity ratio can be taken as 1060 Btu/lb of moisture:

$$\dot{Q}_m = 1060 \frac{\dot{V}}{v}(W_i - W_o). \tag{6.46}$$

The difficulty is found in establishing the amount of infiltration which is uncontrolled. It is affected by wind (stagnation pressure), by temperature difference (stack effect), by types of doors, windows, and walls, and by workmanship. In the **crack method** of estimation, the perimeters of all openings on one side of a building are calculated (all sides for a small residence), and the lengths are multiplied by leakage rates estimated for the types of openings. The lack of possible precision is obvious. An alternative, which also requires judgment, is the **air-change method**. Rates of air changes from infiltration that are used vary from one-half change per hour for inside rooms to two changes per hour for rooms with doors and windows on three sides. While inaccuracies are inevitable in either method, the heat and moisture quantities are usually of a magnitude that cannot be ignored, and therefore must be estimated.

The purposeful introduction of **fresh air** is contrasted with infiltration in that it is designed for and controlled. The calculations of heat and moisture quantities are as in Eqs. (6.45) and (6.46), but $\dot{V}$ is presumably known.

The heat and moisture contributed to a space by **people** depend on the conditions discussed in Section 6.3, Physiological Factors. Again, the average values to be used for different applications and with estimated proportions of men, women, and children have been compiled and tabulated.

The heat equivalent of **lights** and **power** would seem to be a straightforward calculation, but some adjustments must be made. The rated wattage of installed lights (1 watt = 3.413 Btu/hr) must be multiplied by a **use factor** for the time of the load calculation; all lights would be on all day only in some commercial applications. Fluorescent fixtures utilize about 20 % additional power because of the need for ballasts. Ventilated fixtures can often be installed so that much of the heat is carried outside the conditioned space. Electric motor ratings (1 hp = 2544 Btu/hr) must be multiplied by a use factor and also divided by **motor efficiency** (unless the motor is outside the space and the shaft work is done inside the space as in most but not all ducted fans).

Miscellaneous apparatus, particularly in such applications as restaurant air conditioning, can be extremely important parts of the heat load. With hoods and proper venting, the latent heat fraction can usually be practically eliminated, and the heat gain is what is radiated from hot surfaces. The ASHRAE tables and manufacturers' ratings can provide reasonable estimates for most appliances. Ducts and pipes running through a space can contribute appreciably to heating and cooling loads if they are uninsulated.

The transporting of **materials into and out of** conditioned spaces is an important part of the cooling load in such applications as cold-storage vaults and food freezing. This may also be important in hangars and garages for winter heating loads.

6.6 AIR-CONDITIONING PROCESSES

The **total heating or cooling load** for a conditioned space is obtained, as discussed in the previous section, by the summation of both heat and moisture quantities of all the components. This load can be overcome entirely by heat exchangers such as finned coils or wall and ceiling panels to supply or remove the heat quantity, and by humidifiers or dehumidifiers all operating directly in the conditioned space. However, an alternative is to remove air continuously from the space, condition this air by the removal or addition of heat and moisture, clean it by filtering, and return it to the room. Since some fresh air is usually required, any percentage of recirculated and fresh air can be mixed before distribution to the conditioned space. The question is then: At what conditions should moist air be introduced to the room?

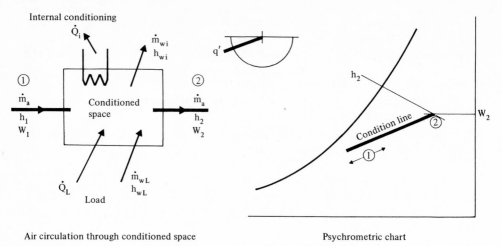

Air circulation through conditioned space   Psychrometric chart

FIG. 6.22 Condition line for a space.

6.6.1 The Condition Line

Satisfactory conditions for supply air are not limited to one point on the psychrometric chart, but it will be shown that they must fall on a particular line on that chart. That line is the **condition line** for the space. It is established by the enthalpy/moisture ratio of the air-conditioning load, where the net heat gain and net moisture gain are used.

Figure 6.22 shows a conditioned space with some internal conditioning and some circulation of air. Let us assume that the heat and moisture loads have been determined as $\dot{Q}_L$ Btu/hr plus the enthalpy of moisture added, and $\dot{m}_{wL}$ lb moisture/hr. To keep a consistent sign convention, we will designate the internal conditioning $\dot{Q}_i$ (heat removal) and $\dot{m}_{wi}$ (dehumidification) both positive as shown on the diagram. The net load to be carried by the introduction of externally conditioned air will be

$$\dot{Q}_{net} = \dot{Q}_L - \dot{Q}_i + \dot{m}_{wL}h_{wL} - \dot{m}_{wi}h_{wi} \text{ Btu/hr} \qquad (6.47)$$

and

$$\dot{m}_{wnet} = \dot{m}_{wL} - \dot{m}_{wi} \text{ lb moisture/hr.} \qquad (6.48)$$

The sign convention of the diagram and of these equations agrees with that of Fig. 6.21 for which it was noted that positive gains constitute a positive cooling load. If there is no attempt at internal air conditioning, the circulation of conditioned air through the space must, obviously, carry the whole load.

To find the conditions of the supply air at section ①, a heat and mass balance must be calculated for the space. The air leaving at ② will have the design conditions of the air in the space:

In	**Out**
$\dot{m}_a h_1 + \dot{Q}_{net}$	$= \dot{m}_a h_2,$
$\dot{m}_a W_1 + \dot{m}_{wnet}$	$= \dot{m}_a W_2.$

Thus we have

$$\frac{\dot{Q}_{net}}{\dot{m}_{wnet}} = \frac{h_2 - h_1}{W_2 - W_1} = q'. \tag{6.49}$$

On a psychrometric chart, the coordinates of which are h and W, the line joining conditions ① and ② has a slope dictated by $\dot{Q}_{net}/\dot{m}_{wnet}$. This slope is q' by the definition of enthalpy/moisture ratio. Air supplied with conditions at any point on a line through ② with a slope given by q' will satisfy the heat and moisture balances. This is the **condition line** for the space. The values of h_1 and W_1 will be dictated by the mass rate of flow of air. Conversely, the air circulation required is specified by h_1 or W_1, and the condition of the supply air must fall on the space-condition line.

It will be recalled that the enthalpy/humidity ratio protractor on the psychrometric charts has another scale, which is sensible heat/total heat. In the discussion of the chart, it was noted that these terms are not relevant as moist air properties. They can be used in air-conditioning calculations, but only with caution. If it is assumed that all moisture released in the space must be condensed out by giving up its latent heat, then a distinction can be made between **sensible heat,** for which there must be a corresponding temperature change in the air, and **total heat,** which is the sum of sensible heat plus heat to condense the moisture and cool it to room conditions. The terms are used justifiably for an **approximate** calculation of the load on a cooling apparatus that must change $\dot{V}$ standard cfm (0.075 lb/cu ft) from outside conditions t_o and W_o to conditions t_i and W_i:

$$\text{Sensible load:} = \dot{V} \times 60 \times 0.244 \times 0.075(t_o - t_i) \text{ Btu/hr} \tag{6.50}$$
$$\text{Latent load:} = \dot{V} \times 60 \times 0.075 + 1076(W_o - W_i) \text{ Btu/hr} \tag{6.51}$$

The average value 1076 includes the latent heat of condensation and "sensible" cooling of the water to t_i. The standard density of air (0.075 lb/cu ft) is used, because it is on this basis that fans and all air-conditioning equipment are generally rated. (For greater accuracy, the specific weight of the dry air alone should be used).

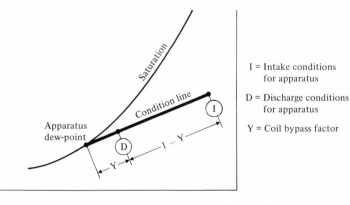

I = Intake conditions for apparatus

D = Discharge conditions for apparatus

Y = Coil bypass factor

Psychrometric chart   **FIG. 6.23** Apparatus dew point.

The condition line for a space establishes another point called the **apparatus dew point**. It is applicable only to air conditioning that involves cooling and dehumidifying. If the air-conditioning apparatus is to bring part of the air to saturation in a dehumidifying process and then mix that with some recirculated air for supply at condition ①, then the saturated air must be at a point determined by the extension of the space-condition line to an intersection with the saturation line as shown in **Fig. 6.23**. Adiabatic mixing of air at the apparatus dew point with air at condition ② must produce conditions on a straight line between the two points, as was proved in Example 6.5. As discussed in connection with Example 6.3 and illustrated in Fig. 6.7, no cooling coil will bring 100% of the air passed through it to saturation. The imperfect heat transfer and imperfect mixing produce the same result as intentional bypassing of the coil and subsequent adiabatic mixing. If intake conditions, the apparatus dew point, and a **bypass factor** are specified for a unit, the final discharge conditions can be determined on the psychrometric chart.

6.6.2 Calculation of Air Quantity

The mass or volume of air to be supplied to a space per minute depends on the net heat load and the allowable temperature difference between room air and supply air. The change of enthalpy of the air during its residence time in the room is the means by which the net heat gain is carried away. Thus

$$\dot{Q}_{net} = \dot{m}_a(h_2 - h_1) \text{ Btu/hr.} \tag{6.52}$$

The volume rate of flow would be

$$\dot{V}_1 = \frac{\dot{m}_a v_1}{60} \text{ cfm.} \tag{6.53}$$

As mentioned previously, fans and other equipment are rated in terms of "standard" air which has a specific weight of 0.075 lb/cu ft:

$$\dot{V}_{std} = \frac{\dot{V}_1}{v_1 \times 0.075} \text{ cfm.} \tag{6.54}$$

The calculation could be made as well on the basis of conditions at the outlet from the space by the substitution of $\dot{V}_2$ and v_2 for $\dot{V}_1$ and v_1 in Eqs. (6.53) and (6.54).

All values are easily determined from the psychrometric chart with sufficient accuracy for most air-conditioning applications if points ① and ② have been established. An **approximation** in the absence of a chart or before conditions are decided on can be obtained simply by the net heat load and the proposed temperature difference:

$$\dot{V}_{std} = \frac{\dot{Q}}{60 \times 0.075 \times 0.244(t_2 - t_1)} \text{ cfm.} \tag{6.55}$$

The **allowable temperature difference** between the supply air and room air depends on the manner of air supply and distribution. When the untempered air is discharged directly into the room and no particular provision is made for mixing the stream and room air, the differential should not exceed 20 F degrees, and when people are close to outlets, it should be less. When a high-velocity aspiration system is used in which the supply air (primary air) induces several times its own volume of room air to mix with it before release to the room, temperature differences of 30 to 35 F degrees can sometimes be tolerated. Of course, the supply air rate must never fall below the required ventilation rate. When the ventilation rather than the heat load determines the air-supply rate, the temperature difference rather than the volume rate of flow is established by Eq. (6.55).

6.6.3 Examples of Air-Conditioning Processes and Adiabatic Mixing

In the following examples of systems for air conditioning, the sources of heat or the means of cooling in the apparatus will not be considered. Heating may be accomplished by steam, hot water, or some other method; cooling may be accomplished by an outside water supply, by the circulation of chilled water or a brine, or by direct expansion coils. All heat-transfer elements would normally be finned coils, and filters would almost invariably be inserted in each air stream. Automatic or manual dampers would usually be included for control in the larger systems. More than one fan may be necessary. All these things do not affect the psychrometric consideration. The space heating or cooling loads shown are net values taking into account any air conditioning internal to the space. Heat and moisture quantities for the apparatus show net direction only; required quantities depend on outside air conditions and ventilation requirements as well as space loads. It is assumed that the heat input of the fans is negligible; actually, it is part of the load, as is friction loss in ducts.

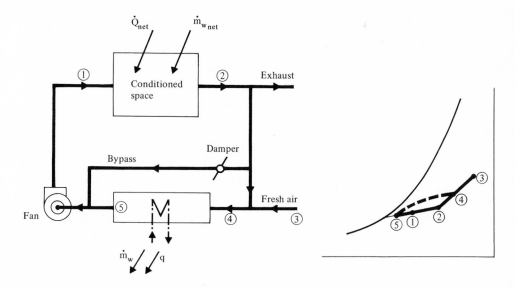

FIG. 6.24 Cooling with bypass.

a) *Cooling with Bypass* (**Fig. 6.24**). The schematic arrangement shown is a common, simple, summer air-conditioning system. The amount of dry air introduced as fresh air must be balanced by the exhaust which may or may not be controlled by a separate exhaust fan. For full load demand, the bypass may be closed. At part load, the bypass allows the delivery rate to the space, and thus the distribution pattern within the space, to remain the same. As in Fig. 6.22, the condition line extends from ② to ①. Adiabatic mixing places ④ on the straight line ②–③ and ① on the straight line ②–⑤.

b) *Absorption Dehumidification with Additional Cooling (A) or Heating (B)* (**Fig. 6.25**). Dehumidification with sorbent materials was discussed in Section 6.4. The system using liquid absorption illustrated in Fig. 6.20 is shown incorporated into an air-conditioning system in Fig. 6.25. Two psychrometric conditions are shown. Chart *A* illustrates the situation in which the cooling necessary in the absorption process must be supplemented by subsequent heat removal. Chart *B* shows the situation in which heating is required. If adiabatic adsorption had been applied at point ④ of chart *B*, the broken line shows that large subsequent cooling would be required. The selection of the type of dehumidification is primarily an economic one unless other factors predominate.

c) *Year-round Air-Conditioning System* (**Fig. 6.26**). The system shown is a relatively simple one for either winter or summer air conditioning. Only the preheat coil, the humidifying spray, and the reheat coil would be used in the winter,

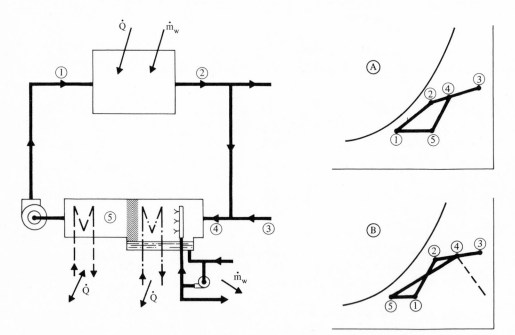

FIG. 6.25 Absorption dehumidification with additional cooling (*A*) or heating (*B*).

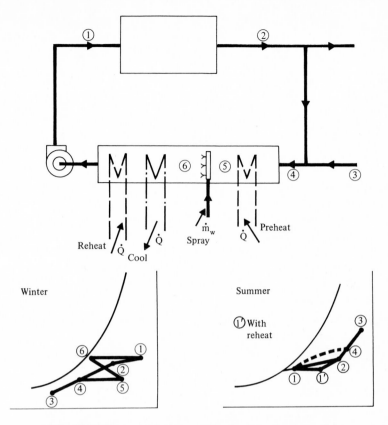

FIG. 6.26 Year-round air conditioning (simple system).

and usually only the cooling coils in the summer. No zoning (cooling one space while heating another) is possible. The preheat coil prevents freezing at the spray in cold weather and controls the amount of moisture pickup. The reheat coil establishes the final dry-bulb temperature.

d) *Evaporative Cooling* (**Fig. 6.27**). In regions where the air, though hot, is quite dry as in some of the southwestern parts of this country, quite effective cooling can be obtained by water spray. The process is that of adiabatic saturation. When maximum cooling is required, recirculation is not possible. A bypass line and heater can be incorporated for year-round operation.

e) *Dehumidification and Reheat with a Refrigeration Cycle* (**Fig. 6.28**). While dehumidification by means of cooling alone may require reheat for suitable supply conditions, it was noted in Section 6.4 that this reheat may be obtained from the condenser of the refrigeration system which provides the cooling. Since the condenser heat transfer is likely to exceed the desired reheat, supplementary condenser cooling water will be required, and either system *A* or system *B* might be used.

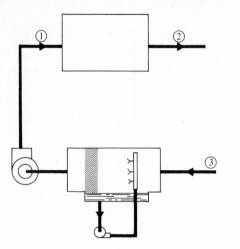

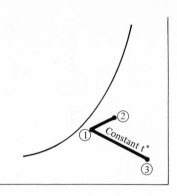

FIG. 6.27 Evaporative cooling (adiabatic saturation).

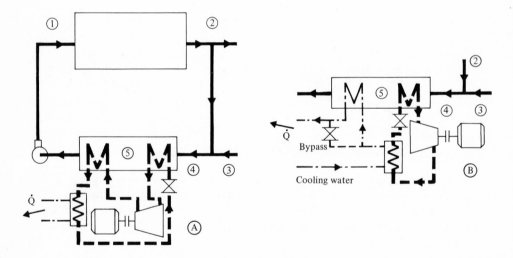

FIG. 6.28 Dehumidification and reheat with refrigeration cycle.

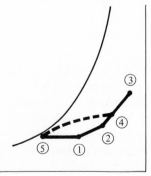

6.7 DISTRIBUTION OF AIR CONDITIONING

The systems described in the following paragraphs are discussed primarily from the point of view of comfort air-conditioning applications. Many of the remarks will apply as well to other applications, but special considerations often outweigh the generalizations attempted here.

Air-conditioning distribution systems can be divided into three major classifications: (a) **all-air systems**, in which sufficient preconditioned air is supplied to the various spaces to carry the whole air-conditioning heat and moisture load and to provide ventilation; (b) **all-water, steam,** or **direct expansion systems**, in which the whole load for each space is carried by the cooling or heating medium supplied to heat exchangers within the spaces, and ventilation, if any, is through wall openings; and (c) **mixed systems** in which at least enough conditioned air is supplied to the space for ventilation and the remainder of the load is carried by local heat exchangers. The relative merits of these systems and their several subclassifications are discussed at length by Jaros [A.28]. Some of the main points will be reviewed here.

6.7.1 Zoning

In almost all applications, various zones of a building have different air-conditioning requirements at any given time. In both summer and winter, the peripheral areas facing east, south, and west will be affected separately by the presence or absence of solar heat gain. Areas immediately below flat roofs may be quite different from the rest of the building in terms of air-conditioning requirements. The wind will also affect one building face differently from the others. Internal rooms (those with no outside walls and surrounded by conditioned spaces) will always require cooling if occupied, and changes in occupancy density may create large demand changes in one part of a building but not in others. Load contributing machines and appliances will often be localized.

Thus buildings can usually be divided into zones that have similar requirements throughout the day. These zones may be vertical in a multistoried building or horizontal as in a two-story building with a flat roof. They may simply divide internal from peripheral areas or occupied from unoccupied areas. To take an extreme example, each room or suite of a hotel or hospital could be considered a separate zone. Of the systems to be discussed, some will lend themselves well to zoning and others will not.

6.7.2 All-Air Systems

To provide a relatively simple system, all the air for conditioning the various spaces in a building might be carried in a **single duct** which is supplied with totally or partially conditioned air from one fan room. If all the air is conditioned in the fan room, then zone control can be accomplished by dampers only. The low static pressures of such systems (1 or 2 in. of water gage) make fine control difficult. Ventilation requirements limit the degree to which conditioning of one space or another can be reduced. Also, dampers closed sufficiently to limit flow, introduce the problem of noise.

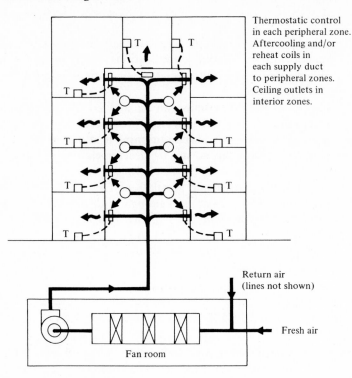

Thermostatic control
in each peripheral zone.
Aftercooling and/or
reheat coils in
each supply duct
to peripheral zones.
Ceiling outlets in
interior zones.

Return air
(lines not shown)

Fresh air

Fan room

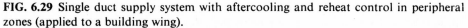

FIG. 6.29 Single duct supply system with aftercooling and reheat control in peripheral
zones (applied to a building wing).

A more flexible and generally more satisfactory all-air system is shown in
Fig. 6.29 where **aftercooling** or **reheating** can be provided at the branch supply to
each peripheral zone. The whole air supply in the installation illustrated is conditioned in the fan room for the requirements of internal spaces, and further conditioning is controlled by individual thermostats in peripheral rooms. Return air
lines (not shown) must go right back to the single fan room. A return air fan separate from the supply fan is almost always required. In general, exhaust and return
air quantities are designed to make up about 85 % of the supply so that exfiltration
is assured.

To reduce return lines and to provide better zone control of air flow in a single
duct system, **booster fans** and additional cooling or heating can be provided at each
zone. Only primary air, 100 % fresh, need be supplied by the principal fan and duct.
Recirculation is accomplished at each zone through the booster fan.

A **cold-deck** and **hot-deck** system can be applied in two ways as shown in **Fig.
6.30**. In both, the air from the supply fan is divided into two plenum chambers
(decks), one containing aftercooling coils and the other reheating coils; primary
conditioning including dehumidification is accomplished ahead of the fan. The
cooled and warmed air can then be supplied separately or mixed in any proportions

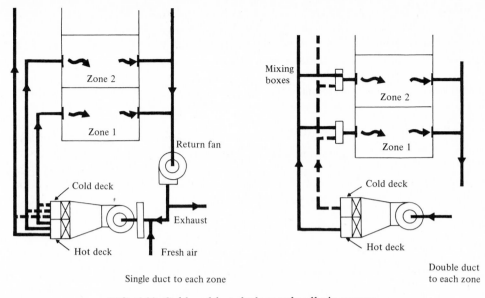

Single duct to each zone   Double duct to each zone

FIG. 6.30 Cold and hot deck supply all-air system.

desired for the various zones. For a relatively small total volume to be conditioned, a **multizone fan room** is designed in which damper controls, in paired outlets from the decks, mix the air for a **single duct** supply **to each zone**. For larger installations, a **double duct** system may be more satisfactory. In this, one duct from the cold deck and one from the hot deck run all the way to each zone, and thermostatically controlled mixing is accomplished at each air-supply outlet. The automatic mixing units can be installed in ceilings or under windows.

Long duct runs, particularly in double duct systems, have led to the use of higher pressures and velocities in supply systems. Instead of values such as 1 to 2 in. of water and 1000 to 1800 fpm in main ducts, 8 to 10 in. of water and 2500 to 5000 fpm are in use. Mixing control is more reliable at these pressures, but aerodynamic noise at such high duct velocities requires much more careful design in turns and outlets and extensive sound trapping. Smaller round ducts usually replace the large rectangular ducts of the low-pressure systems. While the systems may be more expensive, building space is saved.

6.7.3 All-Water Systems

Finned coil heat exchangers can carry the total sensible load in each zone. These coils can be supplied with heated and chilled water, mixed as required for each zone. Water circulation with pumps is much simpler and requires much less space than air circulation. Local air circulation can be accomplished by fan and coil units. When these are located on outside walls, fresh air for ventilation can be drawn in through individual wall openings with dampers (usually not automatic) inside the units controlling the proportion of fresh and recirculated air. Large numbers of wall openings introduce an architectural and a maintenance problem. Air filtering in small units is much less satisfactory than in centralized systems.

If conditioned spaces are not too far from central equipment rooms, direct expansion coils for cooling and steam coils for heating can be used. But the previously described system for mixed water requires only one coil and allows for better graduated control as well as immediate switchover from heating to cooling.

6.7.4 Mixed Systems

Considerable flexibility is provided by the combination of air and water distribution. With the major part of the load being carried by water coils, only primary air, amounting to about 20 % of the total required circulation rate, need be supplied from the main fan rooms and distributed by zones to take care of ventilation and humidity control.

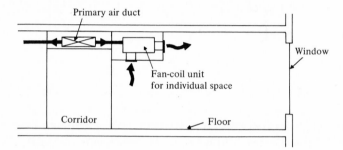

FIG. 6.31 Mixed system: Low pressure air plus heated or chilled water to zone fan-coil unit.

Figure 6.31 shows a low-pressure air supply system coupled with heated or chilled water coils for individual room control and overhead supply. For colder climates, the fan-coil unit would be more effective in the winter if it were installed below a window. The primary air supply could be distributed from below the floor and supplied through the unit, or from a separate wall or ceiling outlet.

High-pressure, high-velocity air supply to peripheral zones is being coupled with **aspirated** coil units developed in recent years. The discharge of air through nozzles at high velocity creates a low pressure and induces a flow of room air in through the face of the unit and over finned coils. The primary and secondary air mix before being released to the room. No local fans are necessary, but baffling of the primary air supply for noise control is a major problem.

The mixed systems can be supplied with water in either a **two-pipe** (one supply and one return) or a **three-pipe** system. The latter supplies hot and cold water in two separate lines to each unit where mixing is thermostatically controlled and one return line is used. Thus zoning before distribution is unnecessary.

In a mixed system, sensible heat can be supplied or removed by means of **radiant ceiling** panels: aluminum panels are usually best for heat transfer and maintenance. Acoustic treatment can be provided by holes that do not appreciably reduce radiant areas, and the same insulating material above the panels can provide sound absorption and heat insulation. Heat transfer per square foot of radiant panel is limited by limited temperature differences. In the summer, surface temperatures must not drop below the dew point; thus circulated water must not in general enter below about 60 F. (The use of polythene sheets below radiant ceiling panels to allow the transmission of long wavelength heat radiation while separating the panels from the room air has been reported [A.36]. This would allow surface temperatures to exist below the dew point and increase cooling rates greatly. However, such arrangements are not yet in general use).

Under certain conditions, it may be possible with a mixed system to provide peripheral spaces with **ventilation air entirely from interior spaces**. Relatively simple air-supply systems may be quite satisfactory for the interior spaces because load demands are often quite steady. The variable requirements of the zones fed by the air that has passed through the interior spaces can be met by finned coils or radiant panels.

PROBLEMS

1. For air at 100 F and standard barometric pressure, calculate the humidity ratio, the degree of saturation, and the dew-point temperature for a reported relative humidity of 60%. Assume that the perfect gas law and the Gibbs-Dalton rule apply.

2. For the conditions of the previous problem, calculate the specific volume and specific enthalpy, both per pound of dry air. Again assume that the perfect gas law and the Gibbs-Dalton rule apply. Check all your answers against psychrometric chart values.

3. For the previous problem, by what percentage would the calculated values of specific volume and specific enthalpy be changed if the coefficients of Table 6.2 were applied? Use linear interpolation for values at 100 F.

4. Use the psychrometric chart only to find the degree of saturation of air at standard barometric pressure, 75 F dry-bulb temperature, and 65 F thermodynamic wet-bulb temperature.

5. Three tons of refrigeration are used to cool 1400 cfm of standard pressure air supplied at 89 F dry-bulb and 65 F wet-bulb temperatures. The coil surface is 54 F. At what relative humidity does air leave the system?

6. An air-conditioning system cools 15,000 cfm of air at standard pressure from 98 F and 60% relative humidity to 63 F and 95% relative humidity. Calculate the refrigeration rate required in tons, and the rate of moisture removal.

7. How much heat and moisture must be supplied to take 3000 cfm of air at standard pressure from 50 F and 30% relative humidity to 80 F and 30% relative humidity? Assume that moisture is sprayed in as liquid at 60 F.

8. An air-conditioning system draws in 2000 cu ft/min of moist air at 95 F and 60% relative humidity. The air is cooled below its dew-point temperature, and then reheated (without addition of moisture) to reach a final condition of 70 F and 50% relative humidity. Assume that cooling and dehumidification are to a relative humidity of 90%, and that the process is at standard atmospheric pressure. Use a psy-

chrometric chart to find

a) the heat removed, Btu per hour;

b) the water vapor removed, pounds per hour;

c) the heat added in the reheat operation, Btu per hour.

9. In an air-conditioning system, 3000 cfm of recirculated air at 75 F and 50% relative humidity are mixed with 9000 cfm of outside air at 95 F and 40% relative humidity. The mixture is cooled to 55 F and 90% relative humidity, and then reheated without addition of moisture to 65 F.

a) What rate of cooling must be provided?

b) How much reheat is required?

c) What are the final values of relative humidity and humidity ratio?

d) If only the outside air were cooled and then tempered with recirculated air before delivery at the identical conditions, to what conditions would the outside air have to be cooled, and what ratio of recirculated to fresh air could be used? Assume again that cooling and dehumidification are to 90% relative humidity.

10. If commercially dry silica gel is used to dehumidify atmospheric air, what relative humidity ratio could be achieved at 80 F if sufficient adsorbent is supplied so that it takes on no more than one tenth of a pound of water per pound of dry silica gel?

11. For a building in Fort Worth, Texas, it is desired to maintain room air at 80 F and 50% relative humidity in the summer. The heat and moisture loads are 150,000 Btu/hr and 50 lb/hr, respectively. The difference between supply and room air temperatures can be as high as 20 degrees F. For ventilation, 2000 cfm of fresh air are required. What recirculation rate is allowable and necessary? What is the net cooling load?

12. The system of Fig. 6.26 is used to heat a building on a day in which the outside conditions are 35 F and 80% relative humidity. Room conditions are to be maintained at 75 F and 60% relative humidity. It is permissible to supply room air at 95 F. For a heat loss of 200,000 Btu/hr, a moisture loss of 40 lb/hr, and a 25% fresh air supply, calculate the following from chart data:

a) The rate of air supply to the room.

b) The preheat rate required.

c) The reheat required.

Assume that the "adiabatic saturation" process brings the preheated air to 95% relative humidity.

13. For the one percent summer design conditions in Las Vegas, Nevada, assume that cooling is accomplished by adiabatic saturation only. If saturation is to 95% relative humidity, what cooling load and what moisture load can be carried per cubic foot of outside air if room conditions can go to 85 F and 70% relative humidity?

REFERENCES

A. *Articles and Separate Publications*

1. "The Air Conditioning of Multi Room Bldgs: High Velocity Dual Duct Systems," *Proceedings of HEVAC Conf.*, *1961*, *IHVE Journal*, Mar., 1963, *IHVE Journal*, Nov., 1963.

2. AIRSEARCH MFG. CO., *Study of the Thermal Processes for Man-In-Space*, Report No. CR-216, NASA, Washington, 1965.

3. ARNOLD, J. H., "The Theory of the Psychrometer," *Physics*, Vol. 4, July, 1933, p. 255; Sept., 1933, p. 334.

4. ASHLEY, C. M., R. HAINES, J. L. THRELKELD, D. D. WILE, and J. A. GOFF, "Dallas Symposium Debates the Ideal Psychrometric Chart," *ASHRAE Journal*, Apr., 1960, pp. 73–76.

5. BELDING, H. S. and T. F. HATCH, "Index for Evaluating Heat Stress in Terms of Resulting Physiological Strains," *ASHRAE Trans.*, Vol. 62, 1956, pp. 213–236.

6. BERNER, R., "Air Conditioning as Applied to Aircraft Missiles and Rockets," *ASHRAE Jl.*, Mar., 1960, pp. 62–68.

7. BLUM, R., "Two-, Three-, Four-Pipe Air Conditioning Systems," *Air Conditioning, Heating and Ventilating*, Apr., 1966, pp. 57–80.

8. BOYAR, R. E., "Room Temperature Dynamics of Radiant Ceiling and Air Conditioning Comfort Systems," *ASHRAE Trans.*, Vol. 69, 1963, pp. 37–45.

9. BRYAN, W. L., "Heat and Mass Transfer in Dehumidifying Extended Surface Coils." *ASHRAE Jl.*, Vol. 4, No. 4, Apr., 1962, pp. 60–3.

10. BULL, L. C., "Central Air Conditioning," *Modern Refrig.*, Vol. 65, No. 768, Mar., 1962, pp. 223–225.

11. CARLSON, G. F., "Dual Temperature Equipment Room Piping Arrangements," *ASHRAE Jl.*, Vol. 8, No. 12, Dec., 1966, pp. 37–43.

12. CERAMI, V. V. and N. S. SHATALOFF, "Quiet High-Velocity Air Distribution," *Air Conditioning, Heating and Ventilating*, Aug., 1964, 10 pp.

13. CROW, L. W., "Derivation of Outdoor Design Temperature from Annual Extremes," Paper prepared for presentation at ASHRAE Semi Annual Meeting, Jan., 1964 (Ref. in *ASHRAE Jl.*, Vol. 6, No. 8, 1964, p. 43).

14. DuBOIS, F. F., "The Temperature of the Human Body in Health and Disease," Ref. [B. 1], pp. 24–40.

15. EAKIN, B. E., and R. T. ELLINGTON, "Gas-Fuel Open Cycle Air Conditioner," *ASHRAE Trans.*, Vol. 66, 1966, pp. 319–338.

16. FAHNESTOCK, M. K., "Comfort and Physiological Responses to Work in an Environment of 75 F and 45 percent RH," *ASHRAE Trans.*, Vol. 69, 1963, pp. 13–23.

17. FANGER, P. O., "Calculation of Thermal Comfort: Introduction of a Basic Comfort Equation," *ASHRAE Paper*, No. 2051, June, 1967.

18. GAGGE, A. P., "Standard Operative Temperature," Reference [B. 1], pp. 544–552.

19. GOFF, J. A., "Standardization of Thermodynamic Properties of Moist Air," *ASHVE Trans.*, Vol. 55, 1949, pp. 459–482.

20. GOFF, J. A., "There should be Thermodynamic Consistency in the Construction and use of Psychrometric Charts," *ASHRAE Jl.*, May, 1960, pp. 70–75.

21. GOFF, J. A., and S. GRATCH, "Thermodynamic Properties of Moist Air," *ASHVE Trans.*, Vol. 51, 1945, pp. 125–157.

22. GOFF, J. A., and S. GRATCH, "Low Pressure Properties of Water in the Range–160 to 212 F," *ASHVE Trans.* Vol. 52, 1946, pp. 95–121.

23. HAINES, G. F., JR. and T. HATCH, "Industrial Heat Exposures—Evaluation and Control," *Heating and Ventilating*, Nov. 1952.

24. HANDEGORD, G. O. and C. E. TILL, "We Propose a New Humidity Standard," *ASHRAE Journal*, June, 1960, pp. 44–49.

25. HANDLER, E. and J. D. HARDY, "Temperature Sensation," *Mechanical Engineering*, Vol. 81, No. 8, Aug., 1959, pp. 69–70.

26. HOLLADAY, W. L., "Improved Methods for Calculation of Design Temperatures," *ASHRAE Trans.* Vol. 68, 1962, pp. 125–140.

27. HOUGHTEN, F. C. and C. P. YAGLOU, "Determining Lines of Equal Comfort," *ASHVE Trans.*, Vol. 29, 1923, pp. 163–176.

28. JAROS, A. L., "A Classification and Analysis of Air Conditioning Distribution Systems," *Air Conditioning, Heating and Ventilating*, Vol. 59, No. 4, Apr., 1962, pp. 79–90.

29. JENNINGS, B. H. and B. GIVONI, "Environment Reactions in the 80 to 105 F Zone," *ASHRAE Trans.*, Vol. 65, 1959, pp. 115–136.

30. JORDAN, R. C., G. A. ERICKSON, and R. R. LEONARD, "Energy Sources and Requirements for Residential Heating," *ASHRAE Trans.*, Vol. 68, 1962, pp. 174–203.

31. KOCH, W., B. H. JENNINGS, and C. M. HUMPHREYS, "Environmental Studies II," *ASHRAE Trans.*, Vol. 66, 1960, pp. 264–287.

32. KOCH, W., "Is Humidity Important in the Temperature Comfort Range?" *ASHRAE Jl.*, Vol. 2, 1960, p. 63.

33. KRANZ, P., "Calculating Human Comfort," *ASHRAE Journal*, Vol. 6, Sept., 1964, pp. 68–77.

34. MACKEY, C. O. and N. R. GRAY, "Heat Gains are not Cooling Loads," *Trans ASHVE*, Vol. 55, 1949, pp. 413–434.

35. McNALL, P. E. and R. G. NEVINS, "Comfort and Academic Achievement in an Air-Conditioned Junior High School." *ASHRAE Paper*, No. 2050, June, 1967.

36. MORSE, R. N. and E. KALETSKY, "New Approach to Radiant Cooling for Human Comfort," *Journal of the Institution of Engineers, Australia*, Vol. 33, No. 6, June, 1961, pp. 181–186.

37. PALMATIER, E. P., "Construction of the Normal Temperature ASHRAE Psychrometric Chart," *Trans ASHRAE*, Vol. 69, 1963, pp. 7–12.

38. PARCZEWSKI, K. I. and R. S. BEVANS, "A New Method of Rating the Quality of the Environment in Heated Spaces," *ASHRAE Jl.*, June, 1965, pp. 80–86.

39. PATTON, W. G., "1965 Passenger-Car Engineering Trends," *SAE Jl.*, Nov., 1964, p. 30.

40. RICKELTON, D., "Psychrometrics of Dual Duct Systems," Part I, *Heating Piping and Air Cond.*, Mar., 1964, pp. 113–117; Part II, Apr., 1964, pp. 147–152.

41. Roberts, J. F., "Development and Analysis of Design Curves from Weather Data for Air Conditioning Studies," *ASHRAE Jl.*, Vol. 8, No. 12, Dec., 1966, pp. 50–53.

42. SHATALOFF, N. S., "Comparative Costs of Dual Duct Air Conditioning Systems," *Air Conditioning, Heating and Ventilating*, June, 1960, 7 pp.

43. SHELTON, S. J., "Dual Duct High Velocity Air Conditioning," *Heating and Air Conditioning*, Oct., 1962, 7 pp.

44. SPANGLER, A. T., "Climate for Peak Production," *Air Conditioning, Heating and Ventilating*, Nov., 1964.

45. STOLL, A. M. and L. C. GREENE, "Radiation Burns," *Mechanical Engineering*, Vol. 81, No. 8, Aug., 1959, pp. 74–76.

46. SZABO, B. S., "Internal-Source Heat Pump Systems," *Air Conditioning, Heating and Ventilating*, June, 1966, pp. 51–82.

47. VERNON, H. M. "The Measurement, in Relation to Human Comfort, of the Radiation Produced by Various Heating Systems," *Inst. Heating Ventilating Engineering Proc.*, Vol. 31, p. 160.

48. WENTZEL, J. D., "An Instrument for the Measurement of the Humidity of Air," *ASHRAE Trans.*, Vol. 63, 1962, pp. 204–220.

49. WHEELER, A. E., "Air-conditioning Concepts for the Patient Care Unit," *ASHRAE Jl.*, Vol. 8, No. 11, Nov., 1966, p. 51.

50. WILE, D. D., "Psychrometry in the Frost Zone," *Refrigeration Engineering*, Vol. 48, Oct., 1944, pp. 291–301.

51. WILLIAMS, L., "Are Present Recommended Winter Outdoor Design Temp. Valid?" *ASHRAE Trans.*, Vol. 68, 1962, pp. 412–425.

52. WINSLOW, C. E., "Man's Heat Exchange with his Thermal Environment," Ref. [B.1], pp. 509–521.

53. Woodcock, A. H., H. L. Thwaites, and J. R. Breckenridge, "Clothed Man," *Mechanical Engineering*, Vol. 81, No. 8, Aug., 1959, pp. 71–74.

B. *Books*

1. AMERICAN INSTITUTE OF PHYSICS, *Temperature—Its Measurement and Control in Science and Industry*. New York: Reinhold, 1941.

2. AMERICAN SOCIETY OF HEATING, REFRIGERATING AND AIR CONDITIONING ENGINEERS, *Hand-book of Fundamentals*. New York: ASHRAE Inc., 1967.

3. AMERICAN SOCIETY OF HEATING, REFRIGERATION AND AIR CONDITIONING ENGINEERS, *Guide and Data Books: Applications*, 1966; *Fundamentals and Equipment*, 1965. New York: ASHRAE, Inc.

4. BROWN, G. G., et al, *Unit Operations*. New York: Wiley, 1950.

5. GOODMAN, W., *Air Conditioning Analysis*. New York: Macmillan, 1943.

6. HAWK, P. B., B. L. OSER, and W. H. SUMMERSON, *Practical Physiological Chemistry* (13th edition). New York: McGraw-Hill, 1954.

7. HOUGEN, O. A., *Chemical Process Principles*, Part I. New York: Wiley, 1954.

8. RABER, B. F. and F. W. HUTCHINSON, *Panel Heating and Cooling Analysis*. New York: Wiley, 1947.

9. SEVERNS, W. H. and J. R. FELLOWS, *Air Conditioning and Refrigeration*. New York: Wiley, 1958.

10. STROCK, C. and W. B. FOXHALL, *Handbook of Air Conditioning Heating and Ventilating*. New York: The Industrial Press, 1959.

11. THRELKELD, J. L., *Thermal Environmental Engineering*. Englewood Cliffs: Prentice-Hall, 1962.

12. WEBB, P. (ed.), *Bioastronautics Data Book*, NASA SP-3006, National Aeronautics and Space Administration, Washington, 1964.

13. WEXLER, A. and R. E. RUSKIN, *Humidity and Moisture—Vol. 1, Principles and Methods of Measuring Humidity in Gases*. New York: Reinhold, 1965.

14. WEXLER, A. and E. J. AMDUR (eds.), *Humidity and Moisture—Vol. 2, Applications*. New York: Reinhold, 1965.

15. WEXLER, A. and W. A. WILDHACK (eds.), *Humidity and Moisture—Vol. 3, Fundamentals and Standards*. New York: Reinhold, 1965.

16. WEXLER, A. and P. N. WINN, JR. (eds.), *Humidity and Moisture—Vol. 4, Principles and Methods of Measuring Moisture in Liquids and Solids*. New York: Reinhold, 1965.

CHAPTER 7

DIRECT ENERGY CONVERSION

7.1 INTRODUCTION

In Chapters 1 through 5 of this book, many power-producing and heat-pumping devices and processes have been discussed, and all have had a common feature: where heat energy was supplied to a system to produce work, a working fluid executed a cycle (open or closed) before work was available; where work was supplied to accomplish refrigeration or heat pumping, a working fluid and cycle were again necessary. The full cycle may not always have been apparent (as, for instance, in the discussion of cooling and heating by means of a vortex tube) but the dependence of a device on, say, a supply of compressed gas, takes other processes for granted to make up at least an open cycle. Where work was produced or required, it may have been in the form of shaft work, but it could always be imagined that that shaft work might come from an electric motor or go to an electric generator.

The systems and processes dealt with in this chapter differ from those so far presented in that an obvious working fluid executing a cycle will not be present. In fact, these devices are characterized by an absence of moving mechanical parts. Energy flow will be found to produce electrical work, or electrical work will be found to produce heat pumping "directly." That is not to say that conversion of energy from one form to another can be accomplished at an efficiency of 100 % in "direct energy conversion" devices. The five principal means of direct energy conversion to be discussed here can be divided into two major classifications. The thermal efficiency or coefficient of performance will be limited by the second law of thermodynamics as it applies to ideal cycle values (for instance, the Carnot cycle efficiency) wherever heat is transferred to the device at one temperature level and from the device at another level; the thermoelectric, thermionic, and magnetohydrodynamic devices fall in this classification. Representing the other classification are the photovoltaic cells and fuel cells, wherein conversion of the primary energy source to heat flow before utilization is not necessary; these latter devices are therefore not limited by the Carnot efficiency for the temperatures at which they operate; they are, in fact, essentially isothermal in their operation.

The Nernst generator which depends on a thermomagnetic effect is discussed separately in Section 7.6.

In any series of operations, the overall efficiency is the product of the efficiencies of the individual operations. It will follow, then, that the elimination of intermediate stages (cycles or processes) in the conversion of energy from one form to another must increase the system's potential efficiency. The actual efficiency obtained by a particular direct conversion device may not at the present time be greater than or even as great as the thermal efficiency of some systems already discussed. However, the fact that the potential efficiency value is less restricted is one of the main reasons for the intensive research in the modes and theories of direct energy conversion and the optimistic predictions for such devices.

A formal presentation of the theories of direct energy conversion might begin with a study of irreversible thermodynamics. This book, including this chapter, is limited to applications of thermodynamic principles, and therefore the theories of irreversible thermodynamics will not be reviewed. Those readers who are already familiar with that relatively new branch of thermodynamics will find no difficulty in identifying some of its generalizations here in particular examples. Nevertheless, the presentation is developed with the realization that most readers will probably lack this sophistication.

The relevant theories of irreversible thermodynamics are developed in other sources. Onsager's reciprocal relations [A.23] formed a basis for a new approach to irreversible thermodynamics. His concepts are incorporated in several recent books on the subject [B.5, B.8, B.9, and B.27]. The utilization of Onsager's relations in the classification of energy-conversion devices was developed by Osterle [A.24] and is presented in Chapter 2 of *Direct Energy Conversion*, by Angrist [B.1].

More directly pertinent to the understanding of direct energy-conversion methods will be some familiarity with the energy levels of electron states in molecular structures. The processes that cause differences in electron energy potential and thus the flow of electrons (or holes) make possible the conversion of the various forms of energy to electrical work. An attempt will be made to describe these phenomena where needed in the discussion of various devices. The reader can find the background theory in books on the fundamentals of moden physics [B.12 and B.32] and semiconductor theory [B.11, B.16, B.19, B.20, B.21, B.23, and B.24]. Angrist presents a rather concise summary of these theories in Chapter 3 of *Direct Energy Conversion* [B.1]. A simple and direct presentation of much useful background is given by Van Vlack, Chapters 2 and 5 [B.35].

The system of units used in this chapter will be different from that used in most of the rest of the book. Most of the published work in direct energy conversion uses the meter-kilogram-second (mks) or centimeter-gram-second (cgs) units more familiar to the physicist and electrical engineer. This and the English engineering system of units are summarized along with conversion factors in Appendix I.

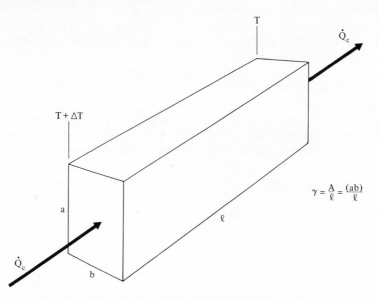

FIG. 7.1 Heat transfer by conduction.

7.2 *THERMOELECTRIC EFFECTS IN SOLIDS (in the absence of an external magnetic field)*

In the analysis of thermoelectric devices, five effects must be considered. In addition to simple **heat conduction** and the **Joule loss**, which accompanies the flow of any electric current against the resistance of the conducting medium, the **Seebeck**, **Peltier**, and **Thomson** effects must be dealt with. It will be seen that the last three phenomena, all resulting from inequalities in electrical potential developed in different ways, are interrelated through relatively simple expressions originally developed by Kelvin, and known as the first and second Kelvin relations.

The **Hall** and **Ettinghausen** (galvanomagnetic) effects and the **Nernst** and **Righi-Leduc** (thermomagnetic) effects will be mentioned later when the effects of external magnetic fields are presented. In Section 7.6, the usefulness of the Hall effect in the determination of semiconductor types will be explained.

7.2.1 *Heat Transfer by Conduction*

The thermodynamic analysis of a thermoelectric device, in which the electrical conducting materials are solid, requires a consideration of heat transfer by conduction. Heat transfer by radiation and convection will be neglected in the work of Sections 2 and 3 of this chapter.

Simple heat conduction in which the heat transfer rate is considered to be directly proportional to the temperature gradient is often referred to as **Fourier** heat flow since it follows the Fourier law of heat conduction:

$$\dot{Q}_c = -\lambda A \frac{dt}{dx}, \tag{7.1a}$$

where λ is the coefficient of thermal conductivity, usually expressed in watts per cm-degree K. For the bar of rectangular cross section shown in **Fig. 7.1**, the heat transfer rate would be

$$\dot{Q}_c = -\lambda \frac{(ab)}{l} \Delta T \text{ watts}$$

$$= -\lambda \gamma \Delta T. \tag{7.1b}$$

It will often be convenient to use the area-to-length ratio γ. It is assumed here that λ does not change with temperature, or alternatively, that the value used is the average over the temperature range ΔT.

7.2.2 Joulean Power Loss

The flow of electric current against any resistance is accompanied by a dissipation of electrical energy, that is, the transformation of electrical to thermal energy. This will raise the temperature of the conducting material unless an equivalent amount of energy is removed by heat transfer. By Ohm's law, $V = IR$. The Joulean (or Joule) heating rate is

$$\dot{Q}_j = IV = I^2 R \text{ watts}. \tag{7.2a}$$

The resistance (R) is determined by the dimensions of the conducting material and the material's **resistivity** (ρ), ohm-centimeter^2/centimeter, or simply ohm-centimeter. The inverse of the resistivity is the **conductivity** (σ); thus

$$\sigma = \frac{1}{\rho} (\text{ohm-cm})^{-1}.$$

For the rectangular bar shown in Fig. 7.1, the electrical resistance is

$$R = \rho \frac{l}{(ab)} = \rho/\gamma \text{ ohm}.$$

The electric current is often conveniently expressed as current density (J), where

$$J = \frac{I}{A} = \frac{I}{(ab)} \text{ amp/cm}^2.$$

Then

$$\dot{Q}_J = I^2 \rho/\gamma \tag{7.2b}$$

$$= J^2 \rho l(ab) \text{ watts}. \tag{7.2c}$$

7.2.3 The Seebeck Effect

The Seebeck thermoelectric effect is familiar to most engineers because of its application in temperature measurement by means of a **thermocouple** or pair of dissimilar conductors, arranged as shown schematically in **Fig. 7.2**. The two different materials making up the thermocouple are labeled P and N and are joined at point ① which is at temperature T_H, the temperature to be determined. The potentiometer circuit is connected through copper leads to conductors P and N at points ② and ③ respectively, both of these junctions being at a reference temperature T_L.

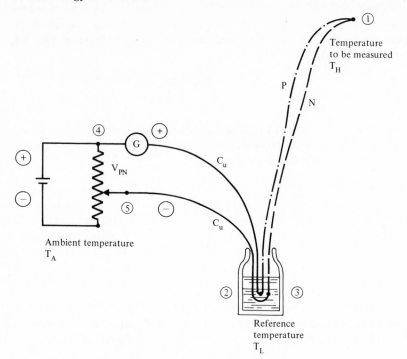

FIG. 7.2 Thermocouple circuit for temperature measurement.

With the potentiometer in the null position (zero current flow through the galvanometer), the open circuit voltage V_{PN}, resulting from temperature difference ($T_H - T_L$) is measured between points ④ and ⑤.

The **Seebeck coefficient (α)**, sometimes called "thermoelectric power," for one material relative to some other material is defined as

$$\alpha = \frac{dV}{dT} \text{ volts/degree.}$$

If, for instance, platinum is taken as a reference as is common for thermocouple materials, then α_P is the slope of the line at some temperature for material P in **Fig. 7.3** where emf is plotted against temperature.

The Seebeck voltage in Fig. 7.2 is the sum of the voltages around the thermocouple circuit:

$$V_{PN} = V_{4-2-1-3-5}$$

$$= \int_{T_A}^{T_L} \alpha_{Cu} \, dT + \int_{T_L}^{T_H} \alpha_P \, dT$$

$$+ \int_{T_H}^{T_L} \alpha_N \, dT + \int_{T_L}^{T_A} \alpha_{Cu} \, dT,$$

$$V_{PN} = \int_{T_L}^{T_H} (\alpha_P - \alpha_N) \, dT. \tag{7.3a}$$

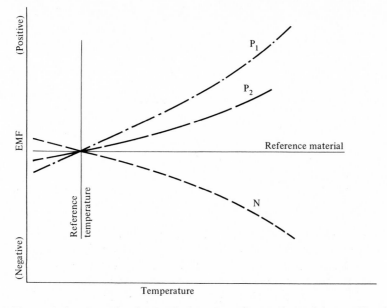

FIG. 7.3 Thermal electromotive force relative to a reference material, usually platinum (P1) for thermocouples, lead (Pb) for thermoelectric devices.

Note that the Seebeck voltage is not affected by either the ambient temperature (T_A) or by the material used for the instrument leads (in this case, copper). Two different materials (say, P_1 and N) are required for the device, and the Seebeck coefficient at a particular temperature for that couple is given by

$$\frac{dV_{PN}}{dT} = (\alpha_P - \alpha_N). \qquad (7.3b)$$

If we refer again to Fig. 7.3, we see that a Seebeck effect would have been produced with the two materials P_1 and P_2, but the emf for a given temperature difference ($T_H - T_L$) would not have been as great as for P_1 and N. Where one Seebeck coefficient is positive and the other negative with respect to the reference material, the couple coefficient is the sum of the numerical values:

$$\frac{dV_{PN}}{dT} = |\alpha_P + \alpha_N| = \alpha_{PN} \text{ volts/degree}. \qquad (7.3c)$$

7.2.4 The Peltier Effect

When an electric current flows from one conducting material to another through a junction as shown in **Fig. 7.4**, energy is brought by the charge carriers to the junction from material A at the left at a rate $\dot{Q}_A$, and energy is carried from the junction to material B at the right at a rate $\dot{Q}_B$. Because the energy level of the charge carriers (see Section 7.3.5) will in general be different in the two materials, $\dot{Q}_A$ will be greater or less than $\dot{Q}_B$. To maintain a constant junction temperature, heat must be

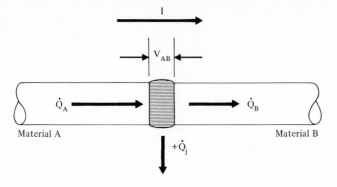

FIG. 7.4 Current flow through the junction of two different materials.

transferred to or from the surroundings as shown. Note that the direction of current flow shown is the conventional current direction which is opposite to the actual direction of the flow of electrons; it may be thought of as the direction of the "hole" flow, where the "hole" is the vacancy left by the removal of an electron. The energy level and thus the amount of energy transported is a function of each material. (See Fermi level, Section 7.4.1.) The Peltier coefficient is defined as

$$\pi = \frac{\dot{Q}}{I} \text{ watts/amp.}$$

The junction itself must have a finite electrical resistance so that current flow through it will result in the dissipation of the usual Joulean power equal to $I^2 R_j$ or IV_{AB}. This process, of course, is not reversible, but is always a conversion from electrical energy to heat. An energy balance for the junction requires that

$$\dot{Q}_j = I(\pi_A - \pi_B) + I^2 R_j$$
$$= I(\pi_{AB}) + I^2 R_j. \tag{7.4}$$

Depending on the relative magnitudes of π_A and π_B, the Peltier effect may be positive or negative. Whether π_A or π_B is larger, the Peltier effect can be reversed by a reversal of the direction of the electric current. However, $\dot{Q}_j$ cannot be as large negatively as it can be positively because the $I^2 R$ term is always positive.

7.2.5 The Thomson Effect

Because of thermal agitation of the charge carriers, it is possible to create a voltage gradient in a homogeneous material when a temperature gradient exists, quite apart from the voltage drop associated with the resistance of the material. In **Fig. 7.5** it is assumed that the voltage gradient resulting from the temperature difference is positive in the same direction as the temperature gradient. The **Thomson coefficient** for the material is

$$\tau = \frac{\Delta V_\tau}{\Delta T} \text{ volts/degree K.}$$

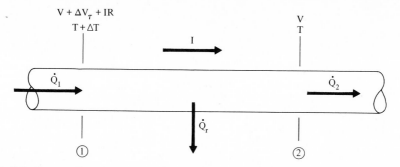

FIG. 7.5 Current flow through a homogeneous *p*-type material with a temperature gradient.

For one electron, the units would be electron-volts per electron per degree K, which is analogous to a specific heat as noted in Section 7.2.6.

If the rate of energy $\dot{Q}_1$, transferred across section ① by the electrical current, is not the same as $\dot{Q}_2$, which is transferred across section ②, an energy balance for the material between sections ① and ② requires that the rate of radial heat loss be

$$\dot{Q}_r = I\Delta V_\tau + I^2R = I\tau\Delta T + I^2R. \tag{7.5}$$

The first term on the right is the **Thomson heat** (or more correctly power), and the second term is the usual Joulean loss.

In a closed circuit, the direction of current flow caused by a temperature difference will depend on the type of material; in a *P*-type material it is opposite to the direction for an *N*-type material. Essentially, the difference depends on whether the current is carried by an excitation of electrons from a donor energy level into the conduction or valence band as is the case for *N*-type materials, or whether the conduction is by holes left vacant as electrons are excited to an acceptor level, which is still below the conduction energy level. (See Section 7.3.5, particularly Fig. 7.21.)

With reference to Fig. 7.5, the current flow (the direction of the hole flow) would be from the higher to the lower temperature, as shown, for a *P*-type material. The same temperature difference would create an opposite voltage difference and a current flow from right to left (opposite to the direction of the electron flow) in an *N*-type material.

7.2.6 Kelvin's (Thomson's) Relations for Thermoelectric Effects

As might be expected, the Seebeck, Peltier, and Thomson effects are not independent phenomena, but are interrelated. Thomson (later Lord Kelvin) predicted the form of the relationships by purely macroscopic thermodynamic reasoning. Because he ignored any irreversible effects, Kelvin's work is not rigorous and would not be accepted today except that (a) the relationships have been repeatedly verified experimentally, and (b) the same relationships have been arrived at through the more rigorous methods of irreversible thermodynamics.

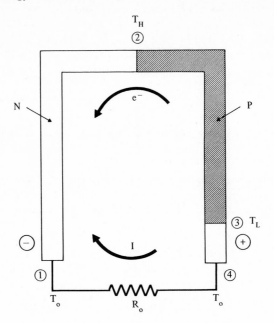

FIG. 7.6 A thermoelectric couple to illustrate Kelvin's relations.

The assumption of complete reversibility and the macroscopic viewpoint will be used here to derive the two Kelvin relationships following the general approach outlined by Jaumot [A.19]. **Figure 7.6** is a simplification of Fig. 7.2, and is more suitable for this discussion; furthermore, it is assumed that a current is flowing through materials P and N so that the couple carries an external load $V_{PN}I = R_0 I^2$ watts.

As a current flows through the circuit, the charge carriers (e) undergo changes in temperature. The Thomson coefficient is analogous to the charge carrier specific heat. An electron flowing from ① to ② must have its temperature increased from T_0 to T_H and absorbs the energy:

$$e \int_{T_0}^{T_H} \tau_P \, dT.$$

The charge carrier at ② flows from material P to material N at temperature T_H. If the entropy of the carrier is different in the two materials, the energy change required, which is proportional to the Peltier coefficient, is analogous to the latent heat. The energy transfer then at temperature T_H is

$$e\pi_{PN(H)}.$$

The same reasoning applies from points ② to ③ and ③ to ④ and at junction ③. Note that at a particular temperature,

$$\pi_{PN} = -\pi_{NP}.$$

We can now apply the first law of thermodynamics to the whole circuit. While there can be heat transfer and work done, there is no net change in the internal energy of the system or circuit. The work done per charge carrier (*e*) is eV_{PN}. Integration around the circuit and the elimination of *e* give us

$$V_{PN} = \pi_{PN(H)} - \pi_{NP(L)} + \int_{T_H}^{T_L} (\tau_N - \tau_P) \, dT. \tag{7.6}$$

The reversibility of the Peltier and Thomson effects would lead one to suspect that the change of entropy in the system is zero by the second law of thermodynamics. Of course, this is not absolutely true because the current flow must be accompanied by irreversible effects. However, as stated earlier, the more rigorous approach of irreversible thermodynamics, which will not be presented here, leads to the same net result. Therefore, as Kelvin did, we shall assume zero entropy change. In general,

$$dS = \left(\frac{dQ}{T}\right)_{\text{rev}}.$$

Therefore

$$dS = \frac{\pi_{PN(H)}}{T_H} + \frac{\pi_{PN(L)}}{T_L} - \int_{T_L}^{T_H} \frac{(\tau_N - \tau_P)}{T} \, dT = 0. \tag{7.7}$$

From these two equations, the Kelvin relationships can be obtained directly. If one of the two temperatures (say, T_L) is assumed constant and each equation is differentiated with respect to the variable temperature (say, $T_H = T$), then from Eq. (7.6), we find that

$$\frac{dV_{PN}}{dT} = \frac{d\pi_{PN(T)}}{dT} - (\tau_N - \tau_P)_T, \tag{7.8}$$

and from Eq. (7.7), we have

$$0 = \frac{d}{dT}\left(\frac{\pi_{PN(T)}}{T}\right) - \frac{(\tau_N - \tau_P)_T}{T}. \tag{7.9}$$

By elimination of $(\tau_N - \tau_P)_T$ from Eq. (7.8), we find that

$$T\left(\frac{dV_{PN}}{dT}\right) = T\left(\frac{d\pi_{PN(T)}}{dT}\right) - T^2\left[T\left(\frac{d\pi_{PN(T)}}{dT}\right) - \pi_{PN(T)}\right]\frac{1}{T^2} = \pi_{PN(T)}. \tag{7.10a}$$

This is generally known as **Kelvin's second relation**.

When this is substituted in Eq. (7.9), we have

$$T\left(\frac{d^2 V_{PN}}{dT^2}\right) = +(\tau_N - \tau_P)_T$$
$$= -(\tau_P - \tau_N)_T, \tag{7.11a}$$

which is generally known as **Kelvin's first relation**. In the preceding equations, the subscript *T* indicates that both π and τ are to be evaluated at temperature *T*.

In Section 7.2.3, the Seebeck voltage for a couple was presented in terms of the Seebeck coefficient (α). The relationships of Eq. (7.3) can be used in the two Kelvin relationships:

$$\frac{dV_{PN}}{dT} = (\alpha_P - \alpha_N) = \alpha_{PN}.$$

Thus an alternative form of Kelvin's second relation is

$$\frac{T(dV_{PN})}{dT} = T\alpha_{PN} = \pi_{PN(T)}. \tag{7.10b}$$

It follows that an alternative form of the first relation is

$$\frac{Td}{dT}\left(\frac{\pi_{PN}}{T}\right) = T\left(\frac{d\alpha_{PN}}{dT}\right) = -(\tau_P - \tau_N). \tag{7.11b}$$

Of course, the Thomson effect can appear in a single conductor, and Eq. (7.11b) can be written for a particular material:

$$T\frac{d\alpha}{dT} = -\tau. \tag{7.11c}$$

This implies that the Thomson effect is present only when there is a change in the Seebeck coefficient with temperature.

7.3 THERMOELECTRIC DEVICES

The term thermoelectric device is used in this chapter to describe a system in which two dissimilar materials are joined electrically and thermally for one of two purposes. When the purpose is the production of electrical power output from the input and removal of thermal energy, the device is a **generator**. When the purpose is to transfer thermal energy from a source to a sink at a higher temperature by the input of electrical power, the device is a **refrigerator** or **heat pump**. Of course, materials other than the two basic ones will be incorporated in the devices in various ways, but the principal effect will result from the arrangement of the two dissimilar legs.

The relationships for the thermoelectric effects in solids discussed in the previous section will now be applied to the analysis of thermoelectric generators and heat pumps.

7.3.1 Generators

A thermoelectric generator is shown schematically in **Fig. 7.7**. Heat transfer to the generator (Q_H) is at temperature T_H, and from the generator (Q_L) is at temperature T_L. Bars B_1, B_2, and B_3 will be assumed to be isothermal at the temperatures shown and to have negligible resistance to flow of heat and electricity compared with the resistances of legs N and P. The legs are of dissimilar materials (one N-type and the other P-type) and have constant rectangular cross sections which are not necessarily equal. The approach will be to assume that the resistivity (ρ), the thermal conductivity (λ), and the Seebeck coefficient (α) for each of the two leg materials will not vary with temperature. (Actually, as will be noted later, this is not so.) As a result,

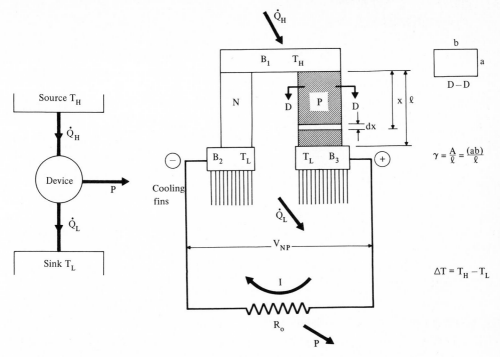

FIG. 7.7 A thermoelectric generator.

the Thomson coefficient as given in Eq. (7.11c),

$$\tau = T\frac{d\alpha}{dT},$$

will be zero in this analysis. This approach is similar to that of Ioffe [B.19], Angrist [B.1], and others.

Consider the differential length of leg P designated dx. Heat will be transferred by conduction from top to bottom, and energy will be dissipated by the electric current flow against the electrical resistance. With the symbols and units introduced in Section 7.2.1 and 7.2.2, an energy balance for a unit of time on a unit cross section of length dx requires that

$$\text{Energy in} + \text{Joulean loss} = \text{Energy out}$$

$$-\lambda\frac{dT}{dx} + J^2\rho dx = -\lambda\left[\frac{dT}{dx} + \frac{d}{dx}\left(\frac{dT}{dx}\right)dx\right]$$

$$\lambda\frac{d^2T}{dx^2} + J^2\rho = 0.$$

When $T = T_H$, $x = 0$, and when $T = T_L$, $x = l$. The resulting expression for the temperature distribution is

$$T = \left[T_H - \left(\frac{x}{l} \right)(T_H - T_L) \right] + \left[\frac{J^2 \rho}{2\lambda} \right] x(l - x). \tag{7.12a}$$

The same solution in terms of the temperature gradient is

$$\frac{dT}{dx} = -\left[\frac{(T_H - T_L)}{l} \right] + \left[\frac{J^2 \rho}{2\lambda} \right](l - 2x). \tag{7.12b}$$

The area-to-length ratio (A/l) is designated by γ, and the overall temperature difference $(T_H - T_L)$ is designated by ΔT. With these symbols in Eq. (7.12b), the rate of heat transfer (from top to bottom) by conduction at the two ends of the leg will be

$$\dot{Q}_{c(x=0)} = -\lambda A \frac{dT}{dx} = \lambda \gamma \Delta T - \frac{I^2 \rho}{2\gamma} \text{ watts}, \tag{7.13a}$$

$$\dot{Q}_{c(x=l)} = \lambda \gamma \Delta T + \frac{I^2 \rho}{2\gamma} \text{ watts}. \tag{7.13b}$$

There are, therefore, two components to these two heat transfer equations: the conduction heat transfer through temperature difference ΔT, and the Joulean energy loss, one half of which goes to either end of the leg.

When both legs of the generator of Fig. 7.7 are considered, the heat is conducted through the two in parallel between the same temperature limits, and the electric current flows through the two in series. The combined coefficient of heat transfer (K) for the two legs is

$$K = \lambda_N \frac{A_N}{l_N} + \lambda_P \frac{A_P}{l_P} = \lambda_N \gamma_N + \lambda_P \gamma_P.$$

The combined internal electrical resistance (R_i) for the two legs is

$$R_i = \rho_N \frac{l_N}{A_N} + \rho_P \frac{l_P}{A_P} = \frac{\rho_N}{\gamma_N} + \frac{\rho_P}{\gamma_P}.$$

The Peltier effect at each end must be added to the combination of the Fourier and Joulean heats in the whole thermocouple arrangement. We shall assume that the electrical resistance at each junction is negligible compared with the leg resistances. From Eqs. (7.4) and (7.10b), we find that

$$\dot{Q}_{j(x=0)} = -I\pi_{PN(H)} = -I\alpha_{PN}T_H \text{ watts}, \tag{7.13c}$$

and

$$\dot{Q}_{j(x=l)} = +I\alpha_{PN}T_L \text{ watts}. \tag{7.13d}$$

The difference in signs results from the direction of current flow which is from N to P at T_H and from P to N at T_L.

When heat conduction, the Joulean loss, and the energy supply or removal to overcome the Peltier effect are combined for the whole generator arrangement, the rates of energy supply $(\dot{Q}_H)$ and of heat removal $(\dot{Q}_L)$ can be calculated. With the directions as indicated on the diagram (Fig. 7.7) taken as positive, we have

$$\dot{Q}_H = I\alpha T_H + K\Delta T - \tfrac{1}{2}I^2 R_i \tag{7.14a}$$

and

$$\dot{Q}_L = I\alpha T_L + K\Delta T + \tfrac{1}{2}I^2 R_i. \tag{7.14b}$$

The subscript is dropped from the Seebeck coefficient since α is the value for the two particular materials involved.

The **useful power** can be obtained in two ways:

a) From the first law of thermodynamics,

$$\text{Useful power} = \dot{Q}_H - \dot{Q}_L = I\alpha\Delta T - I^2 R_i \text{ watts.} \tag{7.14c}$$

b) From an inspection of the diagram,

$$\text{Useful power} = I^2 R_0 \text{ watts.} \tag{7.14d}$$

When these two expressions are equated, the value of the current is found to be

$$I = \frac{\alpha\Delta T}{R_i + R_0} \text{ amp,} \tag{7.14e}$$

which might have been anticipated since $(\alpha\Delta T)$ is the Seebeck voltage and $(R_i + R_0)$ is the total circuit resistance.

The **thermal efficiency** of the thermoelectric generator is simply the ratio of the useful power to the rate of energy supply:

$$\eta_{\text{th}} = \frac{\text{Useful power}}{\dot{Q}_H} = \frac{I^2 R_0}{I\alpha T_H + K\Delta T - \tfrac{1}{2}I^2 R_i}. \tag{7.15a}$$

Each term in this expression can be simplified and made dimensionless if multiplied by $\Delta T/T_H$ and divided by $I^2 R_i$; the equation can be further simplified by the use of Eq. (7.14e):

$$\eta_{\text{th}} = \left[\frac{\Delta T}{T_H}\right]\frac{R_0/R_i}{(1 + R_0/R_i) + (R_i K/\alpha^2)(1/T_H)(1 + R_0/R_i)^2 - \tfrac{1}{2}\Delta T/T_H}. \tag{7.15b}$$

There are two very significant terms in this equation. First, the term in square brackets will be recognized as the **Carnot cycle thermal efficiency** for the temperature range ΔT below a supply temperature T_H. The remainder of the right-hand side thus becomes the relative or reduced efficiency (η_r). Second, the term $\alpha^2/R_i K$, which is inverted in one component of the denominator, has become known as the **figure of merit** (Z) for a particular arrangement:

$$Z = \frac{\alpha^2}{RK} \text{ (degrees } K)^{-1},$$

where the subscript on R is dropped because it is understood that this resistance is exclusive of the external load.

The figure of merit, as we shall see, is the most important parameter indicating the effectiveness of a thermoelectric device operating either as a generator or as a heat pump. While α depends on the materials used, both R and K depend on the device geometry. (To make the parameter dimensionless, some authors include a temperature term making the figure of merit equal $\alpha^2 T/RK$. The more common definition without T will be used in this book, but care must be taken in comparing numerical values from different sources.)

The thermal efficiency will be increased as Z is increased or, for a particular pair of materials, as (RK) is minimized. The product of these two terms is

$$(RK) = (\rho_N/\gamma_N + \rho_P/\gamma_P)(\lambda_N\gamma_N + \lambda_P\gamma_P)$$
$$= \lambda_N\rho_N + \lambda_N\rho_P(\gamma_N/\gamma_P) + \lambda_P\rho_N(\gamma_P/\gamma_N) + \lambda_P\rho_P. \qquad (7.16a)$$

Once the materials have been selected, the variables are γ_N and γ_P, the area-to-length ratios for the two legs. To minimize (RK), we let the differential equal zero:

$$\frac{d(RK)}{d(\gamma_N/\gamma_P)} = \lambda_N\rho_P - \lambda_P\rho_N(\gamma_N/\gamma_P)^{-2} = 0 \qquad (7.16b)$$

$$\frac{\gamma_N}{\gamma_P} = \left(\frac{\lambda_P\rho_N}{\lambda_N\rho_P}\right)^{1/2}, \qquad (7.16c)$$

which reduces Eq. (7.16a) to

$$(RK)_{\min} = [(\rho_N\lambda_N)^{1/2} + (\rho_P\lambda_P)^{1/2}]^2 \qquad (7.16d)$$

It follows then that the **optimum figure of merit** for two materials N and P will be

$$Z_{\text{opt}} = \frac{\alpha_{NP}^2}{[(\rho_N\lambda_N)^{1/2} + (\rho_P\lambda_P)^{1/2}]^2}. \qquad (7.16e)$$

For the whole system of Fig. 7.7, one other variable remains, the ratio R_0/R_i, which may now be optimized for either maximum efficiency or maximum power.

For **maximum efficiency**, the differential of Eq. (7.15b) with respect to (R_0/R_i) is equated to zero. The optimum figure of merit is assumed also for maximum η_{th}. The resulting optimum resistance ratio is

$$(R_0/R_i)_\eta = [1 + Z_{\text{opt}}T_{Av}]^{1/2}; \qquad (7.16f)$$

T_{Av} is simply $(T_L + T_H)/2$. Then (Eq. 6.15b) for η_{th} becomes

$$\eta_{\text{th(max)}} = \left[\frac{\Delta T}{T_H}\right]\frac{(R_0/R_i)_\eta - 1}{(R_0/R_i)_\eta + (T_L/T_H)} = \left[\frac{\Delta T}{T_H}\right]\frac{(1 + ZT_{Av})^{1/2} - 1}{(1 + ZT_{Av})^{1/2} + T_L/T_H}. \qquad (7.16g)$$

On the other hand, if Eq. (7.14d) is differentiated with respect to (R_0/R_i), Eq. (7.14e) being used for the value of I, then the resistance ratio for **maximum power** can be found. The result is

$$(R_0/R_i)_p = 1.0, \qquad (7.16h)$$

and the thermal efficiency at maximum power will be

$$\eta_{\text{th}(p)} = \left[\frac{\Delta T}{T_H}\right]\frac{1}{2 + [4/(Z_{\text{opt}}T_H)]} - \frac{1}{2}\frac{\Delta T}{T_H}. \qquad (7.16i)$$

The resistance ratio requirement for maximum power is clearly incompatible with that for maximum efficiency. However, since Z is relatively small and since the

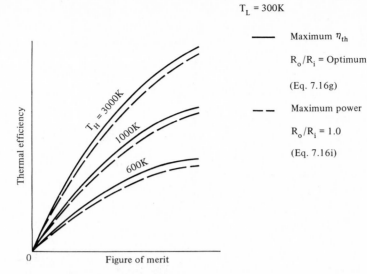

—— Maximum η_{th}

R_o/R_i = Optimum

(Eq. 7.16g)

— — Maximum power

$R_o/R_i = 1.0$

(Eq. 7.16i)

FIG. 7.8 Thermal efficiency of a thermoelectric generator at both maximum efficiency and maximum power conditions.

temperature ratio is more important than the resistance ratio in the thermal efficiency equation, the actual value of thermal efficiency at maximum power output may not be greatly reduced below the maximum possible value. **Figure 7.8** compares Eqs. (7.16g) and (7.16i) on a base of figure of merit.

Materials are available that provide, simultaneously, property values of the order of magnitude listed below in the temperature range 0 to 600 C (273 to 873 K):

$$\alpha = 100 \quad \text{to} \quad 200\mu V/\text{degree K},$$

$$\rho = 0.5 \times 10^{-3} \quad \text{to} \quad 5 \times 10^{-3} \text{ ohm-cm},$$

$$Z = 0 \quad \text{to} \quad 3 \times 10^{-3} \text{ (degree K)}^{-1}.$$

(Note that $1 \mu V = 10^{-6}$ volts.)

Of practical interest is the fact that it has been found more convenient and more reliable to measure α, ρ, and Z than to measure thermal conductivity (λ) [A.27]. For a particular material, λ is determined from the relationship

$$Z = \frac{\alpha^2}{\rho\lambda}.$$

Values for the optimum figure of merit for combinations of materials are then calculated from Eq. (7.16e).

7.3.2 Refrigerators and Heat Pumps

Because the Seebeck, Peltier, and Thomson effects are reversible, the thermoelectric generator arrangement of Fig. 7.7 can be converted to a thermoelectric refrigerator or heat pump when a direct current power source is substituted for the load

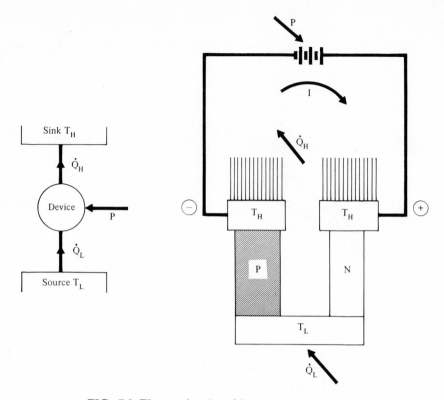

FIG. 7.9 Thermoelectric refrigerator or heat pump.

resistance. The heat pump (or refrigerator) circuit is shown in **Fig. 7.9**. It is inverted only to place the low temperature source at T_L below the high temperature sink at T_H as was usually done in Chapters 4 and 5. The dimensions are as noted in Fig. 7.7 for the generator. Note that the direction of current flow and the direction of net heat flow through the two legs are both the same as before.

The temperature difference applied to the thermocouple arrangement will, of course, produce a Seebeck voltage ($\alpha\Delta T$) in addition to the IR drop through the apparatus. Thus the power required from the source will be

$$P = I\alpha\Delta T + I^2R \text{ watts.} \tag{7.17a}$$

The heat transfer rates and Peltier effects are the same as for the device operating as a generator, except that the current direction and thus the Peltier heat flow are reversed at T_H and T_L, and the conduction heat transfer opposes the direction of desired energy flow. The Joulean heat is still a loss, and will be distributed equally to either end as was found for a generator. Thus the rate of heat removal from the hot junction must be

$$\dot{Q}_H = I\alpha T_H - K\Delta T + \tfrac{1}{2}I^2R \text{ watts.} \tag{7.17b}$$

The rate at which heat can be transferred to the device at the cold junction is

$$\dot{Q}_L = I\alpha T_L - K\Delta T - \tfrac{1}{2}I^2 R. \tag{7.17c}$$

As a **refrigerator**, the coefficient of performance (COP_R) of the device is

$$COP_R = \frac{\dot{Q}_L}{P} = \frac{I\alpha T_L - K\Delta T - \tfrac{1}{2}I^2 R}{I\alpha\Delta T + I^2 R}. \tag{7.17d}$$

Under identical conditions, but considered to be a **heat pump**, the device has a higher coefficient of performance (COP_{HP}) as is true for any heat pumping device:

$$COP_{HP} = \frac{\dot{Q}_H}{P} = \frac{I\alpha T_H - K\Delta T + \tfrac{1}{2}I^2 R}{I\alpha\Delta T + I^2 R}. \tag{7.17e}$$

The expression for power in Eq. (7.17a) can be checked by the application of the first law of thermodynamics to the whole system:

$$P = \dot{Q}_H - \dot{Q}_L \text{ watts.}$$

This also results in the relationship between COP_R and COP_{HP} noted in Chapter 4:

$$COP_{HP} = COP_R + 1.0.$$

The two expressions for the coefficient of performance, Eqs. (7.17d) and (7.17e), while simple, are not in the best form to indicate the significance of either the figure of merit or the ratio (IR/α). The latter parameter assumes an importance in heat-pump analysis that is comparable to that of the resistance ratio (R_0/R_i) in the consideration of the thermoelectric generator. If each term in Eq. (7.17d) is multiplied by (R/α^2), then

$$COP_R = \frac{(IR/\alpha)T_L - (RK/\alpha^2)\Delta T - \tfrac{1}{2}(I^2 R^2/\alpha^2)}{(IR/\alpha)\Delta T + (I^2 R^2/\alpha^2)}.$$

It is now apparent that **for maximum COP** the figure of merit (α^2/RK) should be maximized which, for a particular pair of materials, requires that (RK) be at a minimum. The procedure is the same as in the development of Eq. (7.16d) for a generator, and it is again found that

$$(RK)_{\min} = [(\rho_N \lambda_N)^{1/2} + (\rho_P \lambda_P)^{1/2}]^2.$$

Again, the optimum figure of merit is

$$Z_{opt} = \frac{\alpha_{NP}^2}{[(\rho_N \lambda_N)^{1/2} + (\rho_P \lambda_P)^{1/2}]^2}.$$

The second parameter (IR/α) is optimized when the differential of COP_R with respect to (IR/α) is equated to zero. This result is

$$\left(\frac{IR}{\alpha}\right)_{opt} = \frac{\Delta T}{(1 + ZT_{AV})^{1/2} - 1}, \tag{7.18a}$$

and

$$I_{opt} = \frac{\alpha\Delta T}{R[(1 + ZT_{AV})^{1/2} - 1]}. \tag{7.18b}$$

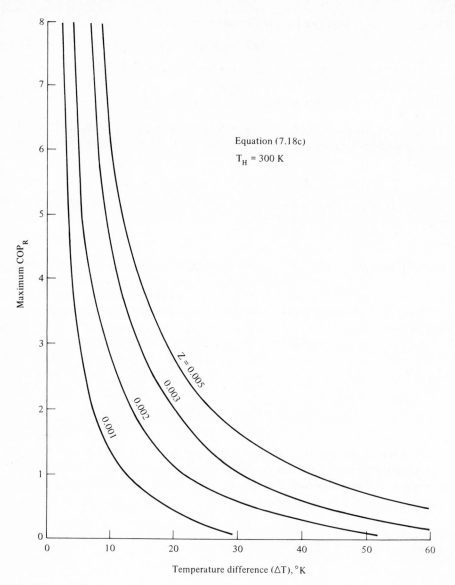

Equation (7.18c)

$T_H = 300$ K

$Z = 0.005$

0.003

0.002

0.001

Temperature difference (ΔT), °K

FIG. 7.10 Maximum COP_R versus temperature difference (figure of merit optimized).

When both the optimum figure of merit and the optimum current flow are used,

$$COP_{R(\max)} = \left[\frac{T_L}{\Delta T}\right] \frac{(1 + ZT_{AV})^{1/2} - T_H/T_L}{(1 + ZT_{AV})^{1/2} + 1}. \qquad (7.18c)$$

This equation is seen to be similar in form to Eq. (7.16g) when it is noted that the ratio (R_0/R_i) is given by Eq. (7.16f). The first term on the right is the coefficient of performance of a Carnot cycle refrigerator operating in the range ΔT above T_L.

An identical procedure applied to Eq. (7.17e) will yield

$$COP_{HP(max)} = \left[\frac{T_H}{\Delta T}\right] \frac{(1 + ZT_{AV})^{1/2} - T_L/T_H}{(1 + ZT_{AV})^{1/2} + 1}. \tag{7.18d}$$

The first term on the right, as might be expected, is the COP for a Carnot cycle heat pump over the same temperature range. The optimum figure of merit and the optimum current flow are the same as for the device operating as a refrigerator. The power requirement at maximum COP conditions will be the value from Eq. (7.17a) when the optimum current is flowing. The optimum applied voltage is simply (P_{opt}/I_{opt}).

Equations (7.18c) and (7.18d) both show that the maximum coefficient of performance of a thermoelectric device operating either as a refrigerator or as a heat pump is a function of only T_H, T_L, and Z. **Figure 7.10** is a plot of COP_R on a base of ΔT for various Z-values and with T_H fixed at a normal ambient value of 300 K.

The prime requirement of a thermoelectric device operating as a refrigerator is not necessarily the coefficient of performance. The temperature range or rate of heat pumping may be more important. The **temperature difference** that can be produced by a thermoelectric refrigerator is given by a rearrangement of Eq. (7.17c):

$$\Delta T = (1/K)(I\alpha T_L - \tfrac{1}{2}I^2 R - \dot{Q}_L).$$

To obtain a minimum T_L, $\dot{Q}_L$ should approach zero (no heat load) and $d\Delta T/dI$ should be zero. Of course, (RK) must be a minimum simultaneously so that the figure of merit is optimized. As a result,

$$I_{(max \, \Delta T)} = \alpha T_L/R \tag{7.19a}$$

and

$$\Delta T_{(max)} = \tfrac{1}{2}ZT_L^2. \tag{7.19b}$$

This quadratic equation can be solved to determine the minimum value of T_L obtainable with a given T_H:

$$T_{L \, (min)} = \frac{(1 + 2ZT_H)^{1/2} - 1}{Z}. \tag{7.19c}$$

A plot of minimum T_L on a base of T_H is shown in **Fig. 7.11** for various values of Z.

When it is required that the refrigeration effect $(\dot{Q}_L)$ should be as large as possible, Eq. (7.17c) must be maximized:

$$\dot{Q}_L = I\alpha T_L - K\Delta T - \tfrac{1}{2}I^2 R.$$

When $d\dot{Q}_L/dI$ equals zero, we have

$$I_{(max \, \dot{Q}_L)} = \alpha T_L/R. \tag{7.20a}$$

This value of current can then produce a cooling rate equal to

$$\dot{Q}_{L(max)} = \frac{(\alpha T_L)^2}{2R} - K\Delta T. \tag{7.20b}$$

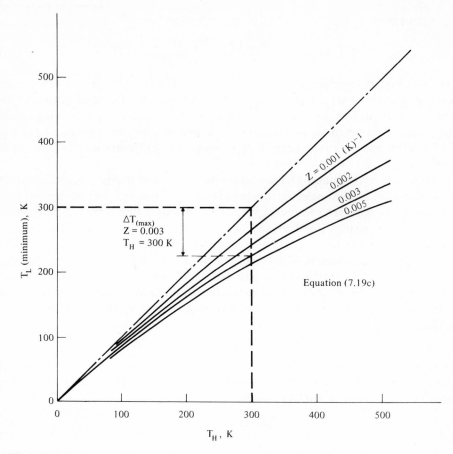

FIG. 7.11 Minimum cold junction temperature obtainable (figure of merit optimized), single stage.

It is interesting that the same expression for current was found for the maximum temperature difference, where one requirement was that $\dot{Q}_L$ should be zero. However, the value of T_L would be different in the two different situations. For either situation, Eq. (7.19a) or (7.20a), the required applied voltage must be

$$V = \alpha\Delta T + IR = \alpha(T_H - T_L) + \alpha T_L = \alpha T_H \qquad (7.21)$$

The power input is then the product (IV). When the purpose is to operate the device as a heat pump rather than as a refrigerator, that is to say, where the purpose is to deliver $\dot{Q}_H$ at T_H rather than to remove $\dot{Q}_L$ at T_L, it would be pointless to allow $\dot{Q}_L$ to approach zero, because then the COP would approach unity. The development of Eq. (7.19b) shows that this occurs when ΔT is $\frac{1}{2}ZT_L^2$. Thus if we have a **heat pump** in which

$$\text{COP}_{HP} > 1.0,$$

then

$$\Delta T < \tfrac{1}{2}ZT_L^2.$$

A coefficient of performance less than unity for a heat pump is not impossible, but it implies that COP_R is less than zero, so that only part of the input power is converted to heat transfer at the upper temperature, while some must be lost by reversing $\dot{Q}_L$ at the lower temperature. The low limit for the coefficient of performance of a device considered as a refrigerator is zero, which is approached as minimum T_L is approached.

7.3.3 Single-Stage Series Operation of a Generator or Heat Pump

The voltage output of a single pair of thermoelectric elements in a generator operating between particular temperature levels is limited by the Seebeck coefficient for the pair, no matter how large the elements might be. At the same time, the power output of a generator or the rate of heat pumping for a refrigerator between particular temperature levels are both affected by the thermal conduction rate and thus the dimensions of the device. Up to this point, only the proportions in terms of the area-to-length ratio (γ) for the elements, but not their size, have been discussed.

In the optimization of figure of merit, the optimum ratio γ_P/γ_N was determined as given by Eq. (7.16c). However, the magnitude of either γ_P or γ_N is not dictated by this; it is determined by the rate of heat transfer required of each couple as given by Eq. (7.1b):

$$\dot{Q}_c = -\lambda\gamma\Delta T.$$

Only the area-to-length ratio is significant, not the length of each element. To minimize weight and conserve thermoelectric material, one would want to keep the length (l) of the two legs small. The size of the legs cannot, however, be reduced to the point at which the contact electrical resistance becomes unduly significant.

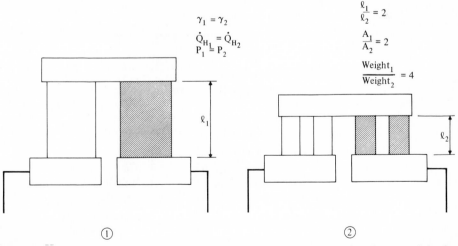

$$\gamma_1 = \gamma_2$$
$$\dot{Q}_{H_1} = \dot{Q}_{H_2}$$
$$P_1 = P_2$$

$$\frac{\ell_1}{\ell_2} = 2$$
$$\frac{A_1}{A_2} = 2$$
$$\frac{Weight_1}{Weight_2} = 4$$

①   ②

FIG. 7.12 Relationship between length and weight for thermoelectric materials for equal power.

As an example of how the weight of material can be greatly reduced, **Fig. 7.12** shows two devices with the same power output or requirement as determined by the fact that γ_1 is equal to γ_2. However, l_1 is twice l_2 and A_1 is twice A_2. Thus the ratio of material weights, which is dependent on the product of the area and the length, is 4:1, the square of the length ratio.

If a practical length (let us say 1 cm or 5 cm) has been selected, and γ is determined by the power and the required rate of heat transfer, the cross-sectional area of the elements is established. This area can be realized, however, in different ways which will determine the internal electrical resistance and the output or required voltage. **Figure 7.13** shows three different methods by which the load-carrying

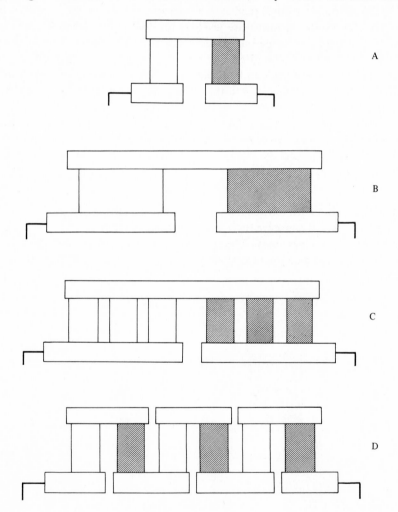

FIG. 7.13 Alternative methods of increasing power of thermoelectric couples in one stage.

capacity or heat-pumping rate of the single thermoelectric couple shown in *A* can be increased threefold. Examples *B* and *C* are essentially the same, because all *N* legs and all *P* legs in *C* are in parallel, both electrically and thermally. Their voltages for a given temperature difference are the same as for *A*, but the current and heat transfer rate are increased in proportion to the increase in cross-sectional area. The heat transfer rate, and thus the power for *D* is the same as for *B* and *C*, but the electrical resistances are in series; the voltage is increased in proportion to the number of couples, but the current remains the same as in *A*.

The arrangement of thermoelectric elements in series as in *D* is known as a **thermopile**. It is used in thermoelectric generators so that a higher voltage output will be available, and in heat-pumping devices so that reasonably high input voltages can be utilized. In temperature measurement, the thermopile arrangement can provide a more easily measured emf signal even from a small temperature difference. However, means are now available to increase or decrease dc voltages through semiconductor voltage converters. The advantage of *D* over *C* may become less important with the improvement of these dc voltage converters.

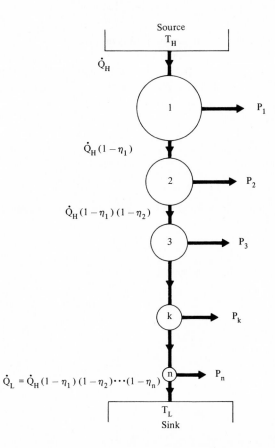

FIG. 7.14 A schematic thermodynamic diagram of a cascaded generator with *n* stages.

7.3.4 Cascaded Multistage Operation

A thermoelectric generator or heat pump may be required to work through a very large temperature range ($\Delta T = T_H - T_L$). For a generator, the main problems are that materials with the best figure of merit values near the lowest temperature might have poor values at higher temperatures or might physically disintegrate or change phase at the highest temperature. For a heat pump, the fact that figure of merit values actually do change with temperature level is important, but an even more serious problem has already been revealed in Eq. (7.19b) and (7.19c): the temperature drop below a given sink temperature is limited for a given pair of materials.

For both the generator and the refrigerator (or heat pump), one solution to these problems is **cascading**. As in Chapters 1 and 4, the term is used where the heat rejection from one part of a system becomes the heat input to another part, which supplements the function of the first. **Figure 7.14** is a schematic representation of a cascaded generator with n stages. The schematic diagram for a heat pump or refrigerator would be essentially the same with all arrows reversed. The diagram implies that the size of the device decreases from one stage to another as the average temperature of the device is lowered, because either a simple "heat balance" or the application of the first law of thermodynamics shows that the heat input to each is less than that for the stage above. **Figure 7.15** presents a closer approach to a real three-stage thermoelectric generator.

In addition to making it feasible to select the best materials at different average temperatures, cascading actually increases the potential maximum thermal efficiency of a generator, as will be demonstrated below. For the system of Fig. 7.14, the first-stage efficiency is

$$\eta_1 = P_1/\dot{Q}_H. \tag{7.22a}$$

The rate of energy supply to the second stage is

$$\dot{Q}_{H2} = \dot{Q}_H - P_1 = \dot{Q}_H(1 - \eta_1). \tag{7.22b}$$

The heat-rejection rate from the kth stage, which is the heat delivery rate to the $k + 1$ stage is

$$\dot{Q}_{Lk} = \dot{Q}_H(1 - \eta_1)(1 - \eta_2)\cdots(1 - \eta_k). \tag{7.22c}$$

Therefore the overall thermal efficiency of the system is

$$\begin{aligned}
\eta_{OA} &= \frac{P_1 + P_2 + \cdots + P_n}{\dot{Q}_H} \\[2mm]
&= \frac{\dot{Q}_H - \dot{Q}_L}{\dot{Q}_H} \\[2mm]
&= \frac{\dot{Q}_H[1 - (1 - \eta_1)(1 - \eta_2)\cdots(1 - \eta_n)]}{\dot{Q}_H} \\[2mm]
&= 1 - \prod_{k=1}^{k=n} (1 - \eta_k) \tag{7.22d}
\end{aligned}$$

in which $\prod$ indicates "the product of terms."

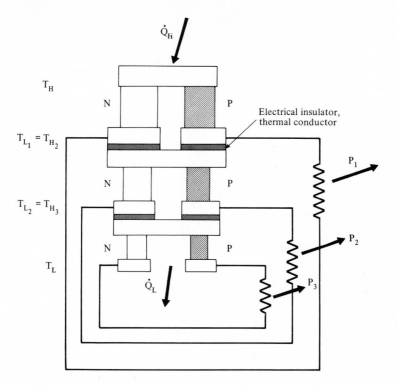

FIG. 7.15 A simple three-stage cascaded thermoelectric generator.

If the efficiency of each stage were the same, say η_s, then

$$\eta_{OA} = 1 - (1 - \eta_s)^n. \tag{7.22e}$$

From this it would seem at first glance that the greater the number of stages, the greater the efficiency. However, it must be recalled that the stage efficiency is limited by the temperature range over which it works, as shown by Eq. (7.16g). Nevertheless, it is logical to assume [A.32, p. 488] that the division of a thermoelectric device into several stages, each operating over only part of the temperature range, will not change the output and overall efficiency when the geometry is not changed from one stage to another. On the other hand, if each stage is optimized for geometry and resistance ratio, each stage efficiency can be improved. Therefore cascading might improve and cannot decrease the efficiency if increased contact resistance is ignored. It is apparent then that the greater the number of optimized stages, the greater the efficiency, and hypothetical **infinite staging** should produce a theoretical maximum value of efficiency over a given temperature range.

If a very large number of stages is assumed, so that T_L approaches T_H for each stage and ΔT is called δT, then Eq. (7.16g) for an optimized stage becomes

$$\eta_{\text{th(max)}} = \left[\frac{\delta T}{T}\right]\frac{(1 + ZT)^{1/2} - 1}{(1 + ZT)^{1/2} + 1}. \tag{7.23a}$$

The **reduced efficiency** (η_r) which is the ratio of the ideal cycle efficiency to the Carnot efficiency, becomes what is known as the **differential efficiency,** since very small temperature differences are assumed; or it is sometimes called the **material efficiency** (η_{mat}), since it is a function of only the material figure of merit at some temperature and the temperature itself:

$$\eta_{mat} = \frac{(1 + ZT)^{1/2} - 1}{(1 + ZT)^{1/2} + 1}. \tag{7.23b}$$

Assuming optimized stages, we find that

$$\eta_k = \eta_{mat}\frac{\delta T}{T_k}$$

$$\dot{Q}_L = \dot{Q}_H\left(1 - \eta_{mat_1}\frac{\delta T}{T_1}\right)\left(1 - \eta_{mat_2}\frac{\delta T}{T_2}\right)\cdots$$

$$= \dot{Q}_H\left(1 - \sum \eta_{mat(k)}\frac{\delta T}{T_k}\right), \tag{7.23c}$$

as long as $\eta_{mat}(\delta T/T)$ is very small relative to unity.* If it is also true that $[(\sum \eta_{mat}(\delta T/T)]$ is very small compared to 1.0, then

$$\dot{Q}_L = \dot{Q}_H \exp\left(-\sum \eta_{mat(k)}\frac{\delta T}{T_k}\right)^\dagger \tag{7.23d}$$

Thus the overall efficiency for infinite staging is

$$\eta_{OA} = \frac{\dot{Q}_H - \dot{Q}_L}{\dot{Q}_H} = 1 - \frac{\dot{Q}_L}{\dot{Q}_H}$$

$$= 1 - \exp\left(-\int_{T_L}^{T_H}\eta_{mat}\frac{dT}{T}\right) = \int_{T_L}^{T_H}\eta_{mat}\frac{dT}{T}. \tag{7.23e}$$

An alternative to the cascading arrangements already discussed is the use of **segmented legs** as shown in **Fig. 7.16.** The material for each segment is chosen for optimum figure of merit and structural stability over its own temperature range. It is not necessary that there be the same number of segments in the positive and negative legs. The heat rejection for an upper segment becomes the heat input to the segment below, so that the heat flow is essentially the same as for cascading. The output voltage is, however, the overall voltage and the segments are in series electrically.

Cascaded operation of thermoelectric devices for refrigeration or heat pumping can be analyzed with essentially the same approach used in the analysis of power generation. For n stages, the coefficient of performance as a refrigerator would be [B.16, p. 468]:

$$\text{COP}_R = \left[\prod_{k=1}^{k=n}\left(1 + \frac{1}{\text{COP}_{R(k)}}\right) - 1\right]^{-1}. \tag{7.24}$$

---

*$(1 - m_1)(1 - m_2)(1 - m_3)\ldots \simeq 1 - m_1 - m_2 - m_3 \ldots = 1 - \sum m$, where $m \ll 1.0$.
†$e^x \approx 1 + x$, where $x \ll 1.0$.

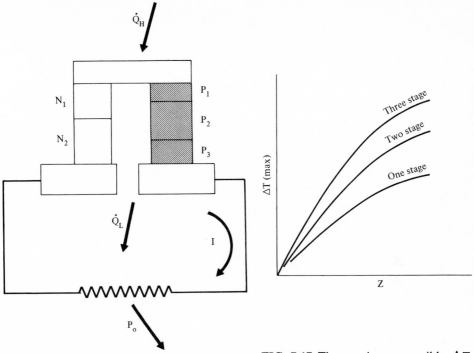

FIG. 7.16 A thermoelectric generator with segmented legs.

FIG. 7.17 The maximum possible ΔT for one-, two-, and three-stage thermoelectric refrigerators.

As noted previously, the advantage of staging for refrigeration is not so much in the increase in coefficient of performance, but in the increase in possible temperature difference. This latter effect is shown clearly in **Fig. 7.17** [B.11, p. 132].

An appreciable improvement in COP by means of cascading is realized only as T_L approaches the minimum achievable T_L, at which condition COP_R approaches zero (See Fig. 7.10). For a particular required temperature difference, the use of more stages allows each stage to operate through a smaller ΔT and therefore at a higher COP.

7.3.5 Materials

The suitability of materials for thermoelectric devices is determined primarily by the figure of merit:

$$Z = \frac{\alpha^2}{\rho\lambda}.$$

The higher this value, the better the material. The values of all three components of Z are interrelated through their mutual dependence on the concentration of charge carriers, electrons or ions, in the materials. The charge carriers for "ionic" conduction may be either positively or negatively charged ions which have either

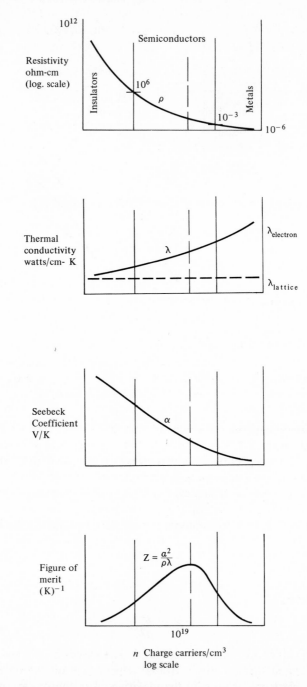

FIG. 7.18 The effect of the concentration of charge carriers on the figure of merit and its components.

a deficiency or a surplus of electrons. For "electronic" conduction, the current may be carried either by the electrons or by the "holes" vacated by electrons.

The unit charge per electron is 1.602×10^{-19} coulomb. Thus the **electrical conductivity** (σ) is given by the product:

$$\sigma = 1.602 \times 10^{-19} \, n\mu$$

$$= (\text{coulomb/carrier})(\text{carriers/cm}^3) \left(\frac{\text{cm/sec}}{\text{V/cm}} \right)$$

$$= \left(\frac{\text{coulomb}}{\text{sec}} \right) \left(\frac{1}{\text{V}} \right) \left(\frac{1}{\text{cm}} \right)$$

$$= \left(\frac{\text{amp}}{\text{V}} \right) \left(\frac{1}{\text{cm}} \right) = (\text{ohm-cm})^{-1},$$

where n is the carrier concentration in the material, μ is the carrier mobility in the material, σ is $1/\rho$, and ρ is the resistivity, ohm-cm.

The **thermal conductivity** (λ watts per cm-degree K) can be thought of as the sum of two separate components. First, there is **electron conduction** (λ_e) by means of the thermally excited electrons free to move in the material; this is a function of the number of charge carriers present. Second, there is **lattice conduction** (λ_e) resulting from thermally produced vibration of the lattice structure of the material; this involves no net transport of matter, and is essentially independent of the number of charge carriers present.

The value of the **Seebeck coefficient** (α volts/degree K) is related to the carrier concentration in a more complicated way than are ρ and λ [B.19 and B.20]. The value of α becomes relatively large in insulators for which ρ is too large for thermoelectric application, and α becomes very small for metals with good conductivity.

The relationships between ρ, λ, α, and Z and the concentration of charge carriers (n) are shown in **Fig. 7.18**. The figure of merit (Z) is seen to reach a maximum at a charge carrier concentration of approximately 10^{19} carriers/cu cm. This is in the range of physical property values arbitrarily designated as the **semiconductor** range.

It is to be expected, then, that the materials that would be most effective in thermoelectric devices would be chemical compounds of pairs of materials falling on either side of the arbitrarily drawn line between "metals" and "nonmetals" in the conventional **periodic table** of elements. In general, this has been found to be so. **Figure 7.19** shows some of the elements that have been found most useful to date. These have been used in various combinations of such compounds as Bi_2Te_3, Bi_2Se_3, $AgSbTe_2$, $SnTe$, $PbTe$, and Sb_2Te_3. Oxides may be useful at high temperatures.

It is now important to see why one material is "positive" and another "negative" thermoelectrically. A very simple example is given by Van Vlack [B.35, p. 113]. Silicon (Si), atomic number 14, has an atomic structure equivalent to the basic neon structure plus two electrons in the lowest level(s) and two in the second level (p) of the third (M) energy shell. Its structure in spectroscopic notation is

$$[s^2 2s^2 2p^6]3s^2 3p^2 = [\text{Neon}]3s^2 3p^2.$$

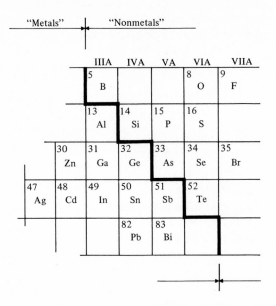

FIG. 7.19 Some elements used in thermoelectric-device semiconductor materials shown in their positions on the periodic table.

If a small percentage of aluminum, atomic number 13, is added to the silicon, the previously regular atomic structure will be lacking one electron in the vicinity of an aluminum atom, because aluminum's structure is

$$[\text{Neon}]\,3s^2\,3p^1.$$

It has only three electrons in the outer or valence shell (M). Any electron from a neighboring silicon atom can move in to fill this vacancy, at the same time leaving a "hole" at its original location [see **Fig. 7.20(a)**]. An applied electric field will encourage electrons to continue to move into the vacated holes always in the general direction from negative toward positive polarity. The net effect is that the "hole," which is in effect a positive charge carrier, is moving from the positive toward the negative side of the applied field. Thus it is the "hole" which has mobility, and silicon with an aluminum impurity is called a *p*-**type semiconductor**.

Let us now consider silicon with a small percentage impurity of phosphorus, atomic number 15. The atomic structure of phosphorus is

$$[\text{Neon}]\,3s^2\,3p^3.$$

It has five electrons in its valence shell. When the phosphorus atom takes its place in the otherwise homogeneous silicon structure, one electron is free to move under the influence of an applied field [see **Fig. 7.20(b)**]. The electron will move toward the positive side of the field. The carrier in this case is the negatively charged electron. Thus silicon with a phosphorus impurity is called an *n*-**type semiconductor**.

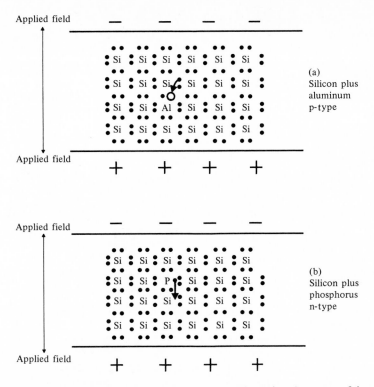

FIG. 7.20 *p*-type and *n*-type semiconductors determined by the type of impurity in a poor conductor (from Van Vlack [B.35]).

Because the materials described became conductors by reason of impurities, they are called **extrinsic semiconductors**. A material such as pure germanium, which has a resistivity of approximately 50 ohm-cm, is an **intrinsic** semiconductor. In its pure form, it falls to the right of the center of the region defined for semiconductors in Fig. 7.18. Pure silicon, another intrinsic semiconductor, has a resistivity of about 6×10^4 ohm-cm, which puts it very close to the insulator region.

What makes a material an insulator, a conductor, or one of the various types of semiconductors is best explained in terms of electron energy levels. For individual atoms, electrons exist at various levels, or in various shells and subshells beyond the nucleus. The core electrons (all except those in the outer shell) remain identified for the most part with the nucleus when a large number of atoms coalesce to form the structure of a solid material. The electrons in the outermost shell (the valance electrons), however, do interact and may be considered to exist in **energy bands**. When sufficient energy is supplied to the electrons of a poor conductor to raise their energy level to the **conduction band**, they are free to move within the material under the influence of an electric field.

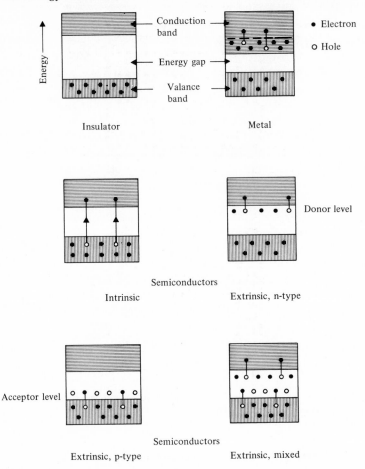

FIG. 7.21 Energy bands in insulators, metals, and semiconductors .

This is shown graphically in **Fig. 7.21**, adapted from Angrist [A.1]. For an **insulator**, the conduction band is separated from the valence band by a wide **energy gap** or **forbidden zone** that is not easily traversed by normal energy input. For a **metal**, the conduction band is partly filled with free electrons bound to no particular nucleus, and thus easily accelerated toward an external positive charge.

An **intrinsic semiconductor** has a relatively small energy gap compared to an insulator, so that valence electrons can be raised by thermal or light energy to the conduction band. Once there, they are free to move as in a metal. Few intrinsic semiconductors have electrical resistivities low enough to be useful in thermoelectricity.

When a poor conductor contains impurities that have energy levels that can appreciably increase conductivity, the combined material is classed as an **extrinsic**

semiconductor. The impurities purposely added to affect electrical conduction are called **dopants**. The type of dopant can determine whether the semiconductor becomes a *p*-type or an *n*-type. In an **n-type** semiconductor, electrons exist at an energy level just below the conduction band to which these **donor** electrons are easily excited. In a **p-type semiconductor**, an energy level within the forbidden zone becomes available to **accept** electrons excited from the valence band. These electrons, however, are not free to move within the material; but the holes left by them in the valence band allow mobility at that level and are thus, in effect, positively charged carriers. A semiconductor can be so doped that it has both donor and acceptor energy levels, and it is then called a **mixed** extrinsic semiconductor. It may be predominantly either positive or negative.

Many examples of extrinsic semiconductors, useful as thermoelements, could be given. Properties of several are given in Reference B.33, Chapter 3, and in Reference B. 1, Appendix C. As an example, one of these is the *n*-type compound 75% Bi_2Te_3 plus $25\% Bi_2Se_3$; a *p*-type compound is $25\% Bi_2Te_3$ plus $75\% Sb_2Te_3$ with about 2% excess Te or Se. For the thermoelectric properties of numerous combinations and the effect of temperature, see Rosi, Hockings, and Lindenblad [A.26].

In the evaluation of materials for thermoelectric applications, it is convenient to have a single, simple, dimensionless parameter indicative of the generator thermal efficiency. Such a parameter can be developed from Eq. (7.23a) for which it was assumed that the ratio T_L/T_H is nearly unity. Then

$$\eta_{th(max)} = \frac{\Delta T}{T} \frac{(1 + ZT)^{1/2} - 1}{(1 + ZT)^{1/2} + 1}.$$

Using the first two terms of the series,

$$(a + x)^n = a^n + na^{n-1}x + \cdots$$

The equation now becomes

$$\eta_{th(max)} \simeq \frac{\Delta T}{T} \frac{(1 + \frac{1}{2}ZT) - 1}{(1 + \frac{1}{2}ZT) + 1} \simeq \frac{Z\Delta T}{4}$$

(since $\frac{1}{2} ZT$ is much less than 2.0). The parameter $\frac{1}{4} Z\Delta T$ is commonly used for the comparison of various materials over particular temperature ranges.

7.4 THERMIONIC GENERATORS

As is the case for many apparently new types of energy conversion devices, the principle of operation of the **thermionic generator** has been known for many years. It is essentially the same as for a thermionic vacuum tube. However, the utilization of thermionic emission for a practical power-producing device has been developed only within the last decade. This rate of development has been phenomenal, so that the thermionic generator is now seen to be practical not only for relatively small power outputs, but also as a possible component if not actually the main part of a central power station.

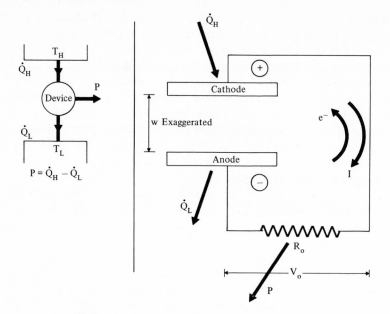

FIG. 7.22 A schematic thermionic generator.

The basic arrangement is shown schematically in **Fig. 7.22**. The accompanying thermodynamic diagram shows that this is analogous to any device which produces work by taking thermal energy in at a high temperature and rejecting some thermal energy at a lower temperature. The energy input $\dot{Q}_H$ drives electrons from the cathode to the anode where energy is rejected at a rate $\dot{Q}_L$. The **electron flow** through the external resistance R_0 is from the anode back to the cathode. That is to say, the **conventional** current flow as shown on the diagram is from the cathode to the anode in the external circuit, which agrees with conventional electric-battery nomenclature. The spacing (w) between the cathode and the anode is usually extremely small, less than 0.001 in. in many cases. When the interspace is essentially a vacuum, this distance must be small for satisfactory operation. When the interspace is filled with a plasma such as ionized cesium vapor at low pressure (see Section 7.4.3), the distance between the cathode and anode can be increased. At the same time, even with such a neutralizing plasma in the interspace, operation is still found to be most satisfactory with very small cathode to anode spacing. The operating characteristics of plasma diodes are more difficult to analyze because of the complicating electron-plasma interactions, including that of heat transfer.

Thermionic generators are essentially low-voltage high-current flow devices. Current densities of 20 to 50 amp/sq cm have been achieved at voltages from 1 to 2 V. Thermal efficiencies of 10 to 20% have been realized, but higher values are certainly possible in the near future.

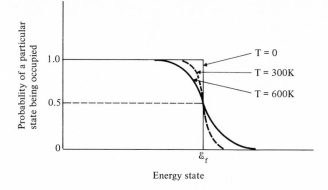

FIG. 7.23 Fermi energy ($\mathscr{E}_f$) (see Reference [B.22]).

7.4.1 Electron Emission

To drive an electron from a solid material requires the addition of sufficient energy to free the electron from that material, in the process overcoming what is sometimes called the surface **work function** (ϕ). The energy achieved by the electron must then be at least at the value ϕ above the **Fermi level** ($\mathscr{E}_f$) of electron energy in the material. The Fermi level of energy can be thought of loosely as the minimum possible electron energy at the outer level of an atomic structure; this is achieved only at absolute zero temperature. More precisely, the Fermi level is by definition "the energy at which the probability of a state being occupied is one half." This is illustrated in **Fig. 7.23** in which the Fermi-Dirac probability (which will not be discussed in detail here) is plotted vertically with electron energy as a base. At zero temperature, the probability of outer-shell electron energy being at the Fermi level is unity. Should energy be added to the material, an electron can be raised to a higher energy level. This, of course, means that the probability of an electron occupying an energy level lower than the Fermi value has been decreased, and the probability of it occupying a level above the Fermi value has been increased. The energy level of the electrons is a reflection of the temperature of the material. Probability curves for two different temperatures above absolute zero are shown on the diagram, but not to scale. The curve shown for the higher of these two temperatures shows lower probabilities of energy levels below the Fermi level and higher probabilities above the Fermi level.

The Fermi level is peculiar to each material because of the unique atomic structure. Cathode materials must have relatively low Fermi levels while anode materials must have fairly high values of this important function. Thus a free electron, which is above the Fermi level in either material, must be at a higher energy level in the cathode than it is in the anode. An electron that has moved from the cathode to the anode must now occupy a higher energy level than it did previously.

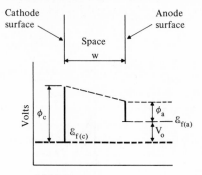

(a) Ideal case
zero retarding potential

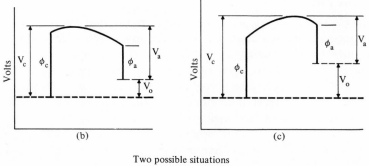

Two possible situations
with retarding potential
in space

FIG. 7.24 Potential gradient in space between anode and cathode.

In **Fig. 7.24**, three possible situations are illustrated: (a) the ideal case in which there is zero **retarding potential** in the interspace; (b) the situation with a retarding potential in the interspace but with the sum of the Fermi level and the work function for the cathode material exceeding the sum for the anode material; and (c) the situation with a retarding interspace potential, but with the sum of the Fermi level and the work function for the anode material being higher than the sum for the cathode material. Regardless of the situation in the interspace and regardless of the values of the work function, the **net voltage difference** between the anode and the cathode, indicated on the diagram by V_0, is the output voltage that can be applied to the load; in each situation it is the difference between the Fermi level of the anode and the Fermi level of the cathode. This does not mean that the work function of the two materials and the interspace retarding potential are not significant in the analysis of a thermionic generator. On the contrary, these are extremely important in the determination of the electric current density at particular cathode and anode temperatures.

The **unit** of energy chosen for illustration in this last diagram is the **volt**. If we are to consider an electron which will move through this voltage difference, we must multiply the voltage by the charge per electron to achieve the unit **electron-volt**. As noted in Appendix I, the unit charge per electron is 1.602×10^{-19} coulombs. Also, one joule is one volt-coulomb. Thus one electron-volt is equal to 1.602×10^{-19} joule.

The significance of temperature and of the material surface work function in determining the current flow in a thermionic device is shown by the **Richardson-Dushman equation** which is the basic equation for electron emission:

$$J = A_1 T^2 \exp\left[-e\phi/kT\right], \tag{7.25}$$

in which J is current density, amp/cm^2; A_1 is a constant for which the theoretical value is 120 amp/cm^2-K^2; T is the material temperature, K; e is the unit charge per electron, 1.60210×10^{-19} coulombs; ϕ is the surface work function in volts; and k is the Boltzmann constant per molecule, 1.38054×10^{-23} joule/K. Since one joule is equivalent to one volt-columb, the exponent is dimensionless. If ϕ were given in electron-volts (which is common) so that e did not appear in the exponent, then the value of k to make the exponent dimensionless would be

$$\frac{k}{e} = \frac{1.38054 \times 10^{-23}}{1.60210 \times 10^{-19}} = 8.6168 \times 10^{-5} \text{ eV/K}.$$

The value of current density (J) given by the Richardson-Dushman equation is the **saturation current density**. If both the cathode and the anode are heated, the net current density from the cathode to the anode will be the difference between the two values ($J_c - J_a$), calculated for the respective materials and temperatures. Should there be a retarding potential barrier in the interspace, then $V_c > \phi_c$ and $V_a > \phi_a$ must be used in place of ϕ_c and ϕ_a, respectively, in the calculation of J_c and J_a. The term J_a is called the **back emission**. This will be illustrated further in Section 7.4.2.

For readers who are unfamiliar with the significance of the **Boltzmann constant** (k), it might be useful to digress at this point to explain why it is sometimes called the "universal gas constant per molecule." The perfect gas law for n moles of gas containing N molecules is

$$pV = nR_U T,$$

in which R_U is the universal gas constant,

$$1545 \text{ ft-lb/lb-mole-R}, \quad \text{or} \quad 8314.3 \text{ joule/kg-mole-K}.$$

Avogadro's hypothesis states that equal volumes of all gases at the same temperature and pressure contain the same number of molecules. **Avogadro's number (N_0)**

is 6.0225×10^{26} molecules/kg-mole. Thus

$$pV = \frac{N}{N_0} R_U T = NkT,$$

$$k = \frac{R_U}{N_0} = \frac{8.3143 \times 10^3 \text{ joules/kg-mole-K}}{6.0225 \times 10^{26} \text{ molecules/kg-mole}}$$
$$= 1.38054 \times 10^{-23} \text{ joules/molecule-K.}$$

For a gas, the Maxwell-Boltzmann statistics show that the **mean translational kinetic energy** of a molecule is

$$\tfrac{1}{2} m \bar{u}^2 = \tfrac{3}{2} kT \text{ joule,}$$

where $\bar{u}^2$ is the mean-square speed (which is not the same as the mean speed squared). This is true regardless of the mass of the molecule. For molecules in a beam, however, for which the flux (number times velocity) is averaged in one direction only (x), high velocity molecules have more significance. Thus for a beam, the average kinetic energy per carrier is

$$KE_{\text{avge}(x)} = 2kT \text{ joule.} \tag{7.26}$$

See, for instance, Reference [B.15], p. 417, or Reference [B.22], p. 59.

7.4.2 Thermodynamic Analysis

To carry out a thermodynamic analysis on the thermionic converter in order to relate its thermal efficiency to its physical characteristics, it is necessary to write an energy balance for both the cathode and the anode. **Figure 7.25** is a more detailed schematic diagram then Fig. 7.22. It shows an interspace retarding potential equivalent to δ volts above the anode work function (ϕ). The current density from the cathode to the anode (J_c) and from the anode to the cathode (J_a) will be given by the expressions

$$J_c = A_1 T_c^2 \exp\left[-eV_c/kT_c\right] \text{ amp/cm}^2, \tag{7.27a}$$

$$J_a = A_1 T_a^2 \exp\left[-eV_a/kT_a\right] \text{ amp/cm}^2. \tag{7.28b}$$

The output voltage across the external resistance (R_0) is

$$V_0 = V_c - V_a = \frac{1}{e}\left[\mathscr{E}_{f(a)} - \mathscr{E}_{f(c)}\right] \text{ volt,}$$

where the Fermi energy levels $(\mathscr{E}_f)$ are given in electron volts.

For simplicity we shall assume that the heat conduction from the cathode to the anode through the interspace and through the lead wires is negligible and that the same is true for the net radiation heat transfer from the cathode to the anode, relative to the larger amounts of energy transported in **electron cooling**. This energy removal is the result of electron flow, and it has two components.

Each electron must overcome the interspace potential $(V_c - \phi_c)$ and the work function (ϕ_c) as it leaves the cathode. In doing this work it carries away the

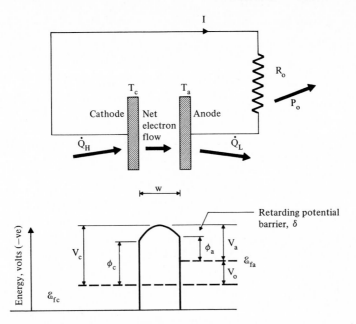

FIG. 7.25 A thermionic generator with interspace retarding potential.

net energy

$$\dot{Q}_{1c} = J_c(V_c - \phi_c + \phi_c) = J_c V_c \text{ watts/cm}^2.$$

Each electron also carries away its own kinetic energy. From Eq. (7.26), this is $2kT_c$ per electron. This component of the energy transfer is

$$\dot{Q}_{2c} = J_c \left(\frac{2kT_c}{e}\right) \text{ watts/cm}^2.$$

Note that k is expressed in joule/K, e is coulombs, and one joule/coulomb is one volt.

The back emission from the anode must similarly carry energy to the cathode. Each electron that returns to the cathode must give up the energy represented by V_c as it joins the cathode material. In addition, it carried the kinetic energy $2kT_a$ with it as it left the anode, and this too must be given up as the electron joins the cathode material.

The net rate of energy supply to the cathode must be the net rate of energy loss:

$$\dot{Q}_H = J_c \left(V_c + \frac{2kT_c}{e}\right) - J_a \left(V_c + \frac{2kT_a}{e}\right) \text{ watts/cm}^2. \qquad (7.29a)$$

The useful output from the device is the product of the voltage and the net current:

$$P_0 = V_0(J_c - J_a) \text{ watts/cm}^2. \qquad (7.29b)$$

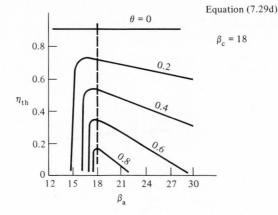

Equation (7.29d)

$\theta = 0$

$\beta_c = 18$

FIG. 7.26 Thermal efficiency of a thermionic generator (heat conduction, radiation, and Joule losses ignored) (from Houston [A.17]).

Thus the thermal efficiency is

$$\eta_{th} = \frac{V_0(J_c - J_a)}{\dot{Q}_H}$$

$$= \frac{(V_c - V_a)(J_c - J_a)}{J_c\left(V_c + \dfrac{2kT_c}{e}\right) - J_a\left(V_c + \dfrac{2kT_a}{e}\right)}. \tag{7.29c}$$

The values of J_c and J_a to be put in the equation are obtained from Eqs. (7.27a) and (7.27b).

To simplify the equation for thermal efficiency, the exponents in Eqs. (7.27a) and (7.27b) will be designated by β, and the temperature ratio T_a/T_c by θ, after the manner of Houston [A.17], so that

$$\beta_c = \frac{eV_c}{kT_c}, \qquad \beta_a = \frac{eV_a}{kT_a}, \qquad \theta = \frac{T_a}{T_c}.$$

Then the expression for thermal efficiency becomes

$$\eta_{th} = \frac{(\beta_c - \theta\beta_a)[1 - \theta^2 \exp(\beta_c - \beta_a)]}{(\beta_c + 2) - \theta^2(\beta_c + 2\theta)\exp(\beta_c - \beta_a)}. \tag{7.29d}$$

This relationship is plotted on **Fig. 7.26** which shows η_{th} as a function of β_a at various values of θ; β_c is constant at the arbitrary value of 18 which is near the middle of the rather narrow range of values reported by Houston.

It is useful to note that at all values of θ, the efficiency curves peak very near the value of β_a equal to β_c. This fact allows a simple expression to be written for **maximum thermal efficiency**:

$$\eta_{th(max)} \simeq [1 - \theta]\frac{\beta}{\beta + 2}\left[\frac{1 - \theta^2}{1 - \theta^2(\beta + 2\theta)/(\beta + 2)}\right]. \tag{7.29e}$$

Note that the first term $[1 - \theta]$ is the Carnot cycle efficiency over the temperature range T_c to T_a. The remainder of the expression is, therefore, the **reduced efficiency** (η_r). Compare Eq. (7.23a) for infinite staging of a thermoelectric generator.

It is interesting to note also that the final bracketed term in Eq. (7.29e) is very nearly unity. Consequently, with β nearly 18, the thermal efficiency is approximately 90% of the Carnot value. In other words,

$$\eta_{r(\text{max})} \simeq 0.90.$$

Another implication of the fact that maximum efficiency occurs when β_a is approximately equal to β_c is that a condition for maximum efficiency is

$$\frac{V_c}{T_c} \simeq \frac{V_a}{T_a}.$$

From the relationship between anode and cathode voltages and anode and cathode temperatures close to maximum efficiency, and the fact that β is close to 18, the values of V_a or T_a can be found approximately in terms of one another:

$$V_a = V_c \frac{T_a}{T_c} = \frac{\beta_c k T_c}{e} \frac{T_a}{T_c} = \frac{\beta_c T_a}{11,606} \simeq \frac{T_a}{645}.$$

In other words, an optimum T_a can be approximated for a given V_a.

7.4.3 Reduction of Interspace Retarding Voltage

The mere presence of electrons in the interspace between the cathode and the anode, once emission has begun, will establish a negative charge that will discourage further emission. Thus the ideal of emission into a **vacuum** without a retarding voltage cannot be achieved. The most direct method of reducing this interspace charge to a minimum is to bring the two electrodes as close together as is practical. Spacings (w) less than 0.001 cm have been used. However, the difficulties of maintaining extremely close spacings at constant values when the cathode and anode are both at very high temperatures can easily be imagined. Furthermore, the temperatures are bound to change with load demand.

The sensitivity of power output to changes in electrode spacing is shown in **Fig. 7.27**, which is adapted from Reference A.34. The actual current density even for the closest spacing is well below the saturation value. A small change in V_c has a large effect on J:

$$J_{\text{sat}} = A_1 T_c^2 \exp\left[-e\phi_c/kTc\right]$$
$$J_c = A_1 T_c^2 \exp\left[-eV_c/kT_c\right] = J_{\text{sat}} \exp\left[-e(V_c - \phi)/kT_c\right]$$
$$= J_{\text{sat}} \exp\left[-11606(V_c - \phi)/T_c\right]. \tag{7.30}$$

A so-called "**low pressure**" diode is a further attempt to reduce the retarding voltage. A vapor at a very low pressure, less than 10^{-4} mm Hg, is supplied to the interspace. The vapor must be that of an easily ionized element. Cesium has been used because of its very low ionization potential. (See Table 7.2.) If that ionization potential is less than the cathode work function, a valence electron from the vapor is attracted and bound to the cathode surface. A positively charged ion is left in

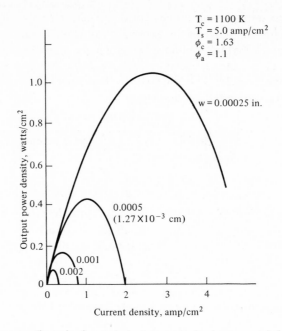

FIG. 7.27 High vacuum thermionic generator power output vs. current density for various spacings between cathode and anode (from Webster [A.34]).

the interspace which tends to encourage electron emission. The whole process of charge reduction in this manner is rather complex, and perhaps not fully understood.

A so-called **"high pressure"** converter will have cesium or some other vapor in the interspace at a pressure exceeding 1 mm Hg. In this situation, the mean free path of electrons and ions is less than the electrode spacing. Despite space charge neutralization, the internal electrical resistance is greatly increased. An analysis is further complicated by appreciable heat conduction through the vapor.

Another suggestion for reduction of retarding voltage is the use of a **magnetic triode** as described in Reference A.16. This allows the anode and cathode to be separated and even to lie in the same plane. A third electrode acts as an accelerator to attract electrons leaving the cathode, while at the same time a cross magnetic field deflects the moving electrons away from that accelerator plate and toward the anode. This is only one of several ingeneous configurations that may improve thermionic generator characteristics in the future.

7.5 MAGNETOHYDRODYNAMIC AND ELECTROGASDYNAMIC CONVERTERS

It will be recalled that in the introduction to this chapter, the **magnetohydrodynamic (MHD) converter** was grouped with thermoelectric and thermionic devices as one of the direct energy conversion methods that depend on the supply of thermal energy at a relatively high temperature, and the rejection of a part of that energy at a lower temperature in order to produce useful work energy. Therefore the thermal efficiency of the MHD converter is limited, as are the efficiencies of the other two, to the Carnot cycle efficiency as a maximum over the temperature range of operation.

The operation of an MHD converter depends on the ability of a fluid medium to conduct electricity. When this conducting medium is moved at high velocity through a magnetic field, a **voltage is induced** so that the device will operate as a generator. If a **voltage is applied** in the direction opposite to that of the induced voltage, the conducting medium can be accelerated across the magnetic field, and the device can then operate as a fluid accelerator. In this sense, the MHD converter is analogous to the more conventional electric generator or motor with its solid metal conductors. It can be operated either as a motor or as a generator depending on whether the input motion is converted to an electric field (a generator) or the input electric field creates motion (a motor).

The fluid medium could be liquid or gaseous. The term magnetohydrodynamics has been retained since the time of the earliest work done in the field, in which the fluid was water. Mercury, too, was an obvious choice for early experiments. Present-day developments in the use of the same principles for power production are almost entirely concerned with gaseous conductors. The terms **magnetogasdynamics** or, even better, **magnetofluidmechanics** have been suggested and would be preferable to MHD. However, this discussion will follow the common practice in using the term magnetohydrodynamics, MHD, regardless of the nature of the fluid.

For a gaseous medium to be a reasonably good conductor, the gas molecules or atoms must be **ionized**. This means that sufficient energy must be added to the gas to free some electrons from their normal molecular or atomic structure in order that they might move through the medium under the influence of an electric field. Such a gas is called a **plasma**. Although each separated or **free electron** carries a **negative charge** and the remaining nucleus and electrons, now called an **ion**, carry a net **positive charge**, the plasma as a whole is electrically neutral. A neutral atom or molecule that gains an electron is a **negatively charged ion**. Ions may carry single, double, or more charges, either positive or negative in polarity.

The **electrogasdynamic (EGD)** converter, to be discussed in Section 7.5.6, is basically different from an MHD converter because the EGD type does not depend on the presence of a magnetic field for electron or ion acceleration. In this type of device, a high-velocity fluid physically carries with it electrons or ions, separately supplied, from one electrode to another within the system. The carrier fluid itself need not be ionized, and therefore operating temperatures can be considerably lower. The transfer of charged particles creates an electric potential which may achieve a very high voltage. This system has been called a gaseous Van de Graaff generator; in that more familiar device, a high voltage is obtained by the transfer of charged ions from one point to another on a moving belt. Like the Van de Graaff generator, the EGD converter could be called a particle accelerator.

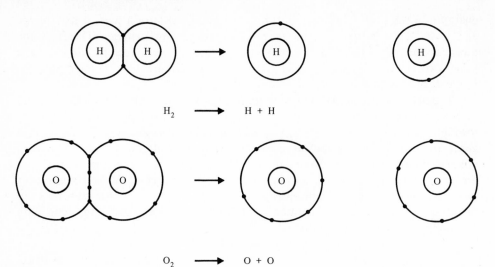

FIG. 7.28 A schematic representation of dissociation of diatomic hydrogen and diatomic oxygen.

7.5.1 Dissociation and Ionization

Before the forces and voltages within an MHD device can be discussed, it will be necessary to review the nature of dissociation and ionization of a gas. A review of atomic and molecular structure of material can be found in any standard book on modern chemistry or physics, and a rather succinct review is to be found in *Elements of Materials Science* by Van Vlack [B. 35]. The phenomena of dissociation and ionization are reviewed by Angrist [B.1] and by Cambel [B.6].

In **Fig. 7.28**, the **dissociations** of the diatomic gases, hydrogen and oxygen, are shown schematically. Upon dissociation, the two separate atoms of either gas assume less stable conditions: in the first case, one electron is left in orbit; in the second case, six electrons exist in the outermost or "valence" shell. As always, the more stable condition (as shown before dissociation) requires less molecular energy. Therefore the conversion to the less stable dissociated condition requires an input of energy. Typical dissociation energies are shown in **Table 7.1**.

Table 7.1

REPRESENTATIVE DISSOCIATION ENERGIES (ELECTRON VOLTS)*

Molecule	$\mathscr{E}_D$	Molecule	$\mathscr{E}_D$
H_2	4.477	NO	6.48
N_2	9.76	OH	4.37
O_2	5.08	CO_2	16.56
CO	11.11		

* Reprinted by permission from Cambel [B. 6].

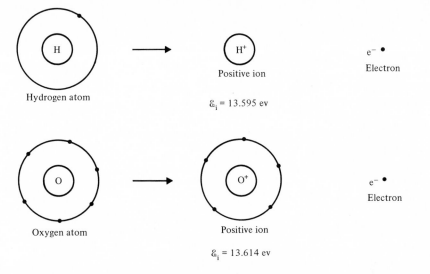

FIG. 7.29 A schematic representation of the ionization of hydrogen and oxygen atoms.

The phenomenon of **ionization** has already been mentioned. It is the separation of one or more electrons from an atomic structure, so that a positive ion is left by the removal of the negative electron or electrons. In **Fig. 7.29**, single ionizations of a hydrogen atom which contains only one electron and of an oxygen atom which contains six electrons in the valence shell are shown schematically. For an atomic structure in which there are more than one electron, further stages of ionization (double, triple, etc.) are possible when other electrons are removed. The positive charge on the remaining ion is proportional to the number of electrons removed. As with dissociation, the condition before ionization is more stable because it requires less energy. An input of energy is therefore required to accomplish ionization. Both the dissociation and ionization processes can be reversed with, of course, a release of the dissociation or ionization energies.

Both oxygen and nitrogen, the major constituents of atmospheric air, are diatomic molecules. With the addition of thermal energy to the gases and an increase in temperature, first dissociation and then ionization will take place. The oxygen molecule, requiring less energy for dissociation, will tend to separate into individual atoms before nitrogen will. This is shown very clearly in **Fig. 7.30**, adapted from Hansen [A.10], which shows the regions of dissociation and ionization for atmospheric oxygen and nitrogen at different pressures and temperatures. Within region *A*, there is practically no dissociation or ionization. Within region *B*, there is some dissociation of oxygen. This is followed by complete dissociation of oxygen and some dissociation of nitrogen within region *C*. Finally, in region *D*, there is dissociated nitrogen, dissociated oxygen, ionized nitrogen, and ionized oxygen

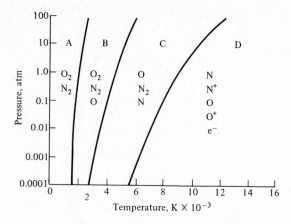

FIG. 7.30 Dissociation and ionization of atmospheric oxygen and nitrogen (from Hansen [A.10]).

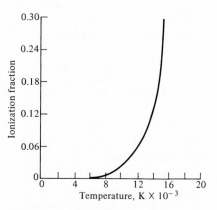

FIG. 7.31 The fraction of molecules thermally ionized in air at 1 atm (from Cambel [B.6]).

plus the free electrons (e^-). Of course, the boundaries between two adjacent areas are not as sharp as might be implied by the lines on the diagram. In addition, some NO may be formed. It can be seen from this that ionization does not take place below rather high values of temperature, well over 5000 K. Both dissociation and ionization are discouraged by increased pressure as shown in the diagram. **Figure 7.31** shows the degree of ionization for atmospheric air at temperatures above 5000 K. Over 12,000 K, ionization of air at 1 atm pressure increases very rapidly. The vertical scale on this figure is the fraction of the atoms of the gas which have been ionized, that is, the number of ionized atoms divided by the total number of atoms originally present. **Table 7.2** shows the energy required for the first two stages of ionization, where known, for the elements. It can be seen that the energy required for the first stage of ionization for the noble gases, which have full outer shells, is quite high because of the stability of this atomic structure. On the other hand, as would be expected, the ionization energies for the alkali metals are relatively low. The energy values for ionization and for dissociation in Tables 7.1 and 7.2 are given in electron volts.

For a gas to be a good conductor of electricity, there must be a significant degree of ionization. Therefore, if unadulterated atmospheric gases are to be used as conductors in an MHD generator or accelerator, the temperatures of operation must be very high. As a matter of fact, these temperatures may be prohibitively high for containment in practical power generation. When gases such as oxygen, nitrogen, and argon are **seeded** with such easily ionized materials as cesium vapor,

Table 7.2

IONIZATION POTENTIALS IN ELECTRON-VOLTS FOR THE FIRST TWO STAGES,
WHERE KNOWN FROM THE HANDBOOK OF PHYSICS*

Z	Element	Stage of ionization		Z	Element	Stage of ionization	
		I	II			I	II
1	Hydrogen	13.595		24	Chromium	6.763	16.49
2	Helium	24.580	54.4000	25	Manganese	7.432	15.64
				26	Iron	7.90	16.18
3	Lithium	5.390	75.619	27	Cobalt	7.86	17.05
4	Beryllium	9.320	18.206	28	Nickel	7.633	18.15
				29	Copper	7.725	20.29
5	Boron	8.296	25.149	30	Zinc	9.391	17.96
6	Carbon	11.264	24.376				
7	Nitrogen	14.54	29.605	31	Gallium	6.00	20.51
8	Oxygen	13.614	35.146	32	Germanium	7.88	15.93
9	Fluorine	17.42	34.98	33	Arsenic	9.81	20.2
10	Neon	21.559	41.07	34	Selenium	9.75	21.5
				35	Bromine	11.84	21.6
11	Sodium	5.138	47.29	36	Krypton	13.99	24.56
12	Magnesium	7.644	15.03				
				37	Rubidium	4.176	27.5
13	Aluminum	5.984	18.823	38	Strontium	5.692	11.027
14	Silicon	8.149	16.34				
15	Phosphorus	10.55	19.65	39	Yttrium	6.5	12.4
16	Sulfur	10.357	23.4	40	Zirconium	6.95	14.03
17	Chlorine	13.01	23.80	41	Niobium	6.77	14.
18	Argon	15.755	27.62	42	Molybdenum	7.06	
				43	Technetium		
19	Potassium	4.339	31.81	44	Ruthenium	(7.5)	(16)
20	Calcium	6.111	11.87	45	Rhodium	7.46	(18)
21	Scandium	6.56	12.89	46	Palladium	8.33	19.42
22	Titanium	6.83	13.63	47	Silver	7.574	21.48
23	Vanadium	6.74	14.2	48	Cadmium	8.991	16.90

* Edited by E. U. Condon and Hugh Odishaw, McGraw-Hill, New York, 1958.   *(continued)*

Table 7.2 (*continued*)

IONIZATION POTENTIALS IN ELECTRON-VOLTS FOR THE FIRST TWO STAGES,
WHERE KNOWN FROM THE HANDBOOK OF PHYSICS*

Z	Element	Stage of ionization		Z	Element	Stage of ionization	
		I	II			I	II
49	Indium	5.785	18.86	70	Ytterbium	6.22	(12)
50	Tin	7.332	14.628	71	Casseopium		
51	Antimony	8.639	16.5	72	Hafnium		(14.8)
52	Tellurium	8.96	(19)	73	Tantalum		
53	Iodine	10.44	19.0	74	Tungsten	7.94	
54	Xenon	12.13	21.2	75	Rhenium	(8)	(13)
				76	Osmium	(8.7)	(15)
55	Cesium	3.893	25.1	77	Iridium	(9.2)	(16)
56	Barium	5.210	10.001	78	Platinum	(8.9)	18.5
				79	Gold	9.23	20.0
57	Lanthanum	5.61	11.43	80	Mercury	10.44	18.8
58	Cerium	6.54		81	Thallium	6.12	20.3
59	Praseodymium	(5.8)		82	Lead	7.42	15.0
60	Neodymium	(6.3)		83	Bismuth	(8.8)	(17)
61	Promethium			84	Polonium	(8.2)	(19)
62	Samarium	(5.6)	(11.4)	85	Astatine		
63	Europium	5.64	11.2	86	Randon	10.75	(20)
64	Gadolinium	6.7		87	Francium		
65	Terbium	6.7		88	Radium	5.27	10.1
66	Dysprosium	(6.8)		89	Actinium		
67	Holmium			90	Thorium		
68	Erbium			91	Protactinium		
69	Thulium			92	Uranium	(4)	

they can be made to have reasonable values of electrical conductivity at temperatures that are still quite high, but not beyond the possibility of use in power plants. See Section 7.5.5.

7.5.2 *Magnetic Flux, Conductor Velocity, Lorentz Force, and Induced Voltage*

It was pointed out at the beginning of Section 7.5 that there is a direct analogy between MHD generators or accelerators and conventional electrical rotating machinery. To introduce a discussion of induced voltages and body forces as they occur in the gaseous-medium generator or accelerator in the presence of a magnetic field, it will be useful to review the laws governing such voltages and forces for solid conductors.

Figure 7.32 summarizes the convention relating to **north** and **south** polarity of an electromagnet and the direction of magnetic lines of flux around a conductor as governed by the **"right-hand screw rule."** Also shown are the net repulsing or attracting forces developed between two parallel conductors as their magnetic

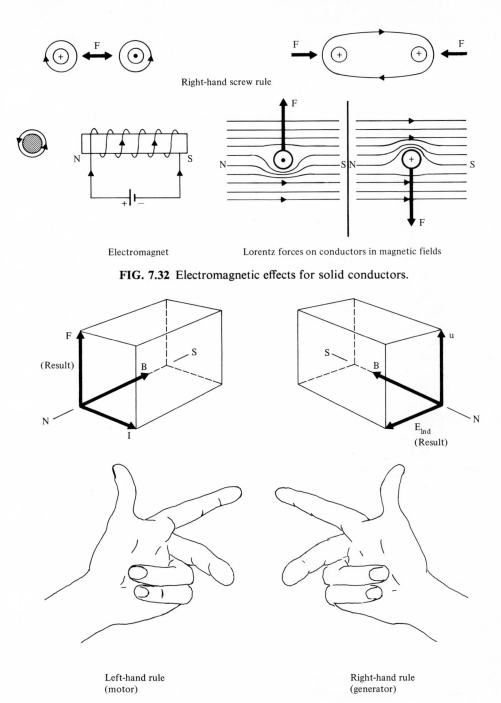

Right-hand screw rule

Electromagnet   Lorentz forces on conductors in magnetic fields

FIG. 7.32 Electromagnetic effects for solid conductors.

Left-hand rule
(motor)

Right-hand rule
(generator)

FIG. 7.33 Left-hand and right-hand rules.

fields interact. The forces reverse direction when the current is reversed in one conductor only. Finally in Fig. 7.32, the direction of the so-called **Lorentz force** is shown acting on a conductor in a magnetic field whose lines of flux are perpendicular to the direction of current flow. The distortion of the magnetic field in the region of the conductor is essentially the result of vectorial addition of the original flux vectors with those of the circular field, produced by the current flow. The direction of the force on the conductor is such as to tend to make it move into a position that would reestablish symmetry of the magnetic flux lines. This is the situation in an electric motor. The oposite effect is not shown, but is well known: If a conductor in a closed circuit is forced to move through the lines of magnetic flux, a current will be induced in a direction that will create a force to oppose the original motion. This is **Lenz's law,** and is the effect found in an electric generator. The current directions shown on the diagram would be the result of motion in directions opposite to the force vectors shown.

The rules governing the mutually perpendicular directions of magnetic flux, current (or induced voltage), and force (or original velocity of conductor motion) are shown in **Fig. 7.33**. These two rules are the **left-hand rule** for a situation analogous to a motor, and the **right-hand rule** for a situation analogous to a generator. In these diagrams, the symbols have the following meanings and units:

B is the magnetic flux density, webers/m²,

I is the electric current, amp;

or alternatively,

I is the electric current density, amp/m²;

F is the force per unit length for a conductor carrying a current I, newtons/m;

or alternatively,

F is the force per unit volume for a flux density J, newtons/m³,

u is the conductor velocity, m/sec; and

E is the voltage gradient, V/m.

Conversion factors and definitions of units are given in Appendix I. Note that

$$
\begin{aligned}
1 \text{ joule} &= 1 \text{ newton-meter} \\
&= 1 \text{ watt-sec} \\
&= 0.7376 \text{ ft-lb,} \\
1 \text{ amp} &= 1 \text{ coulomb/sec,} \\
1 \text{ volt} &= 1 \text{ joule/coulomb,} \\
1 \text{ ohm} &= 1 \text{ volt/amp,} \\
1 \text{ weber} &= 1 \text{ volt-sec} \\
&= 1 \text{ joule/amp,} \\
1 \text{ weber/m}^2 &= 10^4 \text{ gauss} \\
&= 1 \text{ newton/amp-meter,} \\
1 \text{ electron-volt} &= 1.60210 \times 10^{-19} \text{ joule.}
\end{aligned}
$$

An electron volt is the energy acquired by an electron charge (1.60210×10^{-19} coulomb) moving through a potential difference of one volt.

The direction of current flow in all the diagrams discussed above is that for the **conventional current** I. This is the opposite of the direction of electron movement.

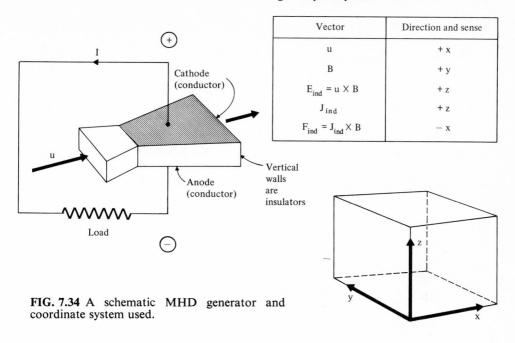

Vector	Direction and sense
u	+ x
B	+ y
$E_{ind} = u \times B$	+ z
J_{ind}	+ z
$F_{ind} = J_{ind} \times B$	− x

Cathode
(conductor)

I

u

Anode
(conductor)

Vertical
walls
are
insulators

Load

FIG. 7.34 A schematic MHD generator and coordinate system used.

7.5.3 *Analysis of an MHD Generator*

Application of the basic principles outlined in Section 7.5.2 to a conducting fluid flowing between two electrodes and across a magnetic field will explain the basic operation of a magnetohydrodynamic generator. Such a device is shown schematically in **Fig. 7.34**. Because this must be shown and analyzed in three dimensions, the coordinate system is shown also; the vectors *x, y,* and *z* on the diagram indicate the positive direction for each.

A high temperature ionized gas, flowing in the positive *x*-direction, enters the converging-diverging nozzle. Clockwise current flow through the windings of an electromagnet will create a north pole at the front face and a south pole at the opposite face of the flow channel so that the magnetic force lines will be in the positive *y*-direction. Application of the right-hand rule (generator) shows that an electromotive force will be generated in the positive *z*-direction. The magnitude of this induced electrical potential is given by the cross product of conductor velocity and magnetic flux density:

$$E_{ind} = [u \times B] \, m/sec \times \text{V-sec}/m^2 = u \times B \text{ V}/m. \tag{7.31a}$$

If the external circuit is completed through a resistance load, the induced voltage will induce a current in the positive *z*-direction. This in turn will create a Lorentz force, which is the cross product of the current and the original magnetic flux. Application of the left-hand rule (motor) shows that this force is in the negative *x*-direction which would retard the flow. Its magnitude is

$$F = [J \times B] \frac{\text{amp}}{m^2} \frac{\text{joule}}{\text{amp-m}^2} = J \times B \frac{\text{newton}}{m^3}. \tag{7.31b}$$

It is sometimes useful to think of this body force F as a pressure gradient in the x-direction with units $(newtons/m^2)/m$.

The upper electrode is called the cathode $(+)$ and the lower electrode the anode $(-)$, following the convention for electric batteries, fuel cells, and thermionic generators: electrons flow from the cathode to the anode within the device, while the conventional current (I) flows from the cathode to the anode in the external circuit.

The total induced voltage cannot be applied to the external load. It must drive the current against both the internal resistance of the flowing gas and the external resistance of the load circuit. The ratio between the actual available voltage (V_0) and the induced or open-circuit voltage is called the **loading factor** (K). For a flow channel with a constant distance (z) between electrodes, we find that

$$K = V_0/E_{ind}z = V_0/uBz = E_0/uB, \qquad (7.31c)$$

where E_0 is the voltage gradient measured at the electrodes in volts/meter.

The current density through the gas within the duct is the product of the conductivity of the gas, $\sigma(ohm\text{-}m)^{-1}$, and the internal voltage drop:

$$J = \sigma(uB - E_0) = \sigma uB(1 - K)\,amp/m^2. \qquad (7.31d)$$

The channel has a constant spacing z meters between electrodes. Each electrode is l meters long and has an average width of b meters. The internal resistance will be

$$R_i = \frac{z}{\sigma lb}\ ohms. \qquad (7.32a)$$

The external resistance is known from the terminal voltage and the current:

$$R_0 = V_0/Jlb = KuBz/\sigma uB(1 - K)lb = KR_i/(1 - K). \qquad (7.32b)$$

A rearrangement of this equation gives us

$$K = 1/[1 + (R_i/R_0)]. \qquad (7.32c)$$

7.5.4 The MHD Accelerator

The load in the schematic diagram of an MHD generator, Fig. 7.34, can be replaced by an external voltage source as shown in **Fig. 7.35**. If the **applied voltage** (E_{appl}) is great enough to exceed the induced voltage $(E_{ind} = uB)$, then the net current flow will be reversed. With a reversal of the direction of the net current flow (J), the net body force (F) will now be reversed to the positive x-direction. The direction and sense of these vectors are tabulated on the diagram.

With F in the positive x-direction, the fluid flow will be accelerated. Thus the device could be used for **propulsion** if the down-stream pressure is low. Alternatively, the work input can tend to raise the down-stream pressure and the device will act as a gas **compressor**.

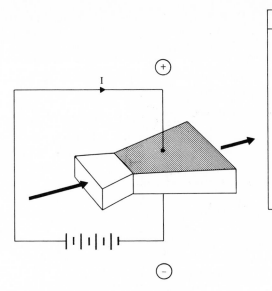

Vector	Direction and sense
u	+ x
B	+ y
E_{ind}	+ z
$E_{appl} > E_{ind}$	− z
J_{net}	− z
$F_{net} = J_{net} \times B$	+ x

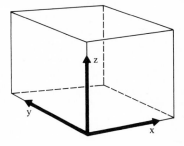

FIG. 7.35 Schematic MHD propulsion.

7.5.5 *Possible Arrangement for Practical Power Generation*

It has already been pointed out that a gaseous medium, even with seeding, has a reasonable conductivity only at high temperature. For this reason, the heat energy rejected in the exhaust gases is so large as to make a simple open-cycle MHD generator very inefficient thermodynamically. Perhaps the most practical means for utilizing this type of device is in the form of a topping plant for a more conventional power cycle.

In **Fig. 7.36**, adapted from Way [A.33], an MHD generator is shown combined with a steam power system. In addition, a regenerator is incorporated in the MHD gas loop to reduce combustion fuel requirements for the upper part of the combined cycle. The accompanying enthalpy-entropy diagram for the gas cycle shows the energy quantities supplied to the gas and rejected from the gas to the vapor cycle below.

The open cycle shown has no provision for reclaiming the seed material which will be lost with the stack gases. For reasons of economy and to avoid air pollution, this may not be tolerable. Closed cycles have also been proposed, particularly for nuclear reactor heat-source applications. Alternatively, continuation of combustion through the generator may eliminate the necessity of seeding for ionization.

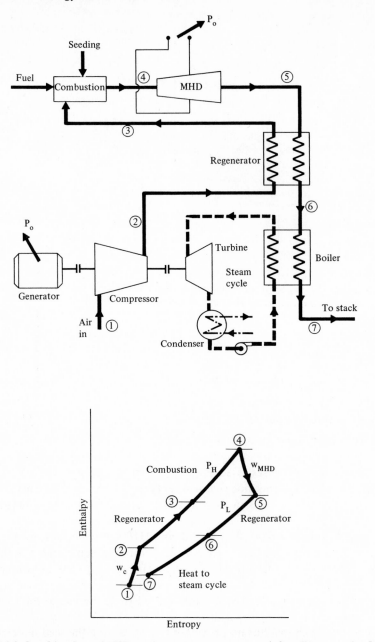

FIG. 7.36 Combined gas MHD and steam power system (after Way, *et al.* [A.33]).

The thermal efficiency of any MHD generator is the ratio of net power output to the rate of energy supply:

$$\eta_{th} = \frac{(\text{Output power}) - (\text{Auxiliary power})}{\text{Energy supply rate}}.$$

The auxiliary power includes that required by the electromagnets. For the large flux densities required, superconducting coils at cryogenic temperatures have been used to keep this drain on net power within practical limits. In an experimental MHD setup, a flux density of 40,000 gauss (4 weber/m²) has been produced.

7.5.6 The EGD Converter

In the introduction to Section 7.5, the **electrogasdynamic** (EGD) generator was described briefly as a particle accelerator in which either positive or negative ions are carried from one electrode to another by a high-velocity gas stream. Such a system is shown schematically in **Fig. 7.37**.

Gas enters the converging-diverging nozzle without appreciable ionization. The **corona** discharge from a pointed electrode provides charged particles, in this case electrons, to the gas stream. The electrons, which were initially attracted to a nearby electrode, are moved physically past this point to a collector electrode at the outlet end of the duct. If all electrons are collected, the gas leaving is electrically neutral. Thus the electron flow through the load circuit, though initiated by a low voltage source, is maintained by the gas stream against a much larger potential. Although the gas-stream velocities are very high in terms of gas flow, the velocities are low for electron movement. Consequently, the device is essentially a low-current, high-voltage generator.

The electrons flowing through the load circuit provide useful power, while those flowing through the auxiliary voltage-source loop, which feeds the corona,

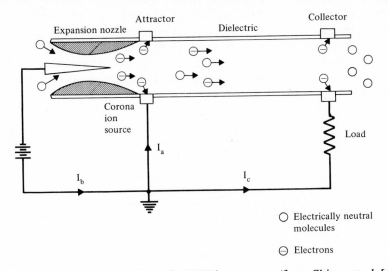

FIG. 7.37 A schematic electrogasdynamic (EGD) generator (from Shiue, *et al.* [A.28]).

constitute a loss. The directions of current flow on the diagram are for the conventional current (I) in each case.

Because entrained particles in the gas flow will aid as ion carriers, the system may prove to be very practical with a coal burning source of heat in which the fly-ash is advantageous. Compared to an MHD generator, the gas temperatures can be low because ionization of the gas stream itself is not required. Consequently, energy losses need not be as great. If a voltage source is put in place of the load so that ion flow is reversed, the system can act as a gas compressor.

7.6 GALVANOMAGNETIC AND THERMOMAGNETIC EFFECTS

Thermoelectric effects in solids in the absence of an external magnetic field have been discussed in Section 7.2. At the beginning of that section, four other effects, found only in the presence of a magnetic field were mentioned. Also mentioned was the usefulness of one of these magnetic effects in the determination of whether a semiconductor material was of the p-type or the n-type.

The four effects found in the presence of a magnetic field are shown in **Fig. 7.38**. They were reviewed in an article by Angrist [A.1]. Two are referred to as **galvanomagnetic** effects, because they result from an electric current flow across a magnetic field. The other two are called **thermomagnetic**, because they are the result of heat transfer, or a thermal current flow across a magnetic field. They are summarized and identified best by reference to the diagrams.

The four effects are interrelated and all depend on the Lorentz force, which results when a charge carrier moves through a magnetic field. The direction of that force was illustrated in Fig. 7.33 by the left-hand rule, but it must be remembered that the direction of current flow (I) was the conventional current direction, which is opposite to the actual movement of electrons where they are the charge carriers.

The **Hall effect** is shown in Fig. 7.38 (a). The Lorentz force on electrons flowing to produce the current in the conventional direction shown will tend to move the electrons, which are negatively charged, to the side marked with a negative sign. If an accumulation of charged electrons is permitted, a **Hall voltage** will be established as shown. The Lorentz body force is the cross product of the field intensity (B) and the charge and its velocity (eu):

$$F = B \times eu \text{ eV/m}. \tag{7.33a}$$

The open-circuit voltage gradient developed in a direction perpendicular to both u and B is

$$E_H = R'BJ \text{ V/m}, \tag{7.33b}$$

where R' is the Hall **coefficient**. When no current flows in the direction of the Hall voltage (open circuit condition), there is equilibrium between the force F and the force resulting from the Hall voltage gradient, E_H:

$$euB = eE_H. \tag{7.33c}$$

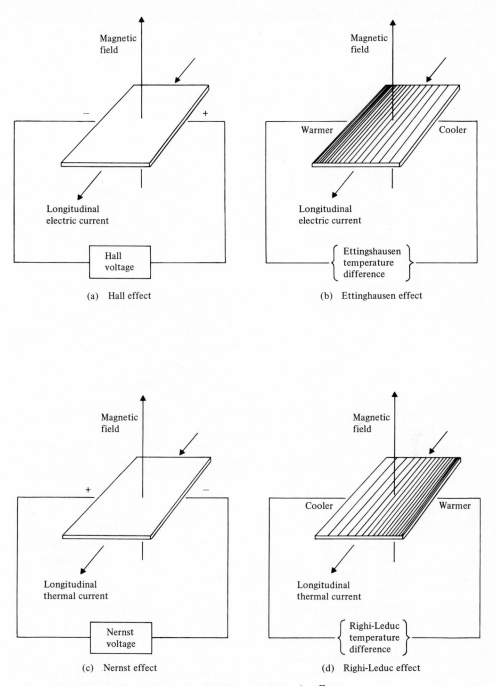

FIG. 7.38 Galvanomagnetic and thermomagnetic effects.

Now the current density, J, in the original direction is the product of the conductivity and the voltage gradient in the same direction. A relationship was given in Section 7.3.5 for conductivity and concentration of charge carriers:

$$J = \sigma E = neu \text{ amp/m}^2. \tag{7.33d}$$

When Eqs. (7.33b, c, and d) are combined, the magnitude of the Hall coefficient is found in basic terms:

$$R' = \frac{E_H}{BJ} = \frac{1}{ne} \text{ (coulomb/m}^3)^{-1}. \tag{7.33e}$$

Thus the Hall coefficient, and consequently the Hall voltage gradient for a given current density and magnetic flux density, vary inversely as the product of the concentration of charge carriers and the charge per carrier. This follows from the dependence of the Lorentz force on carrier velocity. In a semiconductor which has a low density of charge carriers, the Hall effect will be greater for a given current flow than it would be for a conductor, because each carrier must travel at a greater velocity.

The sign of the Hall voltage is reversed if either the direction of the current or the sign of its charge is reversed. If the conventional electric current in Fig. 7.38(a) is from right to left, as shown, because there is an electron flow from left to right, the Lorentz force moves the electrons toward the rear to give that edge a negative polarity. On the other hand, if the same direction of conventional current flow is caused by positively charged carriers (say holes) moving from right to left, both the charge sign and the direction of carrier movement have been changed. The force "moves the holes" to the rear to reverse the polarity of that edge. Thus the sign of the Hall voltage provides a simple means for determining whether a solid semiconductor material is predominantly an *n*-type or a *p*-type conductor.

It is now relatively simple to explain the other three effects. The **Ettinghausen** temperature difference develops because the Lorentz force is proportional to the carrier velocity. Carriers at a higher velocity are exhibiting higher temperature. These have a greater tendency to be moved in the direction of the Lorentz force, and a transverse temperature gradient is established. Like the Hall effect, the Ettinghausen effect is galvanometric in nature. Also like the Hall effect, it is directly proportional to the magnetic field and to the current flow, and inversely proportional to the thickness in the magnetic-field direction. Also, it is reversed by a reversal of the electric current or of the magnetic field.

The thermomagnetic effects are much like the galvanomagnetic effects, because in a solid material, a high percentage of heat conduction is carried by electron movement. The Lorentz forces again act on the electrons and produce a voltage gradient whose direction is established by the same left-hand rule. The **Nernst** voltage, shown in Fig. 7.38(c), results from an electron flow with heat transfer from right to left.

As in the Ettinghausen effect, the faster moving electrons, which are at a higher temperature, are affected more by the Lorentz force as they cross the magnetic field. Thus the higher energy electrons are moved more readily in a direction

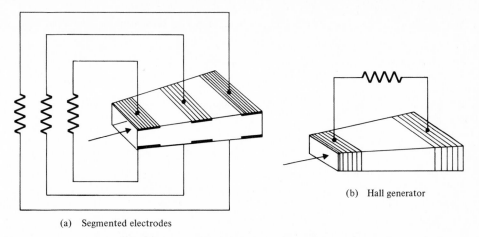

(b)   Hall generator

(a)   Segmented electrodes

FIG. 7.39 Alternative electrode arrangements in an MHD generator: (a) to decrease the Hall effect, and (b) to utilize the Hall effect.

normal to their original path, and the **Righi-Leduc** temperature difference is established as shown in Fig. 7.38(d).

The galvanomagnetic and thermomagnetic effects are likely to become more important than before in engineering devices. In an MHD generator, the Hall effect causes an appreciable reduction in the net electron flow across the gas stream. The magnitude of this reduction in effective current can be decreased by the use of segmented electrodes as shown in **Fig. 7.39**(a). Alternatively, the electrodes can be electrically connected as shown in Fig. 7.39(b) to form what is called a **Hall generator**.

If the Ettinghausen temperature difference can be made significantly large, an effective **Ettinghausen refrigeration** device will be developed. Similarly, the voltage gradient of the Nernst effect promises a means of direct electric power generation from heat transfer in what would be called a **Nernst generator**.

7.7 FUEL CELLS

Before a discussion of fuel cells is undertaken, it might be advisable to review the functioning of **electric batteries** which are more familiar to most readers. The "dry cell" battery shown in **Fig. 7.40** will be used as an example of the type of chemical reaction taking place. However, there are many types of electric batteries just as there are many types of fuel cells. The sign convention for the **cathode** (+) and the **anode** (−) is the same for batteries and fuel cells and for thermoelectric, thermionic, and magnetohydrodynamic generators, as noted in Section 7.5.3: negative ions or electrons flow from the cathode to the anode within the device, so that the conventional current flow is from the cathode to the anode in the external circuit.

The chemical equations, which are noted directly in Fig. 7.40, show the separate reactions at the anode and at the cathode, and also the overall reaction of the whole

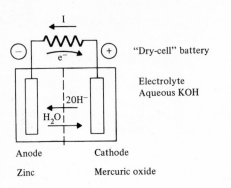

"Dry-cell" battery

Electrolyte
Aqueous KOH

Anode   Cathode

Zinc   Mercuric oxide

Anode reaction: Electrons lost to external circuit
("oxidation")

$$Zn + 2OH^- \rightarrow Zn(OH)_2 + 2e^-$$
$$\rightarrow ZnO + H_2O + 2e^-$$

Cathode reaction: Electrons gained from external circuit
("reduction")

$$HgO + H_2O + 2e^- \longrightarrow Hg + 2OH$$

Cell reaction: Anode and cathode materials eventually
depleted

$$Zn + HgO \longrightarrow ZnO + Hg$$

FIG. 7.40 Typical electric "battery" reactions.

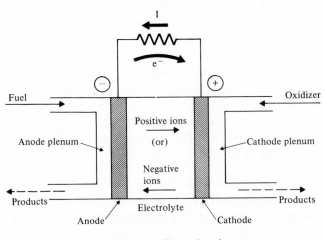

One product exhaust line or the other may
not be required

FIG. 7.41 Fuel-cell operation.

cell. The anode reaction is essentially oxidation of the zinc. This can be imagined as taking place in two steps as shown. The cathode reaction is essentially reduction of the mercuric oxide to mercury. It is typical of batteries that the electrodes and sometimes the electrolyte are chemically changed or exhausted so that eventually the reaction must come to a stop.

A **fuel cell** could be considered as an electric battery in which both the fuel and the oxidizer are continuously replaced. See **Fig. 7.41**. The anode and the cathode material do not normally enter into the chemical reactions although they may act as catalysts. The two electrodes must also serve the function of preventing the movement of nonionized fuel and oxidizer into the electrolyte between the two.

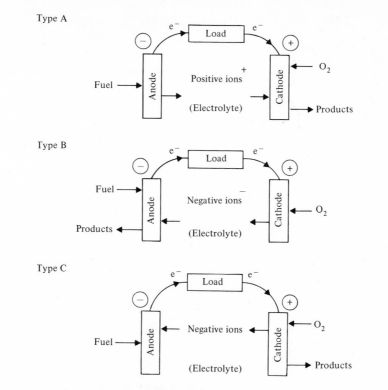

FIG. 7.42 Three different types of fuel-cell reactions.

Fuel cells might be divided into basic categories according to whether the product of the overall reaction must be disposed of in the cathode plenum space or in the anode plenum space, and whether the current flow through the electrolyte is a transfer of negative ions from the cathode to the anode or a transfer of positive ions in the opposite direction. Three types are illustrated in **Fig. 7.42**. More complicated fuel cells may actually operate as combinations of these basic types. Fuel cells are also classed according to the temperatures at which they operate.

There is a significant advantage of the fuel cell over most other types of direct or indirect energy-conversion systems. In the fuel cell, the electrochemical energy is not converted to heat energy, but is transformed directly into electrical energy. Whenever useful work is produced by a system which receives energy by heat transfer at a high temperature and rejects energy at a lower temperature, the thermal efficiency is strictly limited. It cannot be greater than that of a reversible cycle (say Carnot) operating between the the same two temperatures. But the ideal fuel cell operates essentially as an isothermal device, so that its efficiency is not limited in the same manner. Its limit is the ratio of the change in Gibbs free energy to the change in enthalpy for the chemical reaction. The thermodynamic aspects of a fuel cell will be explained more fully in Section 7.7.2.

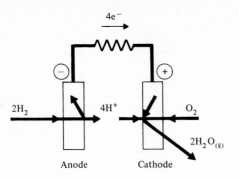

Anode: $2H_2 \longrightarrow 4H^+ + 4e^-$

Cathode: $4e^- + O_2 + 4H^+ \longrightarrow 2H_2O_{(\ell)}$

Cell: $2H_2 + O_2 \rightarrow 2H_2O_{(\ell)}$

FIG. 7.43 Hydrogen-oxygen fuel cell, type A.

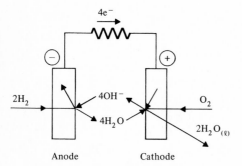

Anode: $2H_2 + 4OH^- \longrightarrow 4H_2O + 4e^-$

Cathode: $2O_2 + 4H_2O + 4e^- \longrightarrow 2HO_2^- + 2OH^- + 2H_2O$
$\longrightarrow 4OH^- + O_2 + 2H_2O_{(\ell)}$
$O_2 + 4H_2O + 4e^- \longrightarrow 4OH^- + 2H_2O_{(\ell)}$

Cell: $2H_2 + O_2 \longrightarrow 2H_2O_{(\ell)}$

FIG. 7.44 Hydrogen-oxygen fuel cell, type C.

7.7.1 *Typical Fuel-Cell Reactions*

In a fuel cell, the type of reaction taking place is determined by the fuel and oxidizer combination, by the composition of the electrolyte, and by the materials of and therefore the catalytic effect of the cathode and anode surfaces.

In **Figs. 7.43, 7.44, 7.45**, and **7.46**, four different fuel-cell reactions are presented. On each figure, the anode, cathode, and overall cell reactions are given, and the sides from which the products are removed are noted. The first two figures of this group are both for hydrogen-oxygen cells. Although the product is water in both examples, the reactions are quite different. In Fig. 7.43, the charge carrier through the electrolyte is a positive hydrogen ion, while in Fig. 7.44, it is a negative hydroxyl ion.

It will be noted that in each reaction presented, four electrons are released to travel through the external circuit for every oxygen molecule supplied for the reaction. This is so because each oxygen molecule (O_2) has 2×6 valence electrons and therefore lacks 2×2 electrons for completion of the outer shells, which would be full with 8 electrons around each oxygen nucleus (O). This fact will be used in the determination of ideal reaction potentials in the following section.

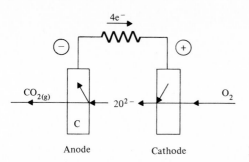

Anode: $C + 2O^{2-} \rightarrow CO_2 + 4e^-$

Cathode: $4e^- + O_2 \longrightarrow 2O^{2-}$

Cell: $C + O_2 \longrightarrow CO_{2(g)}$

FIG. 7.45 Carbon-oxygen fuel cell, type B. (High temperature required for significant reaction rate.)

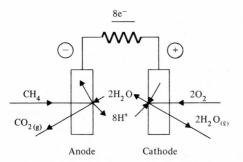

Anode: $CH_4 + \boxed{2H_2O} \rightarrow CO_2 + 8e^- + 8H^+$

Cathode: $8e^- + 8H^+ + 2O_2 \longrightarrow \boxed{2H_2O} + 2H_2O_{(\ell)}$

Cell: $CH_4 + 2O_2 \rightarrow CO_{2(g)} + 2H_2O_{(\ell)}$

FIG. 7.46 Hydrocarbon-oxygen fuel cell.

7.7.2 *The Thermodynamics of Fuel-Cell Reactions*

The fuel cell operates as a steady-state, steady-flow system. Ideally it should operate isothermally. Such a system is illustrated in **Fig. 7.47** in which fluid enters at conditions ① and leaves at conditions ②. By using the first law of thermodynamics applied to an open steady-flow system, we find that

$$H_1 + Q = H_2 + W$$
$$W = (H_1 - H_2) + Q. \tag{7.34a}$$

For ideal isothermal heat transfer, we have

$$Q = T(S_2 - S_1). \tag{7.34b}$$

By definition of the **Gibbs function,** we have

$$G_2 - G_1 = (H_2 - H_1) - (T_2S_2 - T_1S_1)$$
$$= (H_2 - H_1) - T(S_2 - S_1). \tag{7.34c}$$

Therefore the **work** for an isothermal reversible system is

$$W_{(max)} = G_1 - G_2 = -\Delta G. \tag{7.34d}$$

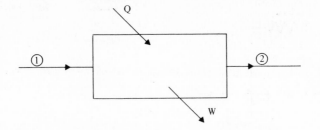

FIG. 7.47 Steady-state steady-flow system of ideal fuel cell.

If the energy supplied is defined as the change in enthalpy of the fluids, the heat transfer term being excluded because it is simply an exchange with the surroundings, then the **thermal efficiency** of an ideal fuel cell (η_i) is given as

$$\eta_i = \frac{\text{Work}}{\text{Energy supplied}} = \frac{\Delta G}{\Delta H}. \tag{7.35a}$$

However,

$$\Delta G = \Delta H - \Delta TS,$$

so that

$$\eta_i = 1 - \frac{T\Delta S}{\Delta H}. \tag{7.35b}$$

We shall see that ΔG and ΔH can be very close in magnitude, so that very high thermal efficiencies are possible. Equation (7.34c) implies a heat transfer from the system when ΔH exceeds ΔG, and the ideal thermal efficiency is less than 100%. In some reactions, ΔG will exceed ΔH. The apparent efficiency over 100% has no real significance, but depends on our arbitrary definition of η_i. The first law of thermodynamics is not violated, because the heat transfer to the system would have to be positive in such a situation to keep the endothermic reaction isothermal. The ideal maximum work is still $-\Delta G$.

The **useful power** output from a fuel cell is, of course, the product of the current flow and voltage (IV) in the external circuit. The value of I depends on the rate of electron release. Each electron traveling through the external circuit carries the charge 1.6021×10^{-19} coulomb. Each molecule involved in the reaction can exchange the number of electrons equivalent to its own valence (j). For instance, a molecule of hydrogen can exchange 2×1 electrons and a molecule of oxygen can exchange 2×2 electrons. By **Avogadro's principle**, a gram mole of each substance contains 6.0225×10^{23} molecules (Avogadro's number). The product of the electronic charge and Avogadro's number gives 96,487 coulombs per gram mole of electrons which is one **Faraday** ($\mathscr{F}$):

$$\mathscr{F} = 96{,}487 \text{ coulombs/g mole of electrons,}$$

$$1 \text{ coulomb} = 1 \text{ amp-sec,}$$

$$\mathscr{F} = 26.8 \text{ amp-hr/g mole of electrons.}$$

Table 7.3

HEATS OF FORMATION AND GIBBS FREE ENERGIES OF FORMATION
IN THE STANDARD STATE (298 K and 1 atm)*

Compound	$\Delta H°$, kcal/mole	$\Delta G°$, kcal/mole	Ion	$\Delta H°$, kcal/mole	$\Delta G°$, kcal/mole
$AgCl(g)$	22.23	16.79	$Ag^+(aq)$	25.31	18.43
$AgCl(s)$	−30.36	−26.22	$Br^-(aq)$	−28.90	−24.57
$CO(g)$	−26.42	−32.81	$Cl^-(aq)$	−40.02	−31.35
$CO_2(g)$	−94.05	−94.26	$CO_3^{--}(aq)$	−161.63	−126.22
$CO_2(aq)$	−98.69	−92.31	$Cu^{++}(aq)$	15.39	15.53
$CH_4(g)$	−17.89	−12.14	H^+	0	0
$C_8H_{18}(g)$	−49.82	4.14	$I^-(aq)$	−13.37	−12.35
$C_8H_{18}(l)$	−59.74	1.77	$K^+(aq)$	−60.04	−67.47
$H_2SO_4(aq)$	−216.9	−177.34	$Li^+(aq)$	−66.54	−70.22
$H_2O(l)$	−68.32	−56.69	$Mg^{++}(aq)$	−110.4	−109.0
$H_2O(g)$	−57.80	−54.64	$Na^+(aq)$	−57.28	−62.59
$LiH(g)$	30.7	25.2	$OH^-(aq)$	−54.96	−37.60
$NaCO_3(s)$	−270.3	−250.4	$Zn^{++}(aq)$	−36.43	−35.18

* From "Selected Values of Chemical Thermodynamic Properties," NBS Circular 500, by
Frederick D. Rossini, Donald D. Wagman, William H. Evans, Samuel Levine, and Irving Jaffe.

The Faraday constant can be expressed in calories or kilocalories per volt-gram
mole of electrons:

$$1 \ IT \ \text{kcal} = 1.162 \ \text{watt-hr},$$

$$\mathscr{F} = 23.06 \ \text{kcal/V-gram mole of electrons.}$$

We return to the fact that a molecule can exchange a number of electrons equivalent to its valence (j). When "$\dot{n}$" gram moles of substance react per second, the
ideal electrical power is

$$P_i = \dot{n}j\mathscr{F}V_i. \tag{7.36a}$$

Now, the ideal power is also the rate of change of the Gibbs function ($\Delta \dot{G}$). When
these two ideal powers are equated, the ideal voltage (V_i) can be determined:

$$V_i = \frac{\Delta \dot{G}}{\dot{n}j\mathscr{F}} = \frac{\Delta G}{nj\mathscr{F}} \ \text{volt}, \tag{7.36b}$$

where ΔG is the total change in the Gibbs function in kilocalories, n is the number
of moles of one of the reacting substances for which the valence per mole is j, and
$\mathscr{F}$ is the Faraday constant in kilocalories per volt-gram mole of electrons.

In **Table 7.3**, values of heats of formation (ΔH) and Gibbs free energy of formation (ΔG) at 298 K and 1 atm pressure (the "standard state") are given. The
values are from NBS Circular 500 by Rossini, et al, in which the superscript ° refers
to standard state. Several examples of fuel cell reactions are given below with values
of ΔH and ΔG taken from this table. Actually, existing fuel cells may not be able to

operate at useful rates at standard-state conditions. Nevertheless, the examples demonstrate the determination of ideal cell efficiencies and voltages.

For the hydrogen-oxygen cells illustrated in Fig. 7.43 and 7.44, the ideal values are as follows:

$$H_{2(g)} + \tfrac{1}{2}O_{2(g)} \rightarrow H_2O_{(l)}$$

$$\Delta G = \sum \Delta G_{(products)} - \sum \Delta G_{(reactants)} = \Delta G_{H_2O_{(l)}} - \Delta G_{H_2(g)} - \tfrac{1}{2}\Delta G_{O_2(g)}$$
$$= -56.69 - 0 - 0.$$

Similarly,

$$\Delta H = -68.32.$$

Then,

$$\eta_i = \frac{\Delta G}{\Delta H} = \frac{56.69}{68.32} = 83.0\%$$

$$V_i = \frac{\Delta G}{nj\mathscr{F}} = \frac{-56.69}{1 \times 2 \times 23.06} = -1.23 \text{ V}.$$

The values for the carbon-oxygen cell of Fig. 7.45 would be

$$\Delta G = -94.26, \qquad \Delta H = -94.05,$$

$$\eta_i \simeq 100\%, \qquad V_i = \frac{-94.26}{1 \times 4 \times 23.06} = -1.02 \text{ V}.$$

As a final example, the methane-oxygen cell of Fig. 7.46 is given:

$$CH_{4(g)} + 2O_{2(g)} \rightarrow CO_{2(g)} + 2H_2O_{(l)},$$

$$\Delta G = [-94.26 - 2 \times 56.69] - [-12.14 - 0] = -195.50,$$

$$\Delta H = [-94.05 - 2 \times 68.32] - [-17.89 - 0] = -212.80,$$

$$\eta_i = \frac{195.50}{212.80} = 92.0\%, \qquad V_i = \frac{-195.50}{8 \times 23.06} = -1.06 \text{ V}.$$

Any of these examples could be extended further to determine the ideal power for an assumed rate of fuel consumption. In the last example, if 1 g-mole of methane could be consumed in 1 min, the ideal power output would be

$$P_i = IV_i = \dot{n}j\mathscr{F}V_i$$
$$= \frac{(1 \times 8 \times 96{,}487) \times 1.06}{60} \simeq 1.36 \times 10^4 \text{ watt}.$$

It is emphasized again that all the above values are ideal. Enthalpy and Gibbs function changes would not be the same at other temperatures, and electrical and heat-transfer losses have been ignored.

7.7.3 Fuel-Cell Power Plant

The intensive research and development effort in fuel cells in recent years has been rewarded by the production of systems which are commercially competitive under certain conditions. Relatively small units are being used in space exploration for auxiliary power. Somewhat larger systems have been announced for moderate-size automotive propulsion.

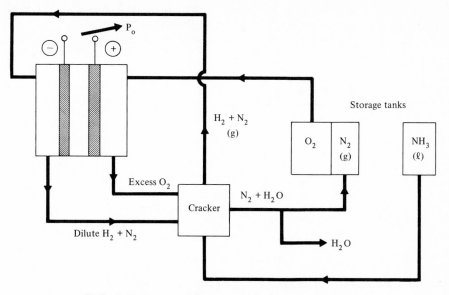

FIG. 7.48 Proposed large scale fuel-cell power plant.

Figure 7.48 is a schematic diagram of a stationary power plant rated at 200 kwatt. Banks of several such plants would make a moderate size central station installation. In a prototype plant, liquid ammonia and liquid oxygen are used. The products are water and nitrogen gas, the latter being used in the arrangement shown to pressurize the oxygen tanks. See reference [A.38]. Figure 7.48 is adapted from a diagram in *European Scientific Notes*, published by the United States Office of Naval Research (ESN-20-9), and is used by permission of ASEA Electric, Inc.

7.8 PHOTOVOLTAIC CELLS

It is possible to convert light energy or other electromagnetic radiation directly to electrical energy in certain materials arranged in a **photovoltaic cell**. The principle has been recognized for many years and is the basis of the common light meter and other measuring and control devices. Only fairly recently has it been practical to utilize the same principle for significant power generation.

In this process, it is not necessary to convert the radiant energy to a high-temperature heat source as in the application of thermoelectric devices to the utilization of solar energy. Nor is it necessary to reject heat at a lower temperature. The conversion process, therefore, is not analogous to thermal power cycles, and consequently, like the fuel cell, is not limited as cycles are in their functional relationship between thermal efficiency and temperatures of operation.

The success of photovoltaic cell development is evidenced by the use of such devices for auxiliary power sources in space exploration in which their reliability over very long periods of time has been convincingly demonstrated. Undoubtedly,

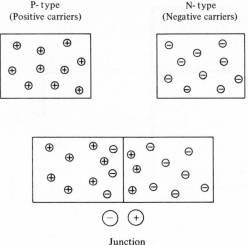

P-type
(Positive carriers)

N-type
(Negative carriers)

Junction
Diffusion of charge carriers

FIG. 7.49 Development of electrostatic potential at junction of *P*-type and *N*-type materials.

the photovoltaic effect will be utilized further in terrestrial applications to harness solar energy and perhaps other radiant energy.

Much of the background information necessary for a general understanding of the phenomena involved has been presented in connection with thermoelectric device materials, Section 7.3.5. As will be seen, positive and negative semiconductor material junctions are used in this field also. The treatment in this section will be essentially descriptive. A detailed thermodynamic analysis would not be closely related to the type of analysis used in connection with other power-producing systems already discussed.

A *p-n* **junction** is shown in **Fig. 7.49**. Because the *p*-type material has bound negatively charged nuclei with mobile "holes," and the *n*-type material has bound positively charged nuclei with mobile electrons, a junction of the two materials produces an electrostatic potential. The mobile electrons from the right will diffuse into the *p*-type material, while the mobile positive charge carriers diffuse from the left into the *n*-type substance. Equilibrium is reached when the electric field so established balances any tendency for further carrier migration.

Parenthetically, we might note that it is exactly this effect at a junction which makes a *p-n* combination an effective electrical diode or "check valve." Connection to an external circuit with a positive potential on the right and a negative potential on the left increases the potential barrier at the junction by driving the respective charge carriers toward the middle. An electron flow from right to left is discouraged. A positive potential on the left and a negative potential on the right, however, encourage charge-carrier movement in the original direction and a conventional electric current flowing from left to right (electrons from right to left) is aided by the presence of the junction.

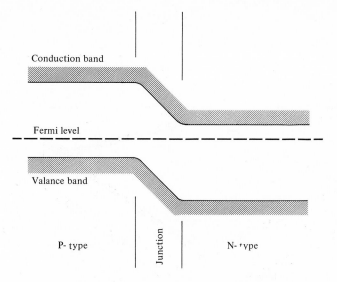

FIG. 7.50 Energy levels in the neighborhood of the junction of *P*- and *N*-type materials.

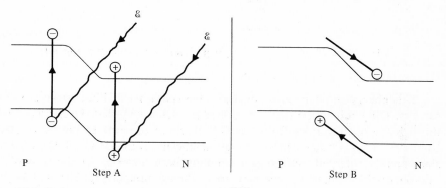

FIG. 7.51 Electron flow across the *P-N* junction with input of energy (ε).

The emf generated by the formation of a junction cannot be used to produce power without some energy input. An electrical connection through an external conductor from one side to the other could not continuously draw on the equilibrium contact potential. Energy levels on either side of the junction are shown in **Fig. 7.50**, where the valence band and conduction-band energy levels of the two materials adjust naturally to produce a continuous Fermi level.

At the junction, however, an input of energy of the right amount to the excess negative charge carriers on the left and to the excess positive charge carriers on the right can drive them across their respective energy gaps (see Fig. 7.21) to the conduction band. This is step *A* shown in **Fig. 7.51**. Electrons are free to move from the higher to the lower energy level from left to right across the junction, while holes at the valence band level move, in effect, from right to left as in step *B*.

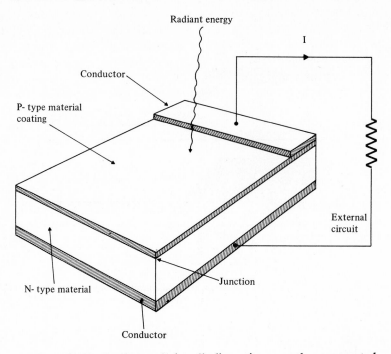

FIG. 7.52 Single photovoltaic cell, dimensions greatly exaggerated.

This energy input is accomplished when photons of light strike the charge carriers. The *p-n* junction must be formed in such a way that the energy can pass through the outer surface from one side to reach the area of junction potential barrier. Such an arrangement for a single **photocell** is shown in **Fig. 7.52** with dimensions greatly enlarged. Silicon is a common base material for the formation of suitable extrinsic *p* and *n* semiconductors. The example of Fig. 7.20 will aid in an understanding of the process: Silicon doped with phosphrous can produce an *n*-type semiconductor. Silicon doped with aluminum (or more commonly with boron which, like aluminum, has only three valence electrons) will make a *p*-type semiconductor. A thin wafer of *n*-type silicon is cut and then coated with a very thin layer of *p*-type impurity. Then the *p*-type layer is removed from all surfaces except the top, and a conductor is bonded to each side of the wafer, as shown. For appreciable power outputs, a large number of individual cells are connected electrically to produce the voltage and current capacity desired.

There are many unavoidable losses. Energy is reflected from the surface. The upper layer (*p*-type in the cell shown) must be thin to transmit radiation, but this increases its internal resistance. Not all radiant energy has a suitable wavelength for useful charge carrier excitation. Not all excited carriers reach and cross the junction. The optimum voltage is unfortunately temperature dependent. All these losses combine to reduce solar cell efficiences to less than 20 %. It is, of course, difficult to manufacture large numbers of identical cell wafers and to connect them identically. Their effectiveness can be seriously deteriorated by particle or high-energy bombardment. It is usual to manufacture the cells from single crystals, because crystal interfaces encourage the recombination of carriers without useful current flow. Still, the simplicity and reliability of the photovoltaic cell ensure that it will find many and diverse applications in the generation of electric power.

PROBLEMS

1. a) If it were true that the human body is limited in thermal efficiency to that of a Carnot cycle, what would its efficiency be?
   b) Is it so limited? Why?

2. A single-stage thermoelectric device is composed of two legs with the following properties and dimensions:

		N	P
α	(V/°K)	-150×10^{-6}	$+200 \times 10^{-6}$
ρ	(ohm-cm)	1.0×10^{-3}	1.2×10^{-3}
Z	$(°K)^{-1}$	2.0×10^{-3}	3.0×10^{-3}
γ	cm²/cm	—	1.0

   a) As a generator, at what value should the external resistance be set to achieve maximum thermal efficiency for operation between 400 K and 700 K?
   b) What would maximum thermal efficiency be?
   c) What should the external resistance be for maximum power output?
   d) What would the thermal efficiency be at maximum power output?

3. Suppose that the value of area-to-length ratio (γ) for the positive leg in the previous problem was changed to 10 cm. Compare the following values for the two different values of area-to-length ratio:
   a) Maximum thermal efficiency.
   b) Current for maximum thermal efficiency.
   c) Power output per couple at maximum thermal efficiency.
   d) Voltage output per couple at maximum thermal efficiency.
   e) Load resistance per couple at maximum thermal efficiency.
   f) Length of each leg.

4. If the area-to-length ratio (γ) for the positive leg of the thermoelectric device of Problem 2 remained 1 cm, how could couples be arranged to give the power output of a device with γ equal to 10 (see Problem 3), maintaining
   a) the same voltage output as for Problem 2?
   b) the same current flow as found for γ equal to 10 in Problem 3?

5. For a thermoelectric generator, show that the maximum power is $(\alpha \Delta T)^2/4R_i$, and write expressions for maximum power densities in terms of watts per unit area (watts/cm²) and watts per unit volume (watts/cm³) of thermoelectric leg material. Note the effect of leg length (l) on each of these two terms.

6. It is proposed that a single-stage thermoelectric device be used to produce a temperature 150 degrees C below a normal room temperature of, say, 300 K. From

your limited knowledge of thermoelectric materials, does this seem possible? Be specific in presenting your reasoning.

7. The same expression for current flow was developed for maximum ΔT and for maximum heat pumping rate for a thermoelectric device operating as a refrigerator. Would the current flow be the same for these two situations, and if not, would it be greater for maximum ΔT or maximum refrigeration effect?

8. A thermoelectric device with material properties the same as listed in Problem 2 is to operate as a refrigerator, rejecting its heat at 300 K.
   a) What is the minimum refrigeration temperature that can be achieved?
   b) What power input is required per couple for operation as in part (a)?
   c) What is the maximum refrigeration effect that can be achieved per couple for one-half the temperature difference found in part (a)?
   d) What are the power input and coefficient of performance (as a refrigerator) for the conditions of part (c)?
   e) For one-half the temperature difference of part (a), what would the maximum coefficient of performance (as a refrigerator) be, and what refrigeration load could be carried per couple at maximum COP_R?
   f) At conditions of part (e), how many thermoelectric couples would be required per ton of refrigeration?

9. To decide whether thermoelectric heat pumping might be economical with an outside ambient temperature of 32 F and a room temperature of 77 F, compare it with direct resistance heating by electricity. Assume required temperature differences of 9 F in the outside heat exchanger and 18 F degrees between room air and heating panels. Also assume the thermoelectric properties listed in Problem 2.

10. Make a rough estimate of the maximum possible ideal thermal efficiency for a thermionic generator operating between cathode and anode temperatures of 2000 K and 1400 K, respectively.

11. Consider a thermionic generator with a cathode work function of 2.2 V and an anode work function of 1.6 V operating at a cathode temperature of 1400 K. For ideal values, assume zero retarding voltage in the interspace.
   a) What should the approximate anode temperature be?
   b) For the temperature, what would the ideal current density be?
   c) Under these conditions, calculate the ideal power output per square centimeter and the thermal efficiency.

12. If a cathode material in a thermionic generator has a Richardson work function of 2.5 V, by what percentage is the current density reduced below the saturation value by a retarding potential in the adjacent space equal to 0.4 V at say 2500 K?

13. The duct of a magnetohydrodynamic converter has a constant spacing between electrodes of 0.4 m. Each electrode area is 0.5 sq m. Ionized gas with a conductivity of 30 $(ohm-m)^{-1}$ flows through at an average velocity of 800 m/sec. A cross magnetic flux of 2.5 webers/sq m is applied, and an external (load) resistance of 0.04 ohms is connected across the electrodes. What is the theoretical power output?

14. For the MHD duct described in the previous problem with the same ionized gas flow, assume that the external resistance is replaced by a voltage source of 900 V opposing the original induced voltage.
   a) What is the rate of work input to the gas?
   b) What is the efficiency of this when compared to the electrical power supplied?

15. If a fuel cell can be made to consume 1 kg-mole/hr of liquid octane (C_8H_{18}) and deliver gaseous carbon dioxide and liquid water, what would be the value for the following:
    a) Ideal thermal efficiency?
    b) Ideal voltage?
    c) Ideal current?
    d) Ideal power?

16. Assume that a hydrogen-oxygen fuel cell delivering liquid water can actually operate with an output voltage that is 65% of its ideal open-circuit voltage. How many grams of oxygen and how many grams of hydrogen will be consumed per kilowatt-hour?

REFERENCES

A. *Articles and Separate Publications*

1. ANGRIST, S. W., "Galvanomagnetic and Thermomagnetic Effects," *Scientific American*, Vol. 205, No. 6, 1961.

2. AUSTIN, L. G., "Fuel Cells," *Scientific American*, Vol. 201, No. 4, 1959.

3. CRANE, W. E., R. L. ROBINSON, and W. U. ROESSLER, "High Power and Endurance —Prime Needs for Spacecraft Auxiliaries," *SAE Jl.*, Aug., 1964.

4. COHN, E. M., "Fuel-Cell Systems," *Mechanical Engineering*, Vol. 88, No. 6, 1966.

5. DAVIS, J. P. and S. BARON, "The Future of Energy Conversion," *Consulting Engineer*, June, 1963.

6. DOMENICALI, C. A., "Irreversible Thermodynamics of Thermoelectricity," *Reviews of Modern Physics*, Vol. 26, No. 2, 1954 (Paper 10 in Reference B. 23).

7. ELLIOTT, J. F., "Photovoltaic Energy Conversion," Chapter 2 of *Direct Energy Conversion*, edited by George W. Sutton (Reference B. 33).

8. GOURDINE, M. C. and D. H. MALCOLM, "Electrogasdynamic Power," *Mechanical Engineering*, Vol. 88, No. 7, 1966.

9. GRUBB, W. T., and L. W. NIEDRACH, "Fuel Cells," Chapter 2 of *Direct Energy Conversion*, edited by George W. Sutton (Reference B. 33).

10. HANSEN, C. F., "Approximations for the Thermodynamic and Transport Properties of High-Temperature Air," NASA TR R-50, *National Aeronautics and Space Administration*, Washington, 1959.

11. HARMAN, T. C., "Multiple Stage Thermoelectric Generation of Power," *Applied Physics*, Vol. 29, No. 10, 1958 (Paper 19 in Reference B. 23).

12. HATSOPOULOS, G. N., "Thermodynamics of Thermionic Engines," Chapter 3 of *Direct Conversion of Heat to Electricity*, edited by Kaye and Welsh (Reference B. 21).

13. HATSOPOULOS, G. N. and J. KAYE, "Analysis and Experimental Results of a Diode Configuration of a Novel Thermoelectron Engine," *Proceedings IRE*, Vol. 46, Sept., 1958 (Chapter 1 in Reference 21).

14. HATSOPOULOS, G. N. and J. H. KEENAN, "Thermodynamics of Thermoelectric Generators," Chapter 15 in *Direct Conversion of Heat to Electricity*, edited by Kaye and Welsh (Reference B. 21).

15. HENDERSON, R. E., "Energetics 7: The Comprehensive View," *Mechanical Engineering*, Vol. 88, No. 12, 1966, pp. 34–40.

16. HERNQUIST, K. G., M. KANEFSKY, and F. H. NORMAN, "Thermionic Energy Converter," *RCA Review*, Vol. XIX, No. 2, 1958 (Paper 24 in Reference B. 23).

17. HOUSTON, J. M., "Theoretical Efficiency of the Thermionic Energy Converter," *Jl. Applied Physics*, Vol. 30, No. 4, 1959 (Paper 21 in Reference B. 23).

18. HOUSTON, J. M., "Thermionic Power," *Mechanical Engineering*, Vol. 88, No. 9, 1966.

19. JAUMOT, F. E., JR., "Thermoelectric Effects," *Proceeding IRE*, Vol. 46, No. 3, 1958 (Paper 5 in Reference B. 23, Chapter 14 in Reference B. 21).

20. JOFFE (also, IOFFE), A. F., "The Revival of Thermoelectricity," *Scientific American*, Vol. 199, No. 5, 1958 (Paper 2 in Reference B. 23).

21. KLEM, R. L. and A. G. F. DINGWALL, "Thermoelectric Power," *Mechanical Engineering*, Vol. 88, Aug., 1966.

22. LIEBHAFSKY, H. A. and D. L. DOUGLAS, "The Fuel Cell," *Mechanical Engineering*, Vol. 81, No. 8, 1959, pp. 64–68.

23. ONSAGER, L., "Reciprocal Relations in Irreversible Processes, I," *Physical Review*, Vol. 37, 1931, pp. 405–426.

24. OSTERLE, J. F., "A Unified Treatment of the Thermodynamics of Steady-State Energy Conversion," *Applied Scientific Research*, Section A, Vol. 12, 1964, pp. 425–434.

25. RAPPAPORT, P., "The Photovoltaic Effect and its Utilization," *RCA Review*, Vol. XX, No. 3, 1959, pp. 373–397 (Paper 31 in Reference B. 23).

26. ROSI, F. D., E. F. HOCKINGS, and N. E. LINDENBLAD, "Semiconducting Power Generation," *RCA Review*, Vol. XXII, 1961, p. 96.

27. ROSI, F. D. and E. G. RAMBERG, "Evaluation and Properties of Materials for Thermoelectric Applications," Chapter 8, *Thermoelectricity*, edited by P. H. Egli (Reference B. 11).

28. SHIUE, J. C., B. KAHN, and H. E. BRANDMAIER, "An Experimental Investigation of the Performance Characteristics of an Electrogasdynamic Power Generator," Paper presented at *ASME International Conference on Energetics*, Rochester, Aug., 1965.

29. SMITH, A. H., "Photovoltaic Power," *Mechanical Engineering*, Vol. 88, No. 10, 1966, p. 38.

30. STOUT, V. L., "Thermionic Emission and Thermionic Power Generation," Chapter 13, *Thermoelectricity*, edited by P. H. Egli (Reference B. 11).

31. TELKES, M., "Solar Thermoelectric Generators," *Jl. Applied Physics*, Vol. 25, No. 6, 1954 (Paper 8 in Reference B. 23).

32. URE, R. W., JR. and R. R. HEIKES, "Theoretical Calculation of Device Performance," Chapter 15 in *Thermoelectricity: Science and Engineering*, edited by Heikes and Ure (Reference B. 16).

33. WAY, S., S. M. DeCORSO, R. L. HUNDSTAD, G. A. KEMENEY, W. STEWART, and W. E. YOUNG, "Experiments with MHD Power Generation," *Trans. ASME*, Vol. 83A (*Jl. Eng. Power*), 1961, p. 397.

34. WEBSTER, H. F., "Calculation of the Performance of a High Vacuum Thermionic Energy Converter," *Jl. Applied Physics*, Vol. 30, No. 4, 1959 (Paper 22 in Reference B. 23).

35. WILSON, V. C., "Conversion of Heat to Electricity by Thermionic Emission," *Jl. Applied Physics*, Vol. 30, No. 4, 1959 (Paper 20 in Reference B. 23).

36. YEAGER, J. F., "Fuel Cells as Energy Converter," Chapter 23 in *Direct Conversion of Heat to Electricity*, edited by Kaye and Welsh (Reference B. 21).

37. ZENER, C., "The Impact of Thermoelectricity upon Science and Technology," Chapter 1, *Thermoelectricity*, edited by P. H. Egli (Reference B. 11).

38. LINDSTROM, O., "The Fuel Cell now Makes its Entry on the Market," *ASEA Jl.*, Vol. 40, No. 6–7, 1967; and "Fuel Cells," *ASEA Jl.*, Vol. 37, No. 1, 1964.

B. *Books*

 1. ANGRIST, S. W., *Direct Energy Conversion*. Boston, Mass.: Allyn and Bacon, 1965.

 2. BOSNJAKOVIC, F., *Technical Thermodynamics*. New York: Holt, Rinehart and Winston, 1965 (Ch. XII).

 3. BRIDGMAN, P. W., *The Thermodynamics of Electrical Phenomena in Metals and a Condensed Collection of Thermodynamic Formulas*. New York: Dover, 1961 (first edition: Macmillian, 1934).

 4. CADOFF, I. B. and E. MILLER (eds), *Thermoelectric Materials and Devices*. New York: Reinhold, 1960.

 5. CALLEN, H. B., *Thermodynamics*. New York: Wiley, 1963.

 6. CAMBEL, A. B., *Plasma Physics and Magnetofluidmechanics*. New York: McGraw-Hill, 1963.

 7. CHANG, S. S. L., *Energy Conversion*. Englewood Cliffs: Prentice-Hall, 1963.

 8. DEGROOT, S. R. and P. MAZUR, *Non-Equilibrium Thermodynamics*. Amsterdam: North-Holland Pub. Co., 1962.

 9. DENBIGH, K. G., *The Thermodynamics of the Steady State*. New York: Wiley (London: Methuen), 1951.

10. DRABBLE, J. R. and H. J. GOLDSMID, *Thermal Conduction in Semiconductors*. New York: Pergamon Press, 1961.

11. EGLI, P. H. (ed.), *Thermoelectricity*. New York: Wiley, 1960.

12. EISBERG, R. M., *Fundamentals of Modern Physics*. New York: Wiley, 1961.

13. *Energy Conversion Systems Reference Handbook, Vol. I*, WADD Techn. Report, 60-699, AD-257357, Office of Techn. Services, U. S. Dept. of Commerce, Washington, 1960.

14. *Energy Conversion Systems Handbook, Vol. IX*, AD-256916, Office Tech. Services, U. S. Dept. of Commerce, Washington, 1960.

15. FAY, J. A., *Molecular Thermodynamics*. Reading, Mass.: Addison-Wesley, 1965.

16. HEIKES, R. R. and R. W. URE, JR., *Thermoelectricity: Science and Engineering*. New York: Interscience Publishers, 1961.

17. HOLT, E. H. and R. E. HASKELL, *Plasma Dynamics (Foundation of)*. New York: Macmillan, 1965.

18. HUGHES, W. F., and F. J. YOUNG, *The Electromagnetodynamics of Fluids*. New York: Wiley, 1966.

19. IOFFE (also JOFFE), A. F., *Physics of Semiconductors*. New York: Academic Press (London: Infosearch), 1960.

20. IOFFE (also JOFFE), A. F., *Semiconductor Thermoelements and Thermoelectric Cooling* (trans. A. Belbtuch). London: Infosearch, 1957.

21. KAYE, J. and J. A. WELSH (eds), *Direct Conversion of Heat to Electricity*. New York: Wiley, (1960).

22. LEE, J. F., F. W. SEARS, D. L. TURCOTTE, *Statistical Thermodynamics*. Reading, Mass.: Addison-Wesley, 1963.

23. LEVINE, S. N. (ed), *Selected Papers on New Techniques for Energy Conversion*. New York: Dover, 1961.

24. MacDONALD, D. K. C., *Thermoelectricity: an introduction to the principles*. New York: Wiley, 1962.

25. MANNAL, C., and N. W. MATHER (eds), *Engineering Aspects of Magnetohydrodynamics*. New York: Columbia University Press, 1962.

26. MITCHELL, W., JR., *Fuel Cells*, Academic Press, New York, 1963.

27. PRIGOGINE, I., *Introduction to Thermodynamics of Irreversible Processes* (second edition). New York: Wiley, 1961.

28. ROSE, R. M., L. A. SHEPARD, and J. WULFF, *Electronic Properties*, Vol. IV, in the *Structure and Properties of Materials*, edited by John Wulff. New York: Wiley, 1966.

29. RUSSELL, C. K., *Elements of Energy Conversion*. New York: Pergamon Press, 1967.

30. SHERCLIFF, J. A., *A Textbook of Magnetohydrodynamics*. New York: Pergamon Press, 1967.

31. SHIVE, J. N., *The Properties, Physics and Design of Semiconductor Devices*. Princeton: D. Van Nostrand, 1959.

32. SPROULL, R. L., *Modern Physics: a textbook for engineers*. New York: Wiley, 1956.

33. SUTTON, G. W. (ed), *Direct Energy Conversion*. New York: McGraw-Hill, 1966.

34. SUTTON, G. W. and A. SHERMAN, *Engineering Magnetohydrodynamics*. New York: McGraw-Hill, 1965.

35. VAN VLACK, L. H., *Elements of Materials Science* (second edition). Reading, Mass.: Addison-Wesley, 1964.

36. YOUNG, G. J. (ed), *Fuel Cells*. New York: Reinhold, 1960.

PHYSICAL CONSTANTS AND CONVERSION FACTORS*

* From *The International System of Units*, NASA SP-7012, 1964, by E. A. Mechtly.

PHYSICAL CONSTANTS

Avogadro's constant	N_0	6.02252×10^{23}	molecule/gram-mole
Faraday's constant	$\mathscr{F}$	9.64870×10^4	coulomb/gram-mole of electrons
Planck's constant	h	6.6256×10^{-34}	joule-sec
Gas constant	R_U	8.3143×10^0	joule/gram-mole-K
Boltzmann's constant	k	1.38054×10^{-23}	joule/molecule-K
Stefan-Boltzmann's constant	σ	5.6697×10^{-8}	watt/meter^2-K^4
Electron-volt	eV	1.60210×10^{-19}	joule/eV

CONVERSION FACTORS

Relationships which are exact (by the National Bureau of Standards definitions) are followed by an asterisk (*). Others are the results of physical measurement, or are only approximate. The first two digits of each conversion factor represent a power of 10. For example, to convert 5 in. to its equivalent in the International System of Units (MKSA), multiply by 0.0254 to obtain 0.1270 m.

To Convert From:	To:	Multiply by:
	(Length)	
angstrom	meter	-10 1.00*
foot	meter	-01 3.048*
inch	meter	-02 2.54*
mile (U.S. statute)	meter	$+03$ 1.609344*
	(Area)	
foot^2	meter^2	-02 9.290304*
inch^2	meter^2	-04 6.4516*
mile^2 (U.S. statute)	meter^2	$+06$ 2.589988110336*
	(Volume)	
foot^3	meter^3	-02 2.8316846592*
gallon (British)	meter^3	-03 4.546087
gallon (U.S. liquid)	meter^3	-03 3.785411784*
liter	meter^3	-03 1.00*
	(Mass)	
gram	kilogram	-03 1.00*
lbm (pound mass, avoirdupois)	kilogram	-01 4.5359237*
slug	kilogram	$+01$ 1.45939029
ton (metric)	kilogram	$+03$ 1.00*
ton (short, 2000 pound)	kilogram	$+02$ 9.0718474*

CONVERSION FACTORS (*continued*)

To Convert From:	To:	Multiply by:
	(Force)	
dyne	newton	−05 1.00*
kilogram force, kgf	newton	+00 9.80665*
pound force lbf (avoirdupois)	newton	+00 4.4482216152605*
	(Pressure)	
atmosphere	newton/meter^2	+05 1.01325*
foot of water (39.2 F)	newton/meter^2	+03 2.98898
inch of mercury (32 F)	newton/meter^2	+03 3.386389
lbf/foot^2	newton/meter^2	+01 4.7880258
lbf/inch^2 (psi)	newton/meter^2	+03 6.8947572
	(Density)	
gram/centimeter^3	kilogram/meter^3	+03 1.00*
lbm/foot^3	kilogram/meter^3	+01 1.6018463
	(Energy)	
British thermal unit (International Steam Table)	joule	+03 1.05504
British thermal unit (thermochemical)	joule	+03 1.054350264488888
calorie (International Steam Table)	joule	+00 4.1868
calorie (thermochemical)	joule	+00 4.184*
electron volt	joule	−19 1.60210
erg	joule	−07 1.00*
foot lbf	joule	+00 1.3558179
joule (international, 1948)	joule	+00 1.000165
kilowatt hour	joule	+06 3.60*
kilowatt hour (international, 1948)	joule	+06 3.60059
watt hour	joule	+03 3.60*
	(Power)	
Btu (thermochemical)/second	watt	+03 1.054350264488888
calorie (thermochemical)/second	watt	+00 4.184*
horsepower (550 foot-lbf/second)	watt	+02 7.4569987
horsepower (electric)	watt	+02 7.46*
	(Electrical)	
ampere (international, 1948)	ampere	−01 9.99835
ampere hour	coulomb	+03 3.60*
coulomb (international, 1948)	coulomb	−01 9.99835

(*continued*)

CONVERSION FACTORS (*continued*)

To Convert From:	To:	Multiply by:
	(Electrical)	
faraday (based on carbon 12)	coulomb	+04 9.64870
faraday (chemical)	coulomb	+04 9.64957
faraday (physical)	coulomb	+00 1.000495
ohm (international, 1948)	ohm	+00 1.000495
gauss	tesla	−04 1.00*
volt (international, 1948)	volt	+00 1.000330
maxwell	weber	−08 1.00*
	(Temperature)	
Celsius (temperature)	Kelvin	$K = C + 273.15$
Rankine (temperature)	Kelvin	$K = (5/9)R$*
Fahrenheit (temperature)	Celsius	$C = (5/9)(F - 32)$
	(Thermal conductivity)	
Btu inch/foot²-second-F	joule/meter-sec-K	+02 5.1887315
	(Viscosity)	
stoke	meter²/second	−04 1.00*
foot²/second	meter²/second	−02 9.290304*
lbf-second/foot²	newton-second/meter²	+01 4.7880258
poise	newton-second/meter²	−01 1.00*
slug/foot-second	newton-second/meter²	+01 4.7880258
	(Miscellaneous)	
degree (angle)	radian	−02 1.7453292519943
Btu (thermochemical)/lbm	joule/kilogram	+03 2.3244444
Btu (thermochemical)/foot² hour	watt/meter²	+00 3.1524808
Btu (thermochemical)/lbm-F	joule/kilogram-C	+03 4.184*
Btu (thermochemical)/lbm-F	calorie (thermochemical)/gram-C	+00 1.00*
calorie(thermochemical)/kilogram-C	joule/kilogram-C	+00 4.184*

CHARTS OF THERMODYNAMIC PROPERTIES

The figures in this appendix have been reprinted by permission from the following sources.

ASME, *Thermodynamic and Transport Properties of Steam,* 1967 (Figure II.1.1); National Bureau of Standards (Figures II.1.2, II.1.3, II.1.5, II.1.6, II.1.7, II.1.8); Butterworth and Co., Ltd. (Figure II.1.4); ASHRAE *Handbook of Fundamentals,* 1967 (Figures II.2.1 and II.6); E. I. Dupont de Nemours, Inc. (Figures II.2.2, II.2.3, II.2.4); *Unit Operations,* by G. G. Brown, *et al.,* John Wiley and Sons, 1950 (Figure II.3.1); Institute of Gas Technology, *Research Bulletin* 14, by Ellington, Kunst, Peck, and Reed (Figure II.3.2); Cornell University, *Bulletin* 30, by Ellenwood, Kulich, and Gray (Figures II.4.1, II.4.2, II.4.3, II.4.4, II.4.5, II.4.6); Combustion Engineering, Inc. (Figure II.5.1); Airesearch Manufacturing Co. (Figure II.5.2).

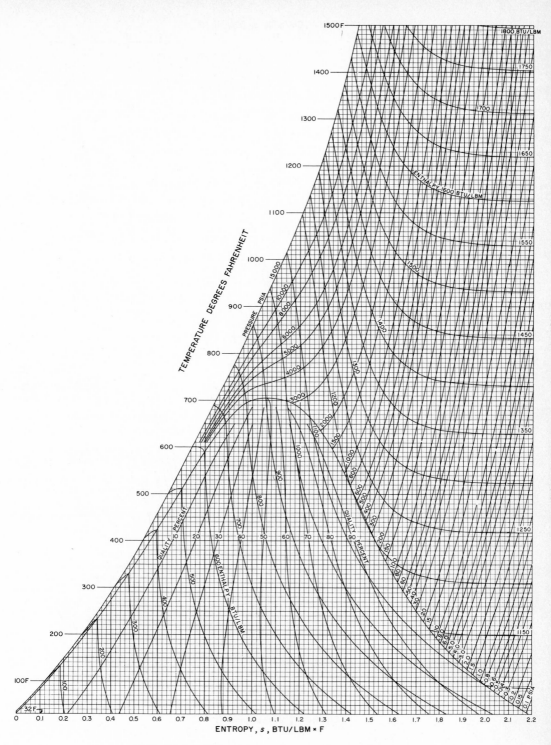

FIG. II.1.1 Temperature-entropy diagram for steam.

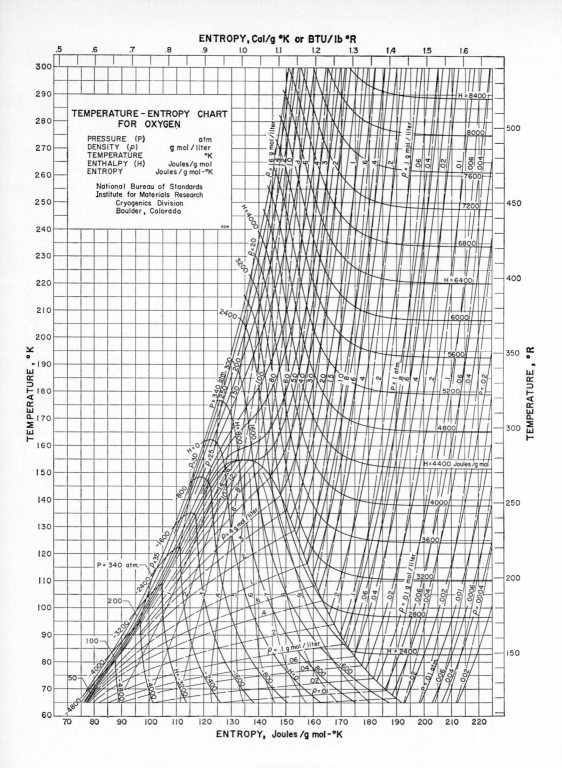

Figure II.1.2

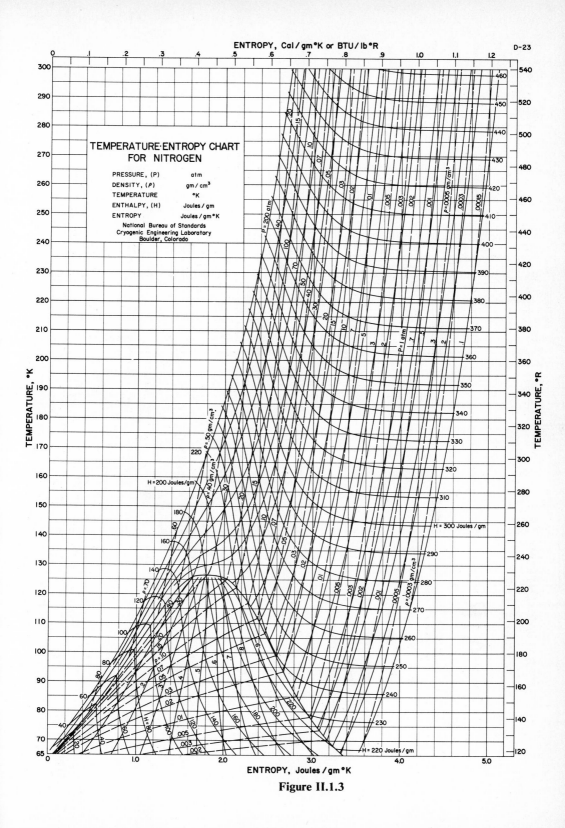

Figure II.1.3

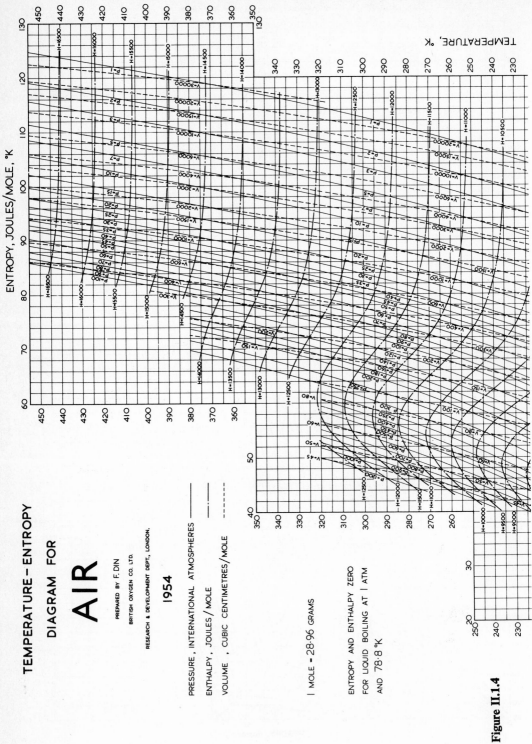

Figure II.1.4

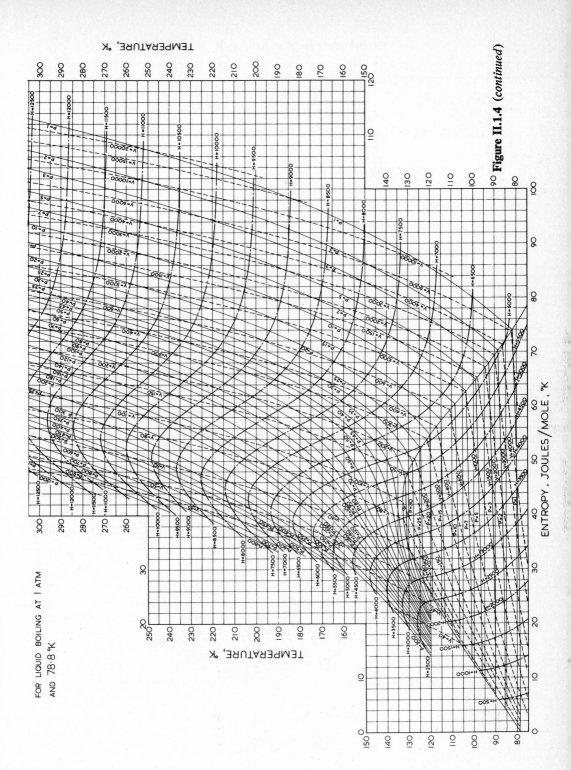

Figure II.1.4 (continued)

FOR LIQUID BOILING AT 1 ATM
AND 78.8 °K

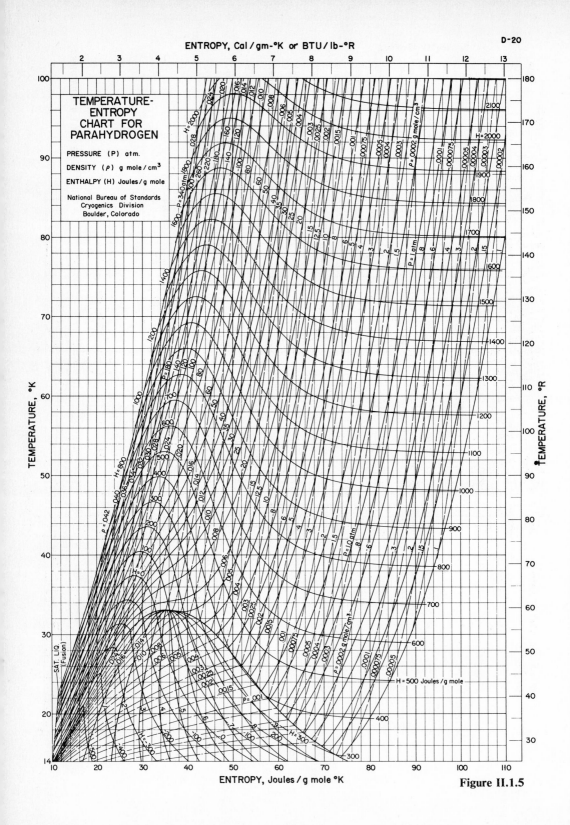

Figure II.1.5

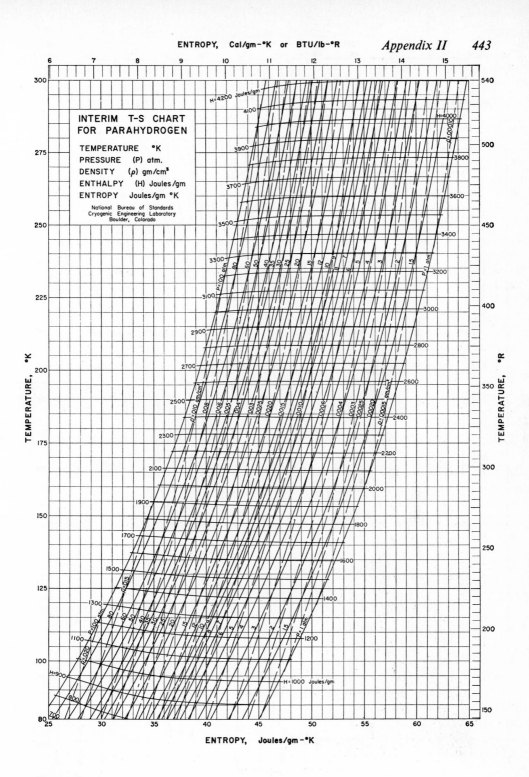

Figure II.1.6

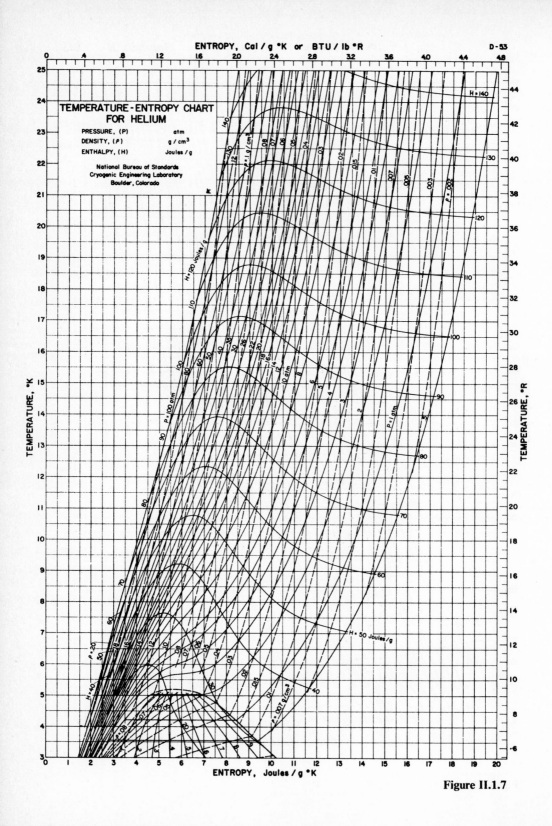

Figure II.1.7

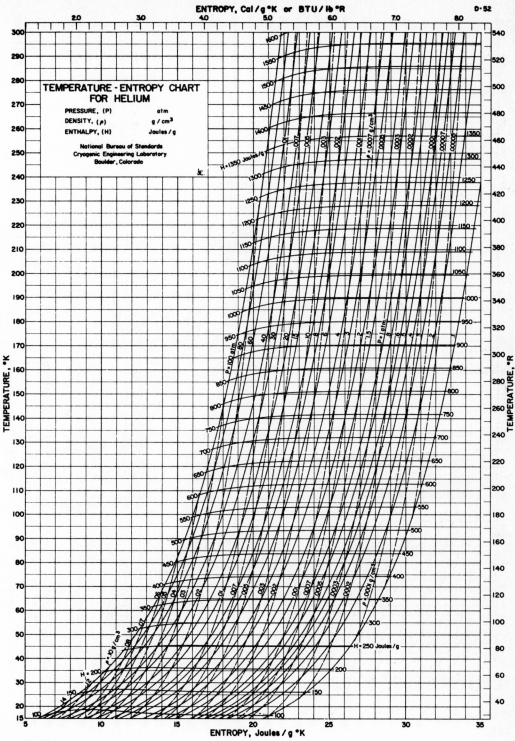

Figure II.1.8

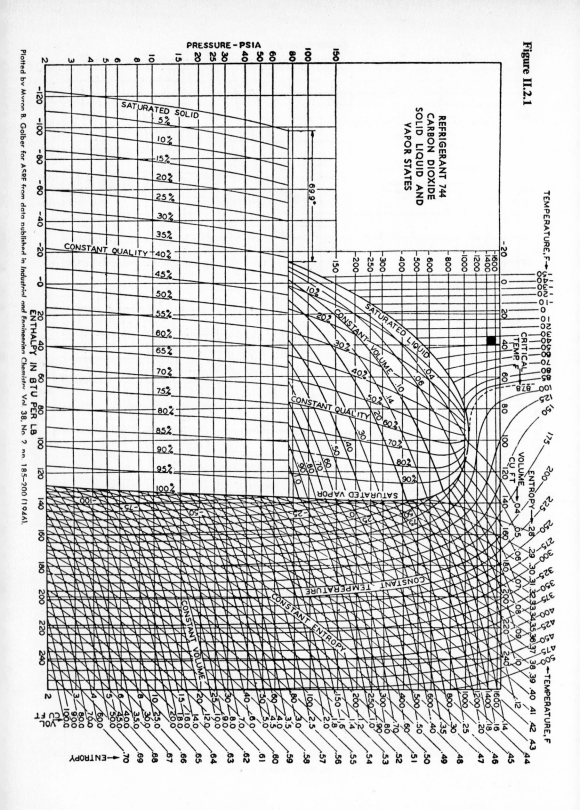

Figure II.2.1

REFRIGERANT 744
CARBON DIOXIDE
SOLID LIQUID AND
VAPOR STATES

Plotted by Myron B. Golber for ASRE from data published in Industrial and Engineering Chemistry Vol 38, No 2 pp 185-200 (1946).

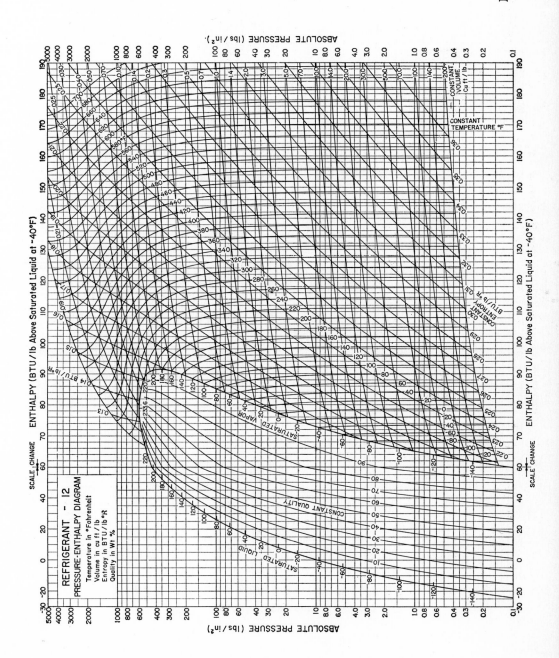

Figure II.2.2

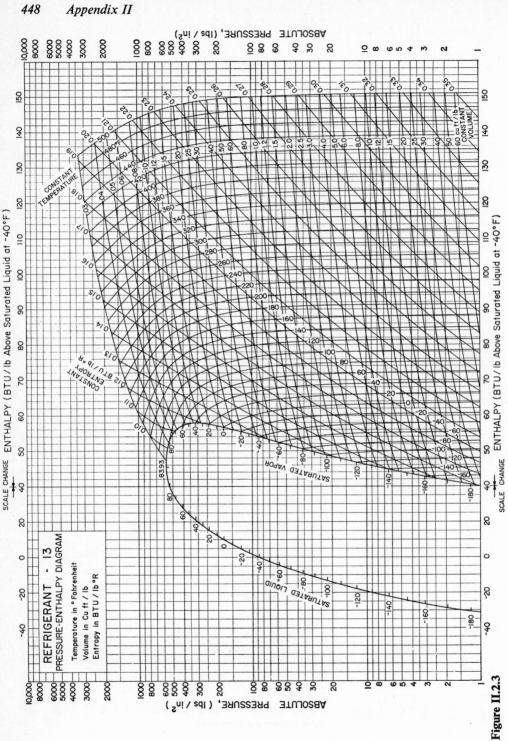

Figure II.2.3

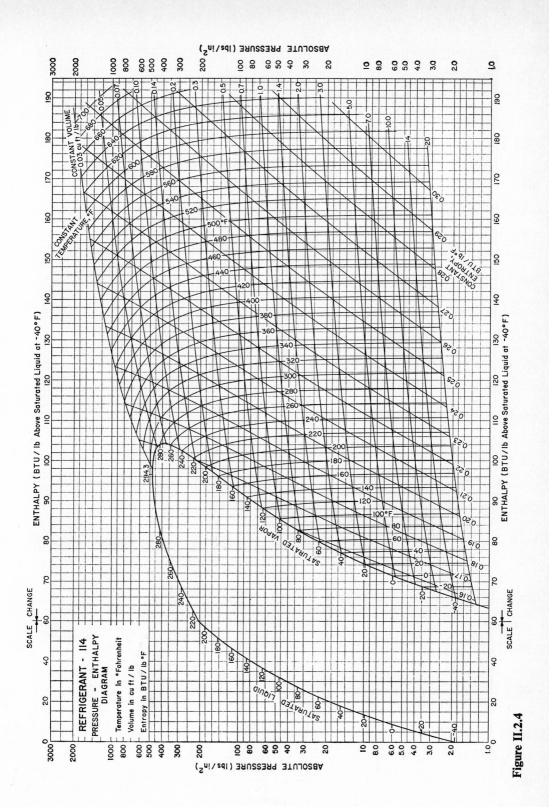

Figure II.2.4

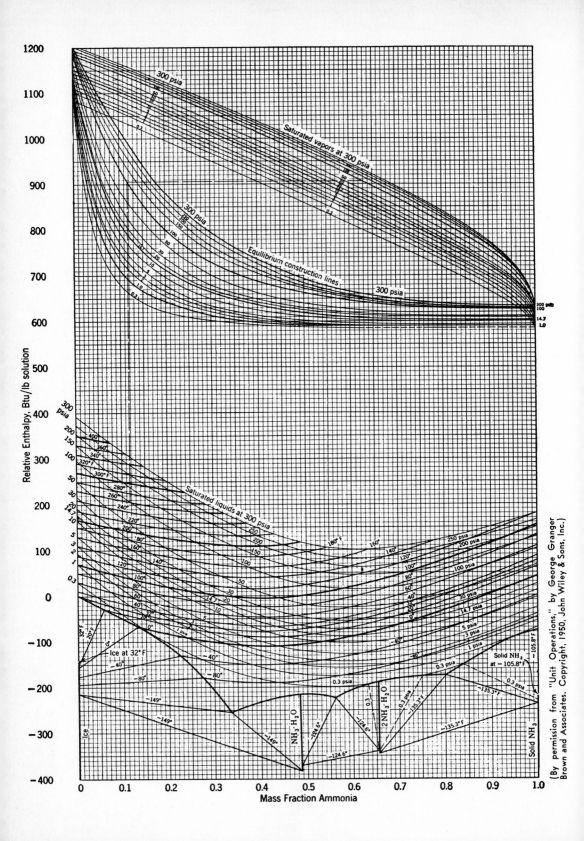

Relative Enthalpy, Btu/lb solution

Mass Fraction Ammonia

1200
1100
1000
900
800
700
600
500
400
300
200
100
0
−100
−200
−300
−400

300 psia

Saturated vapors at 300 psia

Equilibrium construction lines
300 psia

Saturated liquids at 300 psia

Ice at 32°F

Ice

NH₃·H₂O

2NH₃·H₂O

Solid NH₃

Solid NH₃ at −105.8°F

−149°

−124.6°

−135.3°F

0   0.1   0.2   0.3   0.4   0.5   0.6   0.7   0.8   0.9   1.0

(By permission from "Unit Operations," by George Granger Brown and Associates. Copyright, 1950, John Wiley & Sons, Inc.)

◄**FIG. II.3.1** Enthalpy-concentration diagram for ammonia-water combination.

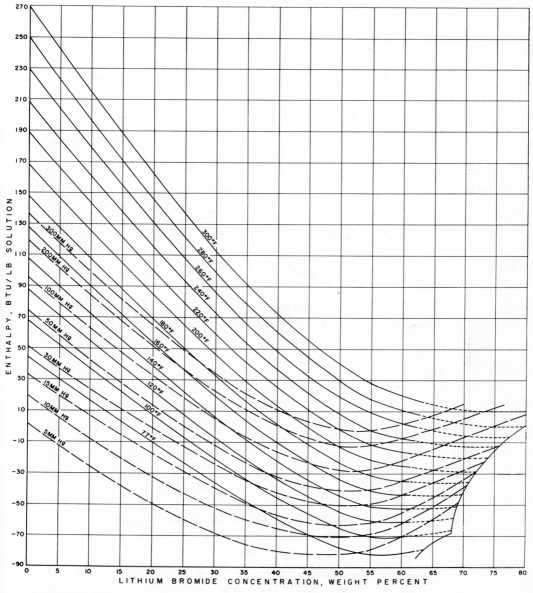

FIG. II.3.2 Enlarged enthalpy-concentration diagram for lithium bromide-water combination.

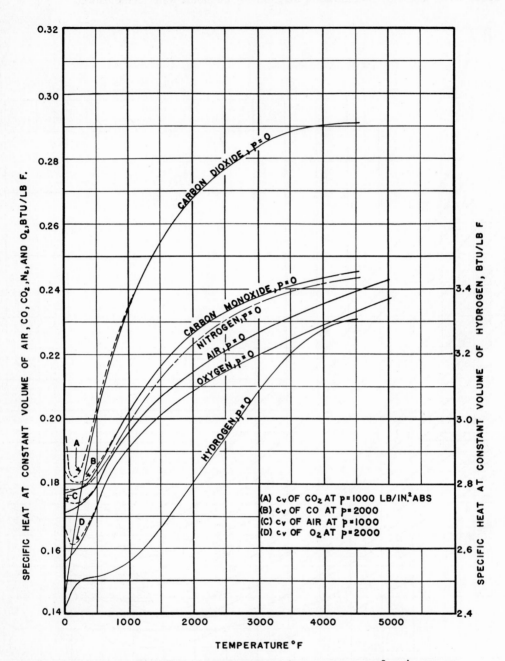

FIG. II.4.1 The effect of temperature and pressure on c_v of various gases.

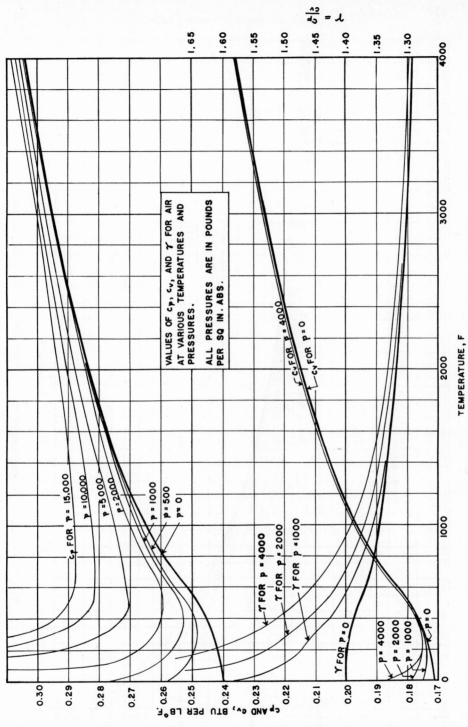

FIG. II.4.2 The effect of temperature on c_p, c_v, and γ of dry air at various pressures.

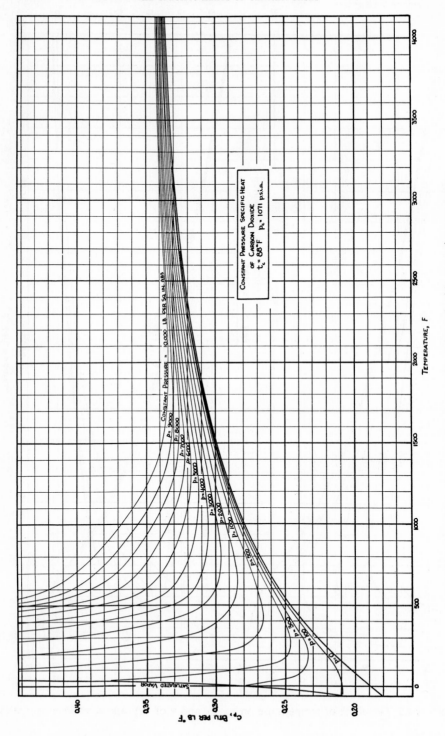

FIG. II.4.3 The effect of temperature on c_p of carbon dioxide at various pressures.

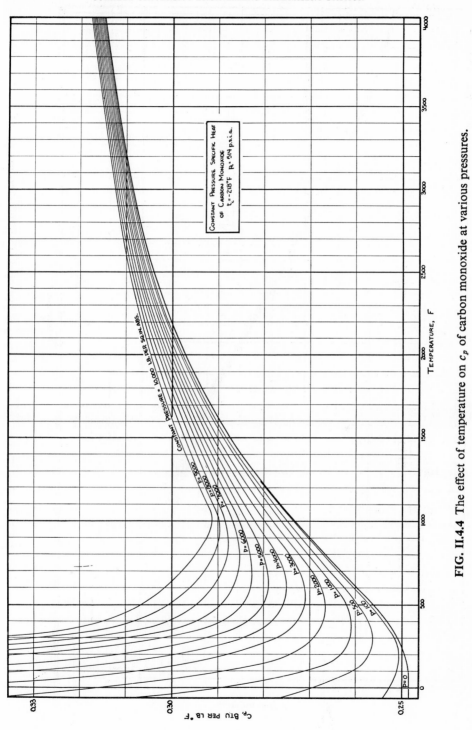

FIG. II.4.4 The effect of temperature on c_p of carbon monoxide at various pressures.

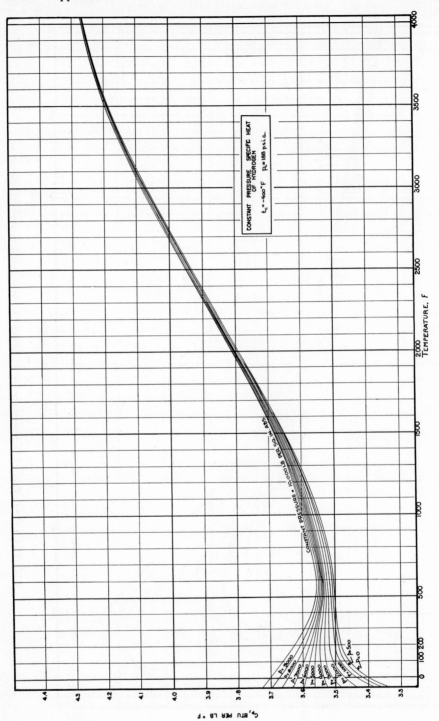

FIG. II.4.5 The effect of temperature on c_p of hydrogen at various pressures.

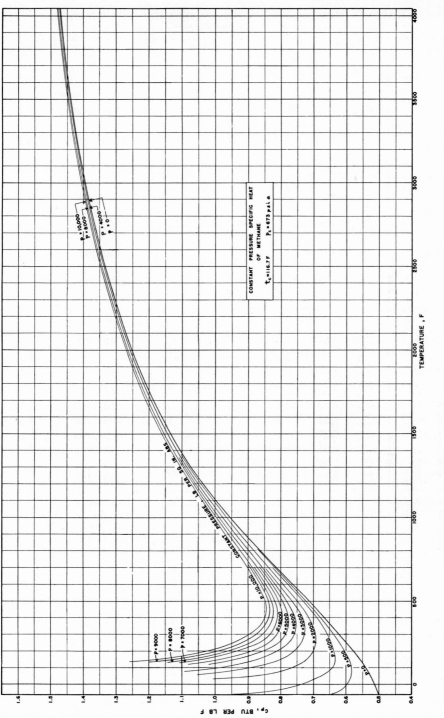

FIG. II.4.6 The effect of temperature on c_p of methane at various pressures.

CONSTANT-PRESSURE SPECIFIC HEATS OF VARIOUS SUBSTANCES
AT ZERO PRESSURE*

Gas or Vapor	Equation, $\bar{C}_{po}$ in Btu/lb mole-R T in degrees Rankine	Range, R	Max. Error, %
O_2	$\bar{C}_{po} = 11.515 - \dfrac{172}{\sqrt{T}} + \dfrac{1530}{T}$	540–5000	1.1
	$= 11.515 - \dfrac{172}{\sqrt{T}} + \dfrac{1530}{T}$ $+ \dfrac{0.05}{1000}(T - 4000)$	5000–9000	0.3
N_2	$\bar{C}_{po} = 9.47 - \dfrac{3.47 \times 10^3}{T} + \dfrac{1.16 \times 10^6}{T^2}$	540–9000	1.7
CO	$\bar{C}_{po} = 9.46 - \dfrac{3.29 \times 10^3}{T} + \dfrac{1.07 \times 10^6}{T^2}$	540–9000	1.1
H_2	$\bar{C}_{po} = 5.76 + \dfrac{0.578}{1000}T + \dfrac{20}{\sqrt{T}}$	540–4000	0.8
	$= 5.76 + \dfrac{0.578}{1000}T + \dfrac{20}{\sqrt{T}}$ $- \dfrac{0.33}{1000}(T - 4000)$	4000–9000	1.4
H_2O	$\bar{C}_{po} = 19.86 - \dfrac{597}{\sqrt{T}} + \dfrac{7500}{T}$	540–5400	1.8
CO_2	$\bar{C}_{po} = 16.2 - \dfrac{6.53 \times 10^3}{T} + \dfrac{1.41 \times 10^6}{T^2}$	540–6300	0.8
CH_4	$\bar{C}_{po} = 4.52 + 0.00737T$	540–1500	1.2
C_2H_4	$\bar{C}_{po} = 4.23 + 0.01177T$	350–1100	1.5
C_2H_6	$\bar{C}_{po} = 4.01 + 0.01636T$	400–1100	1.5
C_3H_8	$\bar{C}_{po} = 2.258 + 0.0320T - 5.43 \times 10^{-6}T^2$	415–2700	1.8
C_4H_{10}	$\bar{C}_{po} = 4.36 + 0.0403T - 6.83 \times 10^{-6}T^2$	540–2700	1.7
C_8H_{18}	$\bar{C}_{po} = 7.92 + 0.0601T$	400–1100	est. 4
$C_{12}H_{26}$	$\bar{C}_{po} = 8.68 + 0.0889T$	400–1100	est. 4

* From *Bulletin No. 2*, Ga. School of Technology, by R. L. Sweigert and M. W. Beardsley, 1938, except C_3H_8 and C_4H_{10}, which are from H. M. Spencer, *J. of Am. Chem. Soc.*, 67, 1859 (1945).

Figure II.4.7

Figure II.6

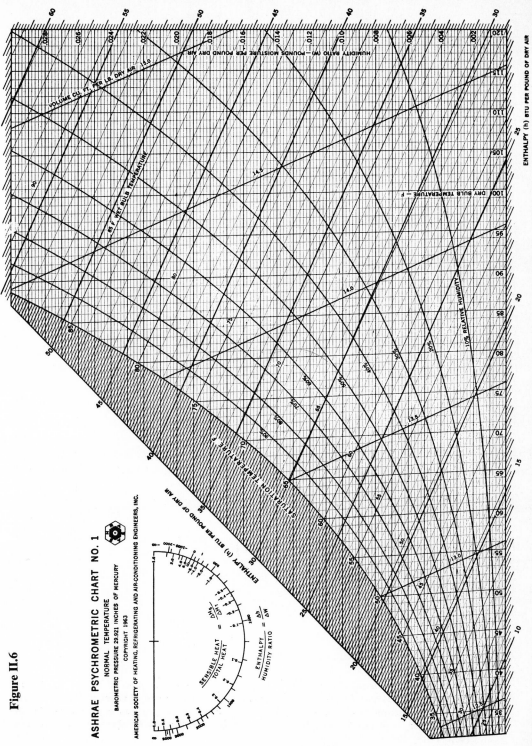

ASHRAE PSYCHROMETRIC CHART NO. 1
NORMAL TEMPERATURE
BAROMETRIC PRESSURE 29.921 INCHES OF MERCURY
COPYRIGHT 1963
AMERICAN SOCIETY OF HEATING, REFRIGERATING AND AIR-CONDITIONING ENGINEERS, INC.

$$\frac{SENSIBLE\ HEAT}{TOTAL\ HEAT} = \frac{\Delta H_S}{\Delta H_T}$$

$$\frac{ENTHALPY}{HUMIDITY\ RATIO} = \frac{\Delta h}{\Delta W}$$

STEAM TABLES*

* From *Thermodynamic and Transport Properties of Steam Comprising Tables and Charts for Steam and Water* ("1967 ASME Steam Tables"). In these Tables, the ASME uses the words "steam" and "water" to mean water in the vapor phase and water in the liquid phase, respectively. Published by the American Society of Mechanical Engineers, New York, 1967.

ASME Table 1
PROPERTIES OF SATURATED STEAM AND SATURATED WATER (Temperature)

Temp. F	Press. psia	Volume, ft³/lbm Water v_f	Evap. v_{fg}	Steam v_g	Enthalpy, Btu/lbm Water h_f	Evap. h_{fg}	Steam h_g	Entropy, Btu/lbm × F Water s_f	Evap. s_{fg}	Steam s_g	Temp. F
705.47	3208.2	0.05078	0.00000	0.05078	906.0	0.0	906.0	1.0612	0.0000	1.0612	705.47
700.0	3094.3	0.03662	0.03857	0.07519	822.4	172.7	995.2	0.9901	0.1490	1.1390	700.0
680.0	2708.6	0.03037	0.08080	0.11117	758.5	310.1	1068.5	0.9365	0.2720	1.2086	680.0
660.0	2365.7	0.02768	0.11663	0.14431	714.9	392.1	1107.0	0.8995	0.3502	1.2498	660.0
640.0	2059.9	0.02595	0.15427	0.18021	679.1	454.6	1133.7	0.8686	0.4134	1.2821	640.0
620.0	1786.9	0.02466	0.19615	0.22081	646.9	506.3	1153.2	0.8403	0.4689	1.3092	620.0
600.0	1543.2	0.02364	0.24384	0.26747	617.1	550.6	1167.7	0.8134	0.5196	1.3330	600.0
580.0	1326.2	0.02279	0.29937	0.32216	589.1	589.9	1179.0	0.7876	0.5673	1.3550	580.0
560.0	1133.38	0.02207	0.36507	0.38714	562.4	625.3	1187.7	0.7625	0.6132	1.3757	560.0
540.0	962.79	0.02146	0.44367	0.46513	536.8	657.5	1194.3	0.7378	0.6577	1.3954	540.0
520.0	812.53	0.02091	0.53864	0.55956	512.0	687.0	1199.0	0.7133	0.7013	1.4146	520.0
500.0	680.86	0.02043	0.65448	0.67492	487.9	714.3	1202.2	0.6890	0.7443	1.4333	500.0
480.0	566.15	0.02000	0.79716	0.81717	464.5	739.6	1204.1	0.6648	0.7871	1.4518	480.0
460.0	466.87	0.01961	0.97463	0.99424	441.5	763.2	1204.8	0.6405	0.8299	1.4704	460.0
440.0	381.54	0.01926	1.19761	1.21687	419.0	785.4	1204.4	0.6161	0.8729	1.4890	440.0
420.0	308.780	0.01894	1.4808	1.4997	396.9	806.2	1203.1	0.5915	0.9165	1.5080	420.0
400.0	247.259	0.01864	1.8444	1.8630	375.1	825.9	1201.0	0.5667	0.9607	1.5274	400.0
380.0	195.729	0.01836	2.3170	2.3353	353.6	844.5	1198.0	0.5416	1.0057	1.5473	380.0
360.0	153.010	0.01811	2.9392	2.9573	332.3	862.1	1194.4	0.5161	1.0517	1.5678	360.0
340.0	117.992	0.01787	3.7699	3.7878	311.3	878.8	1190.1	0.4902	1.0990	1.5892	340.0
320.0	89.643	0.01766	4.8961	4.9138	290.4	894.8	1185.2	0.4640	1.1477	1.6116	320.0
300.0	67.005	0.01745	6.4483	6.4658	269.7	910.0	1179.7	0.4372	1.1979	1.6351	300.0
280.0	49.200	0.01726	8.6267	8.6439	249.2	924.6	1173.8	0.4098	1.2501	1.6599	280.0
260.0	35.427	0.017089	11.745	11.762	228.76	938.6	1167.4	0.3819	1.3043	1.6862	260.0
240.0	24.968	0.016926	16.304	16.321	208.45	952.1	1160.6	0.3533	1.3609	1.7142	240.0

(continued)

ASME Table 1 *(continued)*

PROPERTIES OF SATURATED STEAM AND SATURATED WATER (Temperature)

Temp. F	Press. psia	Volume, ft³/lbm Water v_f	Evap. v_{fg}	Steam v_g	Enthalpy, Btu/lbm Water h_f	Evap. h_{fg}	Steam h_g	Entropy, Btu/lbm × F Water s_f	Evap. s_{fg}	Steam s_g	Temp. F
220.0	17.186	0.016775	23.131	23.148	188.23	965.2	1153.4	0.3241	1.4201	1.7442	220.0
212.0	14.696	0.016719	26.782	26.799	180.17	970.3	1150.5	0.3121	1.4447	1.7568	212.0
200.0	11.526	0.016637	33.622	33.639	168.09	977.9	1146.0	0.2940	1.4824	1.7764	200.0
190.0	9.340	0.016572	40.941	40.957	158.04	984.1	1142.1	0.2787	1.5148	1.7934	190.0
180.0	7.511	0.016510	50.208	50.225	148.00	990.2	1138.2	0.2631	1.5480	1.8111	180.0
170.0	5.9926	0.016451	62.04	62.06	137.97	996.2	1134.2	0.2473	1.5822	1.8295	170.0
160.0	4.7414	0.016395	77.27	77.29	127.96	1002.2	1130.2	0.2313	1.6174	1.8487	160.0
150.0	3.7184	0.016343	97.05	97.07	117.95	1008.2	1126.1	0.2150	1.6536	1.8686	150.0
140.0	2.8892	0.016293	122.98	123.00	107.95	1014.0	1122.0	0.1985	1.6910	1.8895	140.0
130.0	2.2230	0.016247	157.32	157.33	97.96	1019.8	1117.8	0.1817	1.7295	1.9112	130.0
120.0	1.6927	0.016204	203.25	203.26	87.97	1025.6	1113.6	0.1646	1.7693	1.9339	120.0
110.0	1.2750	0.016165	265.37	265.39	77.98	1031.4	1109.3	0.1472	1.8105	1.9577	110.0
100.0	0.94924	0.016130	350.4	350.4	67.999	1037.1	1105.1	0.1295	1.8530	1.9825	100.0
90.0	0.69813	0.016099	468.1	468.1	58.018	1042.7	1100.8	0.1115	1.8970	2.0086	90.0
80.0	0.50683	0.016072	633.3	633.3	48.037	1048.4	1096.4	0.0932	1.9426	2.0359	80.0
70.0	0.36292	0.016050	868.4	868.4	38.052	1054.0	1092.1	0.0745	1.9900	2.0645	70.0
60.0	0.25611	0.016033	1207.6	1207.6	28.060	1059.7	1087.7	0.0555	2.0391	2.0946	60.0
50.0	0.17796	0.016023	1704.8	1704.8	18.054	1065.3	1083.4	0.0361	2.0901	2.1262	50.0
40.0	0.12163	0.016019	2445.8	2445.8	8.027	1071.0	1079.0	0.0162	2.1432	2.1594	40.0
32.018	0.08865	0.016022	3302.4	3302.4	0.0003	1075.5	1075.5	0.0000	2.1872	2.1872	32.018

ASME Table 2
PROPERTIES OF SATURATED STEAM AND SATURATED WATER (Pressure)

Press. psia	Temp. F	Volume, ft³/lbm Water v_f	Evap. v_{fg}	Steam v_g	Enthalpy, Btu/lbm Water h_f	Evap. h_{fg}	Steam h_g	Entropy, Btu/lbm × F Water s_f	Evap. s_{fg}	Steam s_g	Energy, Btu/lbm Water u_f	Steam u_g
3208.2	705.47	0.05078	0.00000	0.05078	906.0	0.0	906.0	1.0612	0.0000	1.0612	875.9	875.9
3200.0	705.08	0.04472	0.01191	0.05663	875.5	56.1	931.6	1.0351	0.0482	1.0832	849.1	898.1
3000.0	695.33	0.03428	0.05073	0.08500	801.8	218.4	1020.3	0.9728	0.1891	1.1619	782.8	973.1
2800.0	684.96	0.03134	0.07171	0.10305	770.7	285.1	1055.8	0.9468	0.2491	1.1958	754.4	1002.4
2600.0	673.91	0.02938	0.09172	0.12110	744.5	337.6	1082.0	0.9247	0.2977	1.2225	730.3	1023.8
2400.0	662.11	0.02790	0.11287	0.14076	719.0	384.8	1103.7	0.9031	0.3430	1.2460	706.6	1041.2
2200.0	649.45	0.02669	0.13603	0.16272	695.5	426.7	1122.2	0.8828	0.3848	1.2676	684.6	1055.9
2000.0	635.80	0.02565	0.16266	0.18831	672.1	466.2	1138.3	0.8625	0.4256	1.2881	662.6	1068.6
1800.0	621.02	0.02472	0.19390	0.21861	648.5	503.8	1152.3	0.8417	0.4662	1.3079	640.3	1079.5
1500.0	604.87	0.02387	0.23159	0.25545	624.2	540.3	1164.5	0.8199	0.5076	1.3274	617.1	1088.9
1400.0	587.07	0.02307	0.27871	0.30178	598.8	576.5	1175.3	0.7966	0.5507	1.3474	592.9	1097.1
1200.0	567.19	0.02232	0.34013	0.36245	571.9	613.0	1184.8	0.7714	0.5969	1.3683	566.9	1104.3
1100.0	556.28	0.02195	0.37863	0.40058	557.5	631.5	1189.1	0.7578	0.6216	1.3794	553.1	1107.5
1000.0	544.58	0.02159	0.42436	0.44596	542.6	650.4	1192.9	0.7434	0.6476	1.3910	538.6	1110.4
900.0	531.95	0.02123	0.47968	0.50091	526.7	669.7	1196.4	0.7279	0.6753	1.4032	523.2	1113.0
800.0	518.21	0.02087	0.54809	0.56896	509.8	689.6	1199.4	0.7111	0.7051	1.4163	506.7	1115.2
700.0	503.08	0.02050	0.63505	0.65556	491.6	710.2	1201.8	0.6928	0.7377	1.4304	488.9	1116.9
600.0	486.20	0.02013	0.74962	0.76975	471.7	732.0	1203.7	0.6723	0.7738	1.4461	469.5	1118.2
500.0	467.01	0.01975	0.90787	0.92762	449.5	755.1	1204.7	0.6490	0.8148	1.4639	447.7	1118.8
400.0	444.60	0.01934	1.14162	1.16095	424.2	780.4	1204.6	0.6217	0.8630	1.4847	422.7	1118.7
300.0	417.35	0.01889	1.52384	1.54274	394.0	808.9	1202.9	0.5882	0.9223	1.5105	392.9	1117.2
250.0	400.97	0.01865	1.82452	1.84317	376.1	825.0	1201.1	0.5679	0.9585	1.5264	375.3	1115.8
200.0	381.80	0.01839	2.26890	2.28728	355.5	842.8	1198.3	0.5438	1.0016	1.5454	354.8	1113.7
180.0	373.08	0.01827	2.5129	2.5312	346.2	850.7	1196.9	0.5328	1.0215	1.5543	345.6	1112.5
140.0	353.04	0.01803	3.2010	3.2190	325.0	868.0	1193.0	0.5071	1.0681	1.5752	324.5	1109.6
120.0	341.27	0.01789	3.7097	3.7275	312.6	877.8	1190.4	0.4919	1.0960	1.5879	312.2	1107.6

(continued)

ASME Table 2 *(continued)*

PROPERTIES OF SATURATED STEAM AND SATURATED WATER (Pressure)

Press. psia	Temp. F	Volume, ft³/lbm Water v_f	Evap. v_{fg}	Steam v_g	Enthalpy, Btu/lbm Water h_f	Evap. h_{fg}	Steam h_g	Entropy, Btu/lbm × F Water s_f	Evap. s_{fg}	Steam s_g	Energy, Btu/lbm Water u_f	Steam u_g
100.0	372.82	0.01774	4.4133	4.4310	298.5	888.6	1187.2	0.4743	1.1284	1.6027	298.2	1105.2
90.0	320.28	0.017659	4.8777	4.8953	290.7	894.6	1185.3	0.4643	1.1470	1.6113	290.4	1103.7
80.0	312.04	0.017573	5.4536	5.4711	282.1	900.9	1183.1	0.4534	1.1675	1.6208	281.9	1102.1
70.0	302.93	0.017482	6.1875	6.2050	272.7	907.8	1180.6	0.4411	1.1905	1.6316	272.5	1100.2
60.0	292.71	0.017383	7.1562	7.1736	262.2	915.4	1177.6	0.4273	1.2167	1.6440	262.0	1098.0
50.0	281.02	0.017274	8.4967	8.5140	250.2	923.9	1174.1	0.4112	1.2474	1.6586	250.1	1095.3
40.0	267.25	0.017151	10.4794	10.4965	236.1	933.6	1169.8	0.3921	1.2844	1.6765	236.0	1092.1
30.0	250.34	0.017009	13.7266	13.7436	218.9	945.2	1164.1	0.3682	1.3313	1.6995	218.8	1087.9
20.0	227.96	0.016834	20.070	20.087	196.27	960.1	1156.3	0.3358	1.3962	1.7320	196.21	1082.0
14.696	212.00	0.016719	26.782	26.799	180.17	970.3	1150.5	0.3121	1.4447	1.7568	180.12	1077.6
10.0	193.21	0.016592	38.404	38.420	161.26	982.1	1143.3	0.2836	1.5043	1.7879	161.23	1072.3
9.0	188.27	0.016561	42.385	42.402	156.30	985.1	1141.4	0.2760	1.5204	1.7964	156.28	1070.8
8.0	182.86	0.016527	47.328	47.345	150.87	988.5	1139.3	0.2676	1.5384	1.8060	150.84	1069.2
7.0	176.84	0.016491	53.634	53.650	144.83	992.1	1136.9	0.2581	1.5587	1.8168	144.81	1067.4
6.0	170.05	0.016451	61.967	61.984	138.03	996.2	1134.2	0.2474	1.5820	1.8294	138.01	1065.4
5.0	162.24	0.016407	73.515	73.532	130.20	1000.9	1131.1	0.2349	1.6094	1.8443	130.18	1063.1
4.0	152.96	0.016358	90.63	90.64	120.92	1006.4	1127.3	0.2199	1.6428	1.8626	120.90	1060.2
3.0	141.47	0.016300	118.71	118.73	109.42	1013.2	1122.6	0.2009	1.6854	1.8864	109.41	1056.7
2.0	126.07	0.016230	173.74	173.76	94.03	1022.1	1116.2	0.1750	1.7450	1.9200	94.03	1051.8
1.0	101.74	0.016136	333.59	333.60	69.73	1036.1	1105.8	0.1326	1.8455	1.9781	69.73	1044.1
0.8	94.38	0.016112	411.67	411.69	62.39	1040.3	1102.6	0.1195	1.8775	1.9970	62.39	1041.7
0.60	85.218	0.016085	540.0	540.1	53.245	1045.5	1098.7	0.1028	1.9186	2.0215	53.243	1038.7
0.40	72.869	0.016056	792.0	792.1	40.917	1052.4	1093.3	0.0799	1.9762	2.0562	40.916	1034.7
0.30	64.484	0.016040	1039.7	1039.7	32.541	1057.1	1089.7	0.0641	2.0168	2.0809	32.540	1032.0
0.20	53.160	0.016025	1526.3	1526.3	21.217	1063.5	1084.7	0.0422	2.0738	2.1160	21.217	1028.3
0.10	35.023	0.016020	2945.5	2945.5	3.026	1073.8	1076.8	0.0061	2.1705	2.1766	3.025	1022.3
0.08865	32.018	0.016022	3302.4	3302.4	0.0003	1075.5	1075.5	0.0000	2.1872	2.1872	0.000	1021.3

ASME Table 3 (continued)

PROPERTIES OF SUPERHEATED STEAM AND COMPRESSED WATER (Temperature and Pressure)

(continued from preceding page)

Temp. F	Volume, v, ft³/lbm			Enthalpy, h, Btu/lbm			Entropy, s, Btu/lbm × F		
100	0.01613	0.01613	0.01613	68.05	68.02	68.01	0.1295	0.1295	0.1295
50	0.01602	0.01602	0.01602	18.11	18.08	18.07	0.0361	0.0361	0.0361
32	0.01602	0.01602	0.01602	0.04	0.01	-0.00	-0.0000	-0.0000	-0.0000

Temp. F	Volume, v, ft³/lbm			Enthalpy, h, Btu/lbm			Entropy, s, Btu/lbm × F		
Press., psia	50.0	100.0	150.0	50.0	100.0	150.0	50.0	100.0	150.0
Tsat	281.02	327.82	358.43	281.02	327.82	358.43	281.02	327.82	358.43
Sat. Steam	8.514	4.431	3.0139	1174.1	1187.2	1194.1	1.6586	1.6027	1.5695
Sat. Water	0.01727	0.01774	0.01809	250.21	298.54	330.65	0.4112	0.4743	0.5141
1500	23.332	11.659	7.7674	1802.9	1802.2	1801.6	2.1561	2.0794	2.0344
1250	20.344	10.160	6.7652	1665.5	1664.5	1663.5	2.0811	2.0042	1.9591
1000	17.350	8.655	5.7568	1533.4	1532.0	1530.5	1.9977	1.9205	1.8751
750	14.344	7.139	4.7366	1406.8	1404.5	1402.2	1.9025	1.8248	1.7787
500	11.306	5.588	3.6799	1284.1	1279.3	1274.3	1.7890	1.7088	1.6602
250	0.01700	0.01700	0.01700	218.63	218.74	218.84	0.3677	0.3676	0.3675
150	0.01634	0.01634	0.01634	118.06	118.19	118.31	0.2150	0.2149	0.2149
100	0.01613	0.01613	0.01612	68.13	68.26	68.39	0.1295	0.1295	0.1294
50	0.01602	0.01602	0.01601	18.20	18.34	18.49	0.0361	0.0360	0.0360
32	0.01602	0.01602	0.01601	0.13	0.29	0.44	-0.0000	-0.0000	-0.0000

Temp. F	Volume, v, ft³/lbm			Enthalpy, h, Btu/lbm			Entropy, s, Btu/lbm × F		
Press., psia	200.0	250.0	300.0	200.0	250.0	300.0	200.0	250.0	300.0
Tsat	381.80	400.97	417.35	381.80	400.97	417.35	381.80	400.97	417.35
Sat. Steam	2.2873	1.8432	1.5427	1198.3	1201.1	1202.9	1.5454	1.5264	1.5105
Sat. Water	0.01839	0.01865	0.01889	355.51	376.14	393.99	0.5438	0.5679	0.5882
1500	5.8219	4.6546	3.8764	1800.9	1800.2	1799.6	2.0025	1.9776	1.9572
1250	5.0679	4.0495	3.3706	1662.5	1661.5	1660.6	1.9270	1.9019	1.8814
1000	4.3077	3.4382	2.8585	1529.1	1527.6	1526.2	1.8426	1.8173	1.7964
750	3.5355	2.8147	2.3342	1399.9	1397.5	1395.2	1.7456	1.7195	1.6980
500	2.7247	2.2462	1.7665	1269.0	1263.5	1257.7	1.6242	1.5951	1.5703

ASME Table 3

PROPERTIES OF SUPERHEATED STEAM AND COMPRESSED WATER (Temperature and Pressure)

Temp. F	Volume, v, ft³/lbm			Enthalpy, h, Btu/lbm			Entropy, s, Btu/lbm × F		
Press., psia	0.50	1.0	2.0	0.50	1.0	2.0	0.5	1.0	2.0
Tsat	79.59	101.74	126.07	79.59	101.74	126.07	79.59	101.74	126.07
Sat. Steam	641.5	333.6	173.76	1096.3	1105.8	1116.2	2.0370	1.9781	1.9200
Sat. Water	0.01607	0.01614	0.01623	47.62	69.73	94.03	0.0925	0.1326	0.1750
1500	—	—	583.65	—	—	1803.5	—	—	2.5112
1250	—	—	509.18	—	—	1666.4	—	—	2.4363
1000	—	—	434.71	—	—	1534.8	—	—	2.3532
750	1441.1	720.5	360.22	1409.0	1409.0	1409.0	2.4115	2.3351	2.2587
500	1143.2	571.5	285.70	1288.6	1288.6	1288.5	2.3001	2.2237	2.1472
250	845.1	422.4	211.03	1173.0	1172.9	1172.6	2.1607	2.0841	2.0074
150	725.7	362.6	180.99	1127.8	1127.6	1127.0	2.0920	2.0153	1.9382
100	665.9	0.01613	0.01613	1105.4	68.00	68.00	2.0537	0.1295	0.1295
50	0.01602	0.01602	0.01602	18.06	18.06	18.06	0.0361	0.0361	0.0361
32	0.01602	0.01602	0.01602	−0.02	−0.02	−0.01	−0.0000	−0.0000	−0.0000

Temp. F	Volume, v, ft³/lbm			Enthalpy, h, Btu/lbm			Entropy, s, Btu/lbm × F		
Press., psia	5.0	10.0	20.0	5.0	10.0	20.0	5.0	10.0	20.0
Tsat	162.24	193.21	227.96	162.24	193.21	227.96	162.24	193.21	227.96
Sat. Steam	73.53	38.42	20.087	1131.1	1143.3	1156.3	1.8443	1.7879	1.7320
Sat. Water	0.01641	0.01659	0.01683	130.20	161.26	196.27	0.2349	0.2836	0.3358
1500	233.45	116.72	58.352	1803.5	1803.4	1803.3	2.4101	2.3337	2.2572
1250	203.66	101.82	50.896	1666.4	1666.3	1666.1	2.3353	2.2589	2.1824
1000	181.10	86.91	43.435	1534.7	1534.6	1534.3	2.2521	2.1757	2.0991
750	144.05	71.99	35.962	1408.8	1408.6	1408.1	2.1576	2.0810	2.0044

PROPERTIES OF SUPERHEATED STEAM AND COMPRESSED WATER (Temperature and Pressure)

P = 400, 500, 600 psia

Temp. F	Volume, v, ft³/lbm			Enthalpy, h, Btu/lbm			Entropy, s, Btu/lbm × F		
	400.0	500.0	600.0	400.0	500.0	600.0	400.0	500.0	600.0
1500	2.9037	2.3200	1.9309	1795.6	1796.9	1798.2	1.9250	1.8998	1.8792
1250	2.5219	2.0128	1.6733	1654.7	1656.6	1658.6	1.8488	1.8233	1.8023
1000	2.1339	1.6992	1.4093	1517.4	1520.3	1523.3	1.7632	1.7371	1.7155
750	1.7333	1.3725	1.1318	1380.4	1385.4	1390.4	1.6634	1.6357	1.6125
500	1.2841	0.9919	0.7944	1215.9	1231.2	1245.1	1.5282	1.4921	1.4590
Sat. Steam	1.1610	0.9276	0.7697	1204.6	1204.7	1203.7	1.4847	1.4639	1.4461
Sat. Water	0.01934	0.01975	0.02013	424.17	449.52	471.70	0.6217	0.6490	0.6723
*T*sat	444.60	467.01	486.20	444.60	467.01	486.20	444.60	467.01	486.20
250	0.01699	0.01699	0.01699	218.94	219.03	219.15	0.3675	0.3674	0.3673
150	0.01633	0.01633	0.01633	118.43	118.55	118.68	0.2148	0.2148	0.2147
100	0.01612	0.01612	0.01612	68.52	68.66	68.79	0.1294	0.1294	0.1294
50	0.01601	0.01601	0.01601	18.63	18.78	18.92	0.0360	0.0360	0.0360
32	0.01601	0.01601	0.01601	0.59	0.74	0.89	−0.0000	−0.0000	0.0000

P = 800, 1200, 1600 psia

Temp. F	Volume, v, ft³/lbm			Enthalpy, h, Btu/lbm			Entropy, s, Btu/lbm × F		
	800.0	1200.0	1600.0	800.0	1200.0	1600.0	800.0	1200.0	1600.0
1500	1.4446	0.9584	0.7153	1792.9	1787.6	1782.3	1.8464	1.7996	1.7657
1250	1.2490	0.8248	0.6127	1650.8	1642.9	1634.9	1.7688	1.7206	1.6852
1000	1.0470	0.6845	0.5031	1511.4	1499.4	1486.9	1.6807	1.6298	1.5916
Sat. Steam	0.5690	0.3624	0.2555	1199.4	1184.8	1164.5	1.4163	1.3683	1.3274
Sat. Water	0.02087	0.02232	0.02387	509.81	571.85	624.20	0.7111	0.7714	0.8199
*T*sat	518.21	567.19	604.87	518.21	567.19	604.87	518.21	567.19	604.87
250	0.01698	0.01697	0.01697	219.36	219.57	219.78	0.3672	0.3670	0.3669
150	0.01632	0.01632	0.01631	118.92	119.16	119.41	0.2146	0.2145	0.2145
100	0.01611	0.01611	0.01610	69.05	69.32	69.58	0.1293	0.1292	0.1292
50	0.01600	0.01600	0.01599	19.21	19.50	19.79	0.0360	0.0360	0.0360
32	0.01600	0.01600	0.01599	1.19	1.50	1.80	0.0000	0.0000	0.0000

(continued)

ASME Table 3 (*continued*)

PROPERTIES OF SUPERHEATED STEAM AND COMPRESSED WATER (Temperature and Pressure)

Temp. F	Volume, v, ft³/lbm			Enthalpy, h, Btu/lbm			Entropy, s, Btu/lbm × F		
	2000.0	3000.0	4000.0	2000.0	3000.0	4000.0	2000.0	3000.0	4000.0
750	0.8303	0.5273	0.3741	1369.8	1346.9	1321.4	1.5742	1.5150	1.4667
500	0.02041	0.02031	0.02023	487.88	487.72	487.60	0.6885	0.6868	0.6851
250	0.01696	0.01693	0.01691	220.19	221.03	221.86	0.3666	0.3660	0.3654
150	0.01630	0.01628	0.01626	119.90	120.88	121.85	0.2143	0.2139	0.2135
100	0.01609	0.01607	0.01605	70.11	71.16	72.21	0.1290	0.1288	0.1286
50	0.01598	0.01596	0.01594	20.36	21.51	22.66	0.0359	0.0359	0.0358
32	0.01598	0.01596	0.01593	2.40	3.60	4.80	0.0001	0.0001	0.0001

Press., psia	2000.0	3000.0	4000.0	2000.0	3000.0	4000.0	2000.0	3000.0	4000.0
Tsat	635.80	695.33		635.80	695.33		635.80	695.33	
Sat. Steam	0.1883	0.0850	—	1138.3	1020.3	—	1.2881	1.1619	—
Sat. Water	0.02565	0.03428	—	672.11	801.84	—	0.8625	0.9728	—

Press., psia	2000.0	3000.0	4000.0	2000.0	3000.0	4000.0	2000.0	3000.0	4000.0
1500	0.5695	0.3753	0.2783	1777.1	1763.8	1750.6	1.7389	1.6888	1.6516
1250	0.4855	0.3159	0.2312	1627.0	1606.8	1586.3	1.6570	1.6031	1.5619
1000	0.3942	0.2484	0.1752	1474.1	1440.2	1403.6	1.5603	1.4976	1.4461
750	0.2805	0.1483	0.0631	1292.6	1197.9	1007.4	1.4231	1.3131	1.1396
500	0.02014	0.01995	0.0198	487.53	487.52	487.7	0.6834	0.6796	0.6760
250	0.01688	0.01682	0.0168	222.70	224.81	226.9	0.3648	0.3634	0.3620
150	0.01624	0.01619	0.0161	122.83	125.28	127.7	0.2131	0.2122	0.2113
100	0.01603	0.01599	0.0159	73.26	75.88	78.5	0.1283	0.1277	0.1271
50	0.01592	0.01587	0.0158	23.80	26.64	29.5	0.0357	0.0355	0.0353
32	0.01591	0.01586	0.0158	5.99	8.95	11.9	0.0002	0.0002	0.0002

PROPERTIES OF SUPERHEATED STEAM AND COMPRESSED WATER (Temperature and Pressure)

Temp. F	Volume, v, ft³/lbm			Enthalpy, h, Btu/lbm			Entropy, s, Btu/lbm × F		
Press., psia	5000.0	10000.0	15000.0	5000.0	10000.0	15000.0	5000.0	10000.0	15000.0
T_{sat}	—	—	—	—	—	—	—	—	—
Sat. Steam	—	—	—	—	—	—	—	—	—
Sat. Water	—	—	—	—	—	—	—	—	—
1500	0.2203	0.1054	0.0690	1737.4	1672.8	1615.9	1.6216	1.5180	1.4491
1250	0.1806	0.0812	0.0516	1565.5	1463.4	1379.4	1.5277	1.4035	1.3197
1000	0.1312	0.0495	0.0337	1364.6	1172.6	1080.6	1.4001	1.2185	1.1302
750	0.0338	0.0251	0.0230	854.9	783.8	769.6	1.0070	0.9270	0.8970
500	0.0196	0.0189	0.0184	488.1	491.8	497.4	0.6726	0.6578	0.6457
250	0.0167	0.0165	0.0162	229.0	239.8	250.7	0.3606	0.3541	0.3481
150	0.0161	0.0159	0.0157	130.2	142.3	154.4	0.2104	0.2061	0.2019
100	0.0159	0.1057	0.0155	81.1	93.9	106.5	0.1265	0.1233	0.1201
50	0.0158	0.0155	0.0153	32.3	45.9	59.1	0.0351	0.0335	0.0314
32	0.0158	0.0155	0.0153	14.8	28.9	42.6	0.0002	-0.0005	-0.0017

ASME Table 9

SPECIFIC HEAT AT CONSTANT PRESSURE OF STEAM AND OF WATER

Temp. F Press., psia	1	6	15	c_p, Btu/lbm $\times$ F 100	1000	6000	15000
Sat. Water	0.998	1.002	1.007	1.039	1.286	—	—
Sat. Steam	0.450	0.466	0.485	0.582	1.191	—	—
1500	0.559	0.559	0.559	0.561	0.580	0.691	0.868
1000	0.515	0.515	0.515	0.519	0.566	1.110	1.306
500	0.472	0.473	0.476	0.508	1.181	1.092	1.017
300	0.458	0.464	0.475	1.029	1.024	1.002	0.973
200	0.453	0.465	1.005	1.005	1.002	0.986	0.963
100	0.998	0.998	0.998	0.998	0.994	0.976	0.955
60	1.000	1.000	1.000	0.999	0.994	0.970	0.942
32	1.007	1.007	1.007	1.006	0.999	0.962	0.904

*AIR TABLES**

* From *Gas Tables* by Joseph H. Keenan and Joseph Kaye, Wiley, New York, 1945.

THERMODYNAMIC PROPERTIES OF AIR AT LOW PRESSURE*

T, R	h, Btu/lbm	P_r	u, Btu/lbm	v_r	ϕ, Btu/lbm R
200	47.67	0.04320	33.96	1714.9	0.36303
220	52.46	0.06026	37.38	1352.5	0.38584
240	57.25	0.08165	40.80	1088.8	0.40666
260	62.03	0.10797	44.21	892.0	0.42582
280	66.82	0.13986	47.63	741.6	0.44356
300	71.61	0.17795	51.04	624.5	0.46007
320	76.40	0.22290	54.46	531.8	0.47550
340	81.18	0.27545	57.87	457.2	0.49002
360	85.97	0.3363	61.29	396.6	0.50369
380	90.75	0.4061	64.70	346.6	0.51663
400	95.53	0.4858	68.11	305.0	0.52890
420	100.32	0.5760	71.52	270.1	0.54058
440	105.11	0.6776	74.93	240.6	0.55172
460	109.90	0.7913	78.36	215.33	0.56235
480	114.69	0.9182	81.77	193.65	0.57255
500	119.48	1.0590	85.20	174.90	0.58233
520	124.27	1.2147	88.62	158.58	0.59173
540	129.06	1.3860	92.04	144.32	0.60078
560	133.86	1.5742	95.47	131.78	0.60950
580	138.66	1.7800	98.90	120.70	0.61793
600	143.47	2.005	102.34	110.88	0.62607
620	148.28	2.249	105.78	102.12	0.63395
640	153.09	2.514	109.21	94.30	0.64159
660	157.92	2.801	112.67	87.27	0.64902
680	162.73	3.111	116.12	80.96	0.65621
700	167.56	3.446	119.58	75.25	0.66321
720	172.39	3.806	123.04	70.07	0.67002
740	177.23	4.193	126.51	65.38	0.67665
760	182.08	4.607	129.99	61.10	0.68312
780	186.94	5.051	133.47	57.20	0.68942
800	191.81	5.526	136.97	53.63	0.69558
820	196.69	6.033	140.47	50.35	0.70160
840	201.56	6.573	143.98	47.34	0.70747
860	206.46	7.149	147.50	44.57	0.71323
880	211.35	7.761	151.02	42.01	0.71886
900	216.26	8.411	154.57	39.64	0.72438
920	221.18	9.102	158.12	37.44	0.72979
940	226.11	9.834	161.68	35.41	0.73509
960	231.06	10.610	165.26	33.52	0.74030
980	236.02	11.430	168.83	31.76	0.74540
1000	240.98	12.298	172.43	30.12	0.75042

* Abridged from Table 1 in *Gas Tables*, by Joseph H. Keenan and Joseph Kaye. Copyright 1948, by Joseph H. Keenan and Joseph Kaye. Published by John Wiley & Sons, Inc., New York.

THERMODYNAMIC PROPERTIES OF AIR AT LOW PRESSURE (*continued*)

T, R	*h*, Btu/lbm	*P*$_r$	*u*, Btu/lbm	*v*$_r$	ϕ, Btu/lbm R
1020	245.97	13.215	176.04	28.59	0.75536
1040	250.95	14.182	179.66	27.17	0.76019
1060	255.96	15.203	183.29	25.82	0.76496
1080	260.97	16.278	186.93	24.58	0.76964
1100	265.99	17.413	190.58	23.40	0.77426
1120	271.03	18.604	194.25	22.30	0.77880
1140	276.08	19.858	197.94	21.27	0.78326
1160	281.14	21.18	201.63	20.293	0.78767
1180	286.21	22.56	205.33	19.377	0.79201
1200	291.30	24.01	209.05	18.514	0.79628
1220	296.41	25.53	212.78	17.700	0.80050
1240	301.52	27.13	216.53	16.932	0.80466
1260	306.65	28.80	220.28	16.205	0.80876
1280	311.79	30.55	224.05	15.518	0.81280
1300	316.94	32.39	227.83	14.868	0.81680
1320	322.11	34.31	231.63	14.253	0.82075
1340	327.29	36.31	235.43	13.670	0.82464
1360	332.48	38.41	239.25	13.118	0.82848
1380	337.68	40.59	243.08	12.593	0.83229
1400	342.90	42.88	246.93	12.095	0.83604
1420	348.14	45.26	250.79	11.622	0.83975
1440	353.37	47.75	254.66	11.172	0.84341
1460	358.63	50.34	258.54	10.743	0.84704
1480	363.89	53.04	262.44	10.336	0.85062
1500	369.17	55.86	266.34	9.948	0.85416
1520	374.47	58.78	270.26	9.578	0.85767
1540	379.77	61.83	274.20	9.226	0.86113
1560	385.08	65.00	278.13	8.890	0.86456
1580	390.40	68.30	282.09	8.569	0.86794
1600	395.74	71.73	286.06	8.263	0.87130
1620	401.09	75.29	290.04	7.971	0.87462
1640	406.45	78.99	294.03	7.691	0.87791
1660	411.82	82.83	298.02	7.424	0.88116
1680	417.20	86.82	302.04	7.168	0.88439
1700	422.59	90.95	306.06	6.924	0.88758
1720	428.00	95.24	310.09	6.690	0.89074
1740	433.41	99.69	314.13	6.465	0.89387
1760	438.83	104.30	318.18	6.251	0.89697
1780	444.26	109.08	322.24	6.045	0.90003
1800	449.71	114.03	326.32	5.847	0.90308
1820	455.17	119.16	330.40	5.658	0.90609
1840	460.63	124.47	334.50	5.476	0.90908
1860	466.12	129.95	338.61	5.302	0.91203
1880	471.60	135.64	342.73	5.134	0.91497

THERMODYNAMIC PROPERTIES OF AIR AT LOW PRESSURE (*continued*)

T, R	h, Btu/lbm	P_r	u, Btu/lbm	v_r	ϕ, Btu/lbm R
1900	477.09	141.51	346.85	4.974	0.91788
1920	482.60	147.59	350.98	4.819	0.92076
1940	488.12	153.87	355.12	4.670	0.92362
1960	493.64	160.37	359.28	4.527	0.92645
1980	499.17	167.07	363.43	4.390	0.92926
2000	504.71	174.00	367.61	4.258	0.93205
2020	510.26	181.16	371.79	4.130	0.93481
2040	515.82	188.54	375.98	4.008	0.93756
2060	521.39	196.16	380.18	3.890	0.94026
2080	526.97	204.02	384.39	3.777	0.94296
2100	532.55	212.1	388.60	3.667	0.94564
2120	538.15	220.5	392.83	3.561	0.94829
2140	543.74	229.1	397.05	3.460	0.95092
2160	549.35	238.0	401.29	3.362	0.95352
2180	554.97	247.2	405.53	3.267	0.95611
2200	560.59	256.6	409.78	3.176	0.95868
2220	566.23	266.3	414.05	3.088	0.96123
2240	571.86	276.3	418.31	3.003	0.96376
2260	577.51	286.6	422.59	2.921	0.96626
2280	583.16	297.2	426.87	2.841	0.96876
2300	588.82	308.1	431.16	2.765	0.97123
2320	594.49	319.4	435.46	2.691	0.97369
2340	600.16	330.9	439.76	2.619	0.97611
2360	605.84	342.8	444.07	2.550	0.97853
2380	611.53	355.0	448.38	2.483	0.98092
2400	617.22	367.6	452.70	2.419	0.98331

SYMBOLS

SYMBOLS

The interpretations given here for symbols are usually those intended throughout the whole book. Where a symbol has a certain meaning in only one part of the book, the chapter number is noted for that particular usage. The symbols and abbreviations that have been chosen are those most commonly used in the field of work under discussion, except where that would have caused confusion. In a few instances, one symbol has different meanings and one term is represented by different symbols in separate chapters; however, such overlapping has been kept to a minimum.

Those symbols used in only a limited portion of the book and most subscripts are not defined here. Their meanings should be clear from the context in which they are used.

Abbreviations whose meanings may not be immediately clear to some readers are defined along with other symbols. While it may be pedagogically important in an introductory textbook to avoid any possible ambiguity in symbols and units, that should not be necessary here. An example is the use in this book, where it seems appropriate, of "psia" for pounds per square inch absolute; a text on the fundamentals of thermodynamics might adhere to lb_f/in^2 abs.

A	area
	Richardson's constant (Chapter 7)
a	acceleration
	speed of sound
B	magnetic flux density (Chapter 7)
Btu	British thermal unit
bhp	brake horsepower
bmep	brake mean effective pressure
C	degrees Celsius

Cal kilocalorie (1000 cal)

cal calorie

c_p specific heat at constant pressure

c_v specific heat at constant volume

cfs cubic feet per second (also cfm: per minute, and cfh: per hour)

d differential (exact)

E voltage gradient (Chapter 7)

e base of natural logarithms
    elementary charge (Chapter 7)

emf electromotive force

eV electron volt

F degrees Fahrenheit

F force

f (subscript) force
    (subscript) saturated liquid

fps feet per second (also fpm: per minute)

G Gibbs function

g Gibbs function per unit mass
    acceleration of gravity
    (subscript) saturated vapor

g_c dimensional constant in $F = ma/g_c$

H enthalpy
    (subscript) higher temperature

h enthalpy per unit mass (specific enthalpy)
    Plank's constant (Chapter 7)

I electric current

ihp indicated horsepower

imep indicated mean effective pressure

J Joule's equivalent (1 Btu = J ft-lb$_f$)
    electric current density (Chapter 7)

K degrees Kelvin

K combined thermal conductivity coefficient (Chapter 7)
    any constant

k thermal conductivity coefficient (Chapter 6)
    Boltzmann's constant

K (prefix) kilo, 1000

L length
    (subscript) lower temperature

lb pound (either force or mass)

lb$_f$ pound force

lb$_m$ pound mass

M Mach number
    momentum

M_w molecular weight

m mass

mph   miles per hour

N_o Avogadro's number

n charge carrier concentration (Chapter 7)
    number of moles

P power

P_b brake mean effective pressure

P_i indicated mean effective pressure

p pressure

psf   pounds force per square foot

psia   pounds force per square inch, absolute

psig   pounds force per square inch, gauge

Q heat transferred (*Note*: $\dot{V}$ is used for volume rate of flow.)

q heat transferred per unit mass

R   degrees Rankine

R individual gas constant (per unit mass)
    electrical resistance

R_U universal gas constant (per mole)

rps   revolutions per second (also rpm: per minute)

S entropy
    scale factor

s entropy per unit mass (specific entropy)

T temperature, absolute, Rankine or Kelvin degrees

t temperature, Fahrenheit or Celsius degrees

V volume
    voltage

$\bar{V}$ velocity

v volume per unit mass (specific volume, inverse of density ρ)

U internal energy

u internal energy per unit mass (specific internal energy)
    velocity (Chatper 7)

W work
    humidity ratio

$\bar{W}$ relative velocity

w work per unit mass

x dryness fraction in a mixture of liquid or solid and vapor; concentration fraction in a solution

z elevation above some datum

Z figure of merit

x, y, z mutually perpendicular coordinates

Greek Letters

α (alpha)   angle
             Seebeck coefficient

β (beta)   angle

γ (gamma)   ratio c_p/c_v
             area-to-length ratio (Chapter 7)

Δ (Delta)   finite difference

δ (delta)   infinitestimal difference

ε (epsilon)   emissivity

η (eta)   efficiency

λ (lambda)   thermal conductivity coefficient

μ (mu)   degree of saturation (Chapter 6)
             charge carrier mobility (Chapter 7)

Π (Pi)   product

π (pi)   ratio of circumference-to-diameter of a circle
             Peltier coefficient

ρ (rho)   density (inverse of specific volume v)
             electrical resistivity (inverse of conductivity σ)

Σ (Sigma)   sum

σ (sigma)   electrical conductivity (inverse of resistivity ρ)

τ (tau)   Thompson coefficient

ϕ (phi)   relative humidity (Chapter 6)
             a thermodynamic property defined by Eq. (1.32c)

ω (omega)   rotational speed, radians per second

Special Symbols

$\bigcirc$ refers to designated point or section on a diagram

$\cdot$ (dot above a letter) time derivative, rate

$\times$ multiplication, particularly a cross product

$đ$ inexact differential

∂ partial differential

$\mathscr{E}$ energy (Chapter 7)

$\mathscr{F}$ Faraday's constant (Chapter 7)

l length

$\ln$ natural or Napierian logarithm

[*Note:* Boldface italic letters in the text, e.g. *a*, correspond to circled letters representing points on diagrams.]

Index

Index